Genomics: Concepts and Applications

Genomics: Concepts and Applications

Edited by Caleb Elliot

SYRAWOOD
PUBLISHING HOUSE
New York

Published by Syrawood Publishing House,
750 Third Avenue, 9th Floor,
New York, NY 10017, USA
www.syrawoodpublishinghouse.com

Genomics: Concepts and Applications
Edited by Caleb Elliot

International Standard Book Number: 978-1-64740-092-7 (Hardback)

Cataloging-in-Publication Data

Genomics : concepts and applications / edited by Caleb Elliot.
p. cm.
Includes bibliographical references and index.
ISBN 978-1-64740-092-7
1. Genomics. 2. Genomes. 3. Genetics. I. Elliot, Caleb.
QH447 .G46 2022
572.86--dc23

TABLE OF CONTENTS

PREFACE

Genomics is a field of biology that studies the genome. It involves the analysis and sequencing of genes to analyze the function and structure of the complete genome. A revolutionary development in genomics has been the Human Genome Project under which the complete genome sequencing of the human body was achieved. This was possible due to an amalgamation of efficient sequencing technologies and utilization of bioinformatics resources. The genome analysis of an organism usually consists of three stages- DNA sequencing, representation of the original chromosome using knowledge of the assembly of the DNA sequence, and the analysis and annotation of that representation. Some of the areas of research in genomics are functional genomics, structural genomics, metagenomics, epigenomics, etc. Such studies contribute to the advancement of biotechnology, medicine, bioengineering and conservation. The objective of this book is to give a general view of the different areas of genomics and its applications. The various studies that are constantly contributing towards advancing technologies and evolution of this field are examined in detail. This book is a collective contribution of a renowned group of international experts.

The information contained in this book is the result of intensive hard work done by researchers in this field. All due efforts have been made to make this book serve as a complete guiding source for students and researchers. The topics in this book have been comprehensively explained to help readers understand the growing trends in the field.

I would like to thank the entire group of writers who made sincere efforts in this book and my family who supported me in my efforts of working on this book. I take this opportunity to thank all those who have been a guiding force throughout my life.

Editor

1

Identification of Novel Causal FBN1 Mutations in Pedigrees of Marfan Syndrome

Yueli Wang, Xiaoyan Li, Rongjuan Li, Ya Yang, and Jie Du

Department of Echocardiography, Beijing Anzhen Hospital, Capital Medical University and Beijing Institute of Heart, Lung and Blood Vessel Diseases, Beijing, China

Correspondence should be addressed to Jie Du; jiedubj@126.com

Academic Editor: Byung-Hoon Jeong

Marfan syndrome (MFS) is an autosomal dominant genetic disorder of the connective tissue, typically characteristic of cardiovascular manifestations, valve prolapse, left ventricle enlargement, and cardiac failure. Fibrillin-1 *(FBN1)* is the causative gene in the pathogenesis of MFS. Patients with different *FBN1* mutations often present more considerable phenotypic variation. In the present study, three affected MFS pedigrees were collected for genetic analysis. Using next-generation sequencing (NGS) technologies, 3 novel frameshift pathogenic mutations which are cosegregated with affected subjects in 3 pedigrees were identified. These novel mutations provide important diagnostic and therapeutic insights for precision medicine in MFS, especially regarding the lethal cardiovascular events.

1. Introduction

Marfan syndrome (MFS; OMIM 154700) is a common autosomal dominant connective tissue disorder; the incidence rate was estimated to be at least 1 per 10,000 individuals without any racial, geographical, or occupational predilection. This disorder affects cardiovascular, ocular, and skeletal systems, as well as skin, lung, and central nervous systems. Mortality and morbidity are mainly determined by the development of cardiovascular events, such as heart failure, aortic aneurysm, and subsequent aortic dissection [1].

MFS was caused by fibrillin-1 (FBN1) gene mutations (NM_000138), which is located on chromosome 15q21.1 and had 65 exons. Fibrillin-1, a kind of extracellular matrix glycoprotein, was reported as an important calcium-binding microfibrillar structural molecule and a regulator of TGF-β signaling [2]. It consists of 47 epidermal growth factor-like (EGF) domains, 7 transforming growth factor β-1-binding protein-like (TB) domains, and a heterozygous domain. To date, about 3000 mutations [3] have been detected in *FBN1* and mainly classified into 3 types including missense mutations, in-frame deletions, and nonsense mutations due to frameshift leading to premature termination codons (PTC) [4, 5], even though progresses had been made for the detection of causal mutations in MFS.

Studies have demonstrated a higher probability of cardiovascular events in patients with mutations reducing the amount of *FBN1* (classified as HI mutation). These genotypes cause severity and worse prognosis, including increased risk for aortic surgery, aortic dissection, and mortality. The genotype–phenotype effect was recognized as an important factor when making a diagnosis of MFS and later predictable phenotype severity as well as clinical decisions [6, 7]. Some have argued that truncating may be associated with a milder disease course, which has long been questioned [8, 9]. For precision medicine of prognosis and treatment of MFS patients, more studies on the genotype–phenotype effect are warranted.

Here, 3 pedigrees with MFS were selected to identify novel mutations and assess genotype–cardiovascular phenotype effects. 3 novel, not yet reported, pathogenic frameshift mutations in *FBN1* were identified, which may contribute to providing diagnostic values in precision medicine of the genotype–cardiovascular phenotype relationship.

2. Material and Methods

2.1. Patients and Clinical Data. In this study, 3 ethnic Han Chinese families were recruited in Beijing Anzhen Hospital and diagnosed as MFS according to the revised Ghent nosology [10] based on their reported family history, clinical features, and essential echocardiography examinations. The study was approved by the ethics committee of Beijing Anzhen Hospital and performed according to the tenets of the Declaration of Helsinki. 3 pedigrees are shown in Figure 1. Detailed clinical information of all patients is listed in Table 1.

2.2. Genetic Testing Panel for Marfan Syndrome and Next-Generation Sequencing. A specific target sequencing panel was designed for the maximum coverage mutation of MFS. The target genes include FBN1, TGFBR1, and TGFBR2. Probes were designed and synthesized according to the Roche NimbleGen SeqCap EZ Choice manuscript. Genomic DNA was nebulized before adaptor ligation. Size selection was carried out using AMPure XP beads with a target size of 200 to 300 bp. Whole-genome shotgun libraries were prepared using the Illumina TruSeq Paired-End Prep Kit (Illumina Inc., San Diego, CA, USA) according to the manufacturer's protocol. The three DNA libraries were pooled together and hybridized to NimbleGen SeqCap EZ panel to enrich the target region.

Samples were sequenced on HiSeq 2500 (Illumina) following the manufacturer's recommendations, in paired-end mode for 100 cycles.

2.3. Sanger Resequencing. The mutations detected with NGS were confirmed by Sanger sequencing using the BigDye Terminator v3.1 Cycle Sequencing kit (Applied Biosystems; Thermo Fisher Scientific Inc.), followed by capillary electrophoresis on an 3500xL Genetic Analyzer (Applied Biosystems; Thermo Fisher Scientific Inc.). The identified mutations were further targeted for Sanger sequencing (cascade testing) in other family members from these 3 pedigrees, respectively.

2.4. Function Prediction of Gene Mutation Function. A forecasting software, MutationTaster, was used to predict the effect of the functional importance of variation loci identified in 3 pedigrees. MutationTaster provides a probability for the alteration to be either a causative mutation or a benign polymorphism [11].

3. Results

3.1. Clinical Findings of the Pedigree

3.1.1. Concerning Family 1. We analyzed a four-generation Marfan syndrome family composed of 9 affected members including 4 males and 5 females. Target gene sequencing was used to detect the proband and his 8-year-old daughter, both manifesting a marfanoid aortic sinus and tall stature. The proband's wife (27-year-old), an unaffected family member, had no clinical features of MFS. Clinical data of the patients are shown in Table 1. The proband (29-year-old), with a 187 cm height and all lanky as well as leggy body, presented to our hospital due to chest tightness and shortness of breath. Transthoracic echocardiography (Figure 2(a)) revealed aortic root dilatation (aortic diameter at the sinuses of Valsalva: 45 mm), mild aortic valve regurgitation, and normal left ventricular end diastolic diameter (LVED). These recommendations do not meet the formal criteria for prophylactic aortic surgery, so conservative therapy is suggested [12]. However, unexpected type A acute aortic dissection developed 2 months later (Figure 2(b)). The proband's mother was described to have suffered from sudden death at the age of 35. As for other relations, individuals I: 1, II: 1, II: 3, and II: 7 passed away. Individuals III: 4, III: 8, and III: 12 suffered from aortic aneurysm/dissection; clinical data are shown in Supplementary Table 1. Therefore, we speculated that inherited mutation in the family is responsible for a young, rapid-progressing, and extremely dangerous aortic dissection phenotype.

3.1.2. Concerning Family 2. 6 of 12 members in this 3-generation family were diagnosed as having MFS. Clinical data of the patients are shown in Table 1. The proband, with an average height, complained of paroxysmal chest pain and heart murmurs. TTE evaluation (Figure 2(c)) presented severe aortic root dilatation (aortic diameter at the sinuses of Valsalva 60 mm), severe aortic valve regurgitation, moderate mitral insufficiency, enlarged LVED, and poor left ventricular ejection fraction (LVEF: 35%), indicating that prophylactic aortic surgery was urgently needed according to the 2014 ESC Guidelines [12]. Since the dilatation of the aortic root was causing aggravation and the aortic valve developed severe regurgitation, the patient underwent a surgical operation for aortic valve and mitral valve replacement. The proband's 14-year-old son (Individual III: 2) (Table 1) not only displayed similar cardiovascular indicators (including marfanoid aortic sinus, enlarged LVED, and moderate mitral valve prolapse as well as mild tricuspid valve prolapse) as his father, but also was 185 cm tall and highly myopic, just like family 1 presented.

The young brother (individual II: 6) of the proband exhibited similar clinical and echocardiographic features (Table 1 and Figure 2(d)), dilatation of the aortic sinus to 67 mm with marked aortic valve regurgitation, and moderate mitral insufficiency. Individual II: 6 underwent the same surgery as the proband, but with a poor prognosis (Figure 2(d)). Individuals I: 1 and II: 1 died due to cardiac events. Taken together, family 2 have different features on cardiovascular disorders rather than early aortic dissection.

3.1.3. Concerning Family 3. The proband (individual II: 2) was a 29-year-old woman who presented typical marfanoid cardiovascular features, facial features, and arachnodactyly, but without ectopia lentis. Echocardiography detected a dilative aortic sinus, with a diameter of 43 mm. Her 5-year-old daughter (individual III: 1) complained of pectus excavatum and was the first to be diagnosed with MFS in Beijing

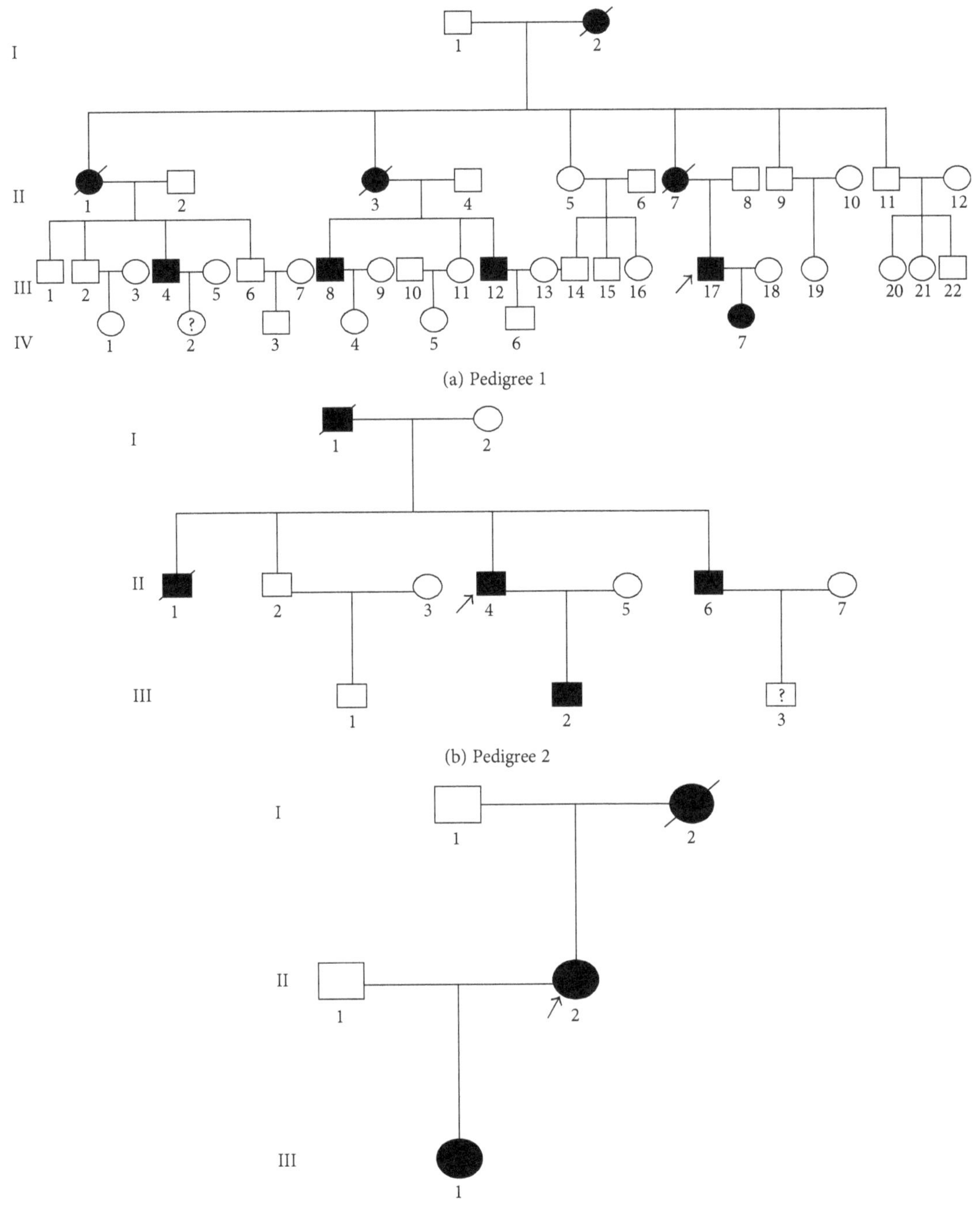

FIGURE 1: The proband pedigrees with MFS. +/− represents the heterozygous type; −/− represents the wild type. "Male" and "female" are indicated by squares and circles, respectively, and the filling symbols represent individuals affected with Marfan syndrome. The arrow shows the proband.

Children's Hospital 2 years before. The mother of the proband had a sudden death at the age of 45.

Further comparing the similarities and differences among these 3 families showed that there is no evidence of skeletal morbidity (including arm span to height > 1.05, scoliosis, and joint hypermobility) in either of the 3 families. Therefore, it is possible to propose that MFS is variable both among and within affected families in severity of cardiovascular manifestations, due to type of mutation.

3.2. Mutation Identification and Bioinformatic Analysis. In 3 families diagnosed with MFS, we identified independent heterozygous frameshift mutations of FBN1 (Figure 3(a)). None of the mutations existed in known databases (UMD-FBN1, HGMD, ClinVar, UCSC common SNP, db SNP, and the 1000-genome project) or in published articles.

A novel frameshift deletion c.4282delC (p.Arg1428AlafsX47) in CDS34 (chr15:48764802) was identified in family 1 individuals III: 17 and IV: 7 (Figure 3(b)). According to

Table 1: Clinical detail of the patients with FBN1 mutation in this study.

Individuals	Pedigree 1		Pedigree 2			Pedigree 3	
	III: 17	IV: 7	II: 6	II: 6	III: 2	II: 2	III: 1
Age (years)	33	10	45	41	14	29	5
Gender	Male	Female	Male	Male	Male	Female	Female
Height (cm)	187	151	173	175	185	182	130
Ocular system							
Ectopia lentis	–	–	+	–	–	–	–
Myopia	+	+	–	–	+	–	–
Cardiovascular system		–					
Diameter of aortic root (mm)	45	30	60	67	36	43	*
Aortic dissection	+	–	–	–	–	–	*
Mitral valve prolapse	–	–	–	+	+	–	*
Tricuspid valve prolapse	–	–	–	–	+	–	*
LVED (mm)	53	41	72	81	58	49	*
LVEF (%)	60	62	50	35	54	66	*
Systemic features							
Thumb sign	+	+	–	–	+	+	*
Wrist sign	+	+	–	–	+	+	*
Thin body	+	+	+	+	+	+	*
Arachnodactyly	+	+	–	–	+	+	*
Pectus excavatum; scoliosis	–	–	–	–	–	–	+

* represents unknown information.

Sakai et al. reported in 2006, a mutation c.4283dup (p.Cys1429LeufsX2) was observed in a Marfan patient from Japan [13]. Further analysis indicated that this mutation leads to truncation of the original 2871-amino acid full-length protein to a 1474-residue protein. This mutation is located between the epidermal growth factor- (EGF-) like domain, according to UniProtKB [14]. We used two algorithmic tools, MutationTaster and Swiss-Model, to examine the impact of this mutation. In the conservation test by MutationTaster [15], the affected residue in family 1 was revealed to be evolutionarily conserved, which implied that the alternation of amino acids in this position is likely to damage protein function. Furthermore, the mutant tertiary structure generated by Swiss-Model predicts that the domain where the mutation is located (1028–1474) differs from that of the wild-type protein (1028–1527) (Figure 4(a)). Therefore, this frameshift mutation was identified as being rare, probably disease causing, which might affect protein function by influencing the structure of the FBN1 protein or the binding between calcium and the cbEGF domain.

In family 2 individuals II: 3 and IV: 2, one de novo frameshift mutation, c.7_8insTC (p.Arg3LeufsX16), in exon 2 of FBN1 was identified (Figure 3(c)). This event resulted in the insertion of 2 base pairs, leading to a change from amino acids 3 to 18 and a deletion of big fragments from amino acids 19 to 2871, which could likely damage protein function overwhelmingly. Furthermore, MutationTaster showed this mutation to be pathogenic.

The sequencing results from family 3 individual II: 2 revealed a heterozygous frameshift mutation c.2192 delC (p.Pro731LeufsX41) in CDS18 of FBN1 (Figure 3(d)), which caused a translation of the protein to stop at the 771st amino acid residue, by the premature termination codon (PTC). Refer to UniProtKB; this mutation is also located between the cb EGF-like domain. The disease-causing potential of this variation was predicted automatically by MutationTaster. Swiss-Model predicts that the domain where the variant is located (723–771) is significantly different from that of the wild-type protein (723–846) (Figure 4(b)). These results indicate that c.2192 delC led to a truncation of the protein, which is associated with cardiovascular phenotype thoracic aortic aneurysms and dissections.

4. Discussion

Most patients with MFS will ultimately develop cardiovascular outcomes (aortic aneurysm, valve prolapse as well as LV dysfunction, etc.) and might further progress into aortic dissection, owing to the clinical variability in FBN1 gene mutations. Thereby, uncovering novel causal mutations and genotype–cardiovascular phenotype correlations would facilitate genetic counselling and allow early symptomatic treatment, in order to improve the prognosis of patients eventually [16].

In the present study, we selected 3 families with MFS and identified novel FBN1 mutations. Intriguingly, 3 novel frameshift PTC mutations caused typical and variable cardiovascular complications in affected family members, from severe to mild.

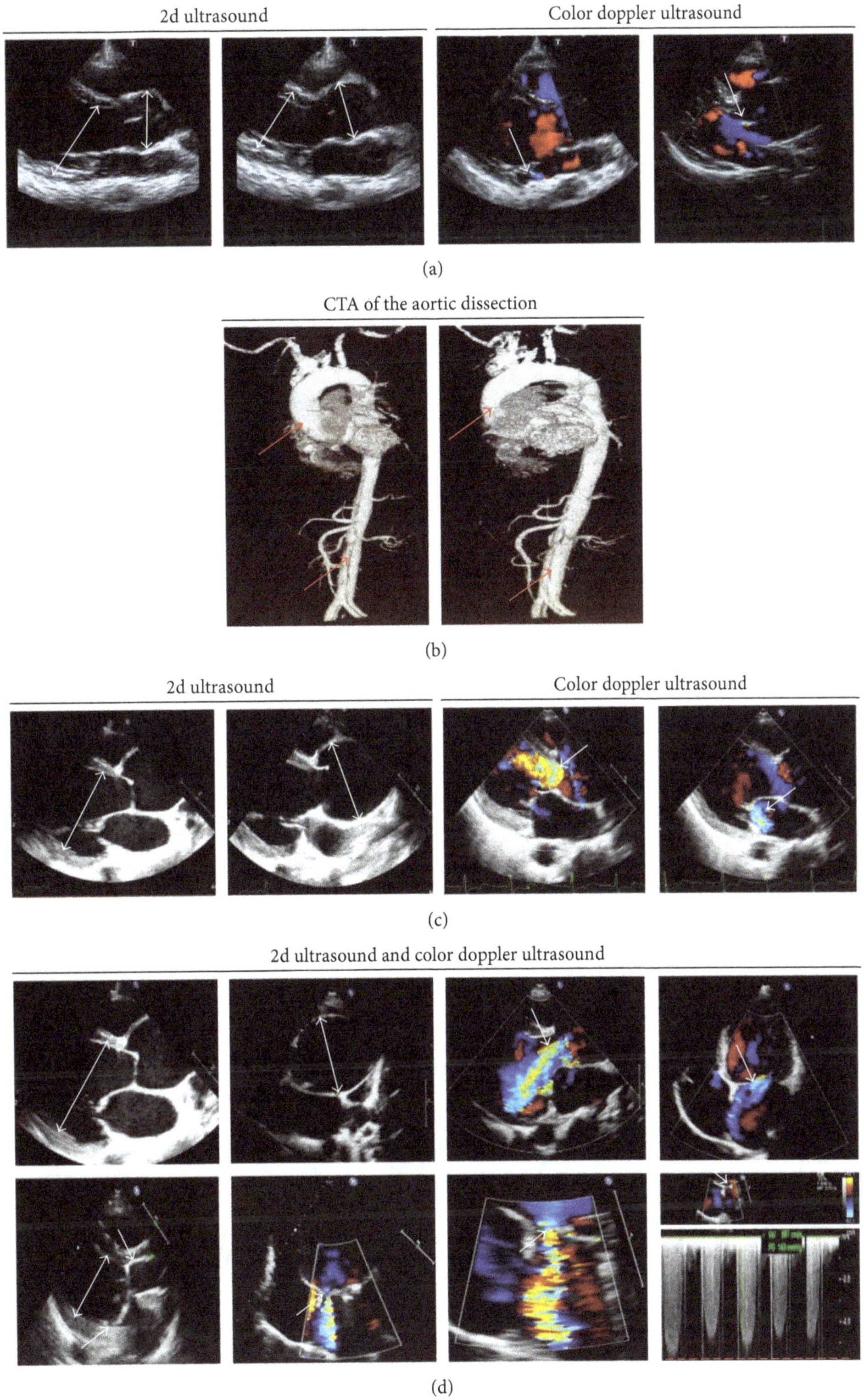

Figure 2: Echocardiography and computed tomography angiography findings for patients. (a) Transthoracic echocardiography at first diagnosis, showing a slight aortic root dilatation (arrow), mild aortic regurgitation (AR, arrow), and mild mitral regurgitation (MR, arrow) of proband in pedigree 1. (b) Aortography results after first checkup at 2 months. Transthoracic echocardiography at first diagnosis of proband in pedigree 2, showing a remarkable aortic root dilatation (arrow), massive aortic regurgitation, and mild mitral regurgitation (arrow). (d) Preoperative (top) and postoperative (bottom) transthoracic echocardiography of individual II: 6 in pedigree 2. Preoperative echocardiographic characteristics (top) showed a remarkable aortic root dilatation (arrow), massive aortic regurgitation, and mitral regurgitation (arrow). Transthoracic echocardiography after the first cardiac surgery (bottom) showed a moderate mitral periprosthetic leak (arrow) and still enlarged left ventricle despite a replaced artificial double valve (DVR).

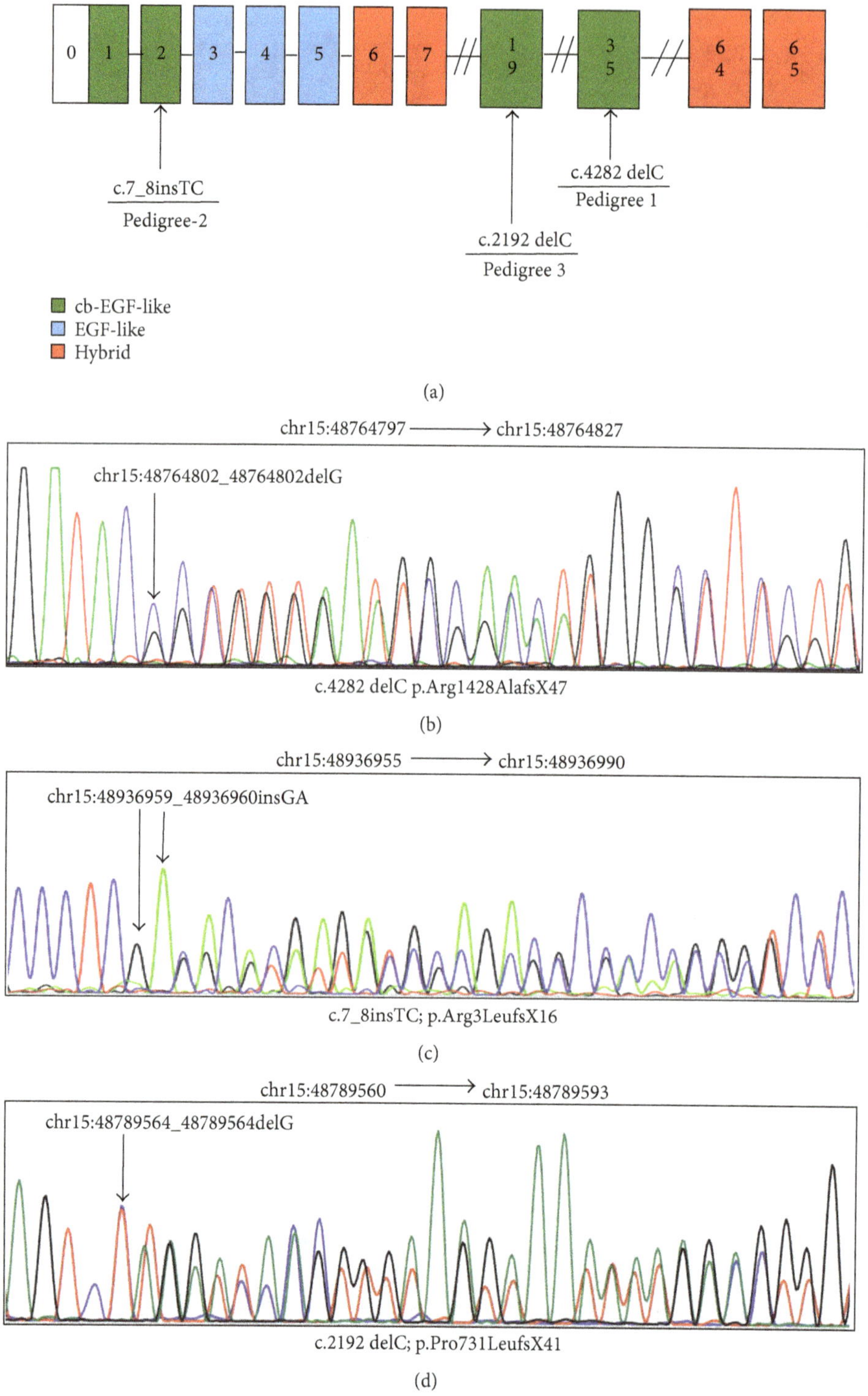

FIGURE 3: Localization of mutations in FBN1 and results from qualitative analysis. (a) Schematic presentation of FBN1 with the localization of the 3 mutations investigated in this study indicated. (b) Fragment of FBN1 cDNA sequence in a patient with c.4282 delC. (c) Fragment of FBN1 cDNA sequence in a patient with c.7_8insTC. (d) Fragment of FBN1 cDNA sequence in a patient with c.2192 delC.

Several genotype–phenotype correlations in cardiovascular involvement have been established. The early observation by Aoyama et al. showed that mutations leading to a very low deposition of the fibrillin-1 protein were associated with shortened event-free survival and more severe cardiovascular complications [17]. In addition, some previous publications have demonstrated a strong association of MFS patients with truncating FBN1 mutations with cardiovascular events [6, 18]. Moreover, recent reports discovered that patients with premature termination codon (classifiable as HI) have a worse prognosis, with increased risk for aortic surgery, aortic complications, and mortality compared with DN mutation

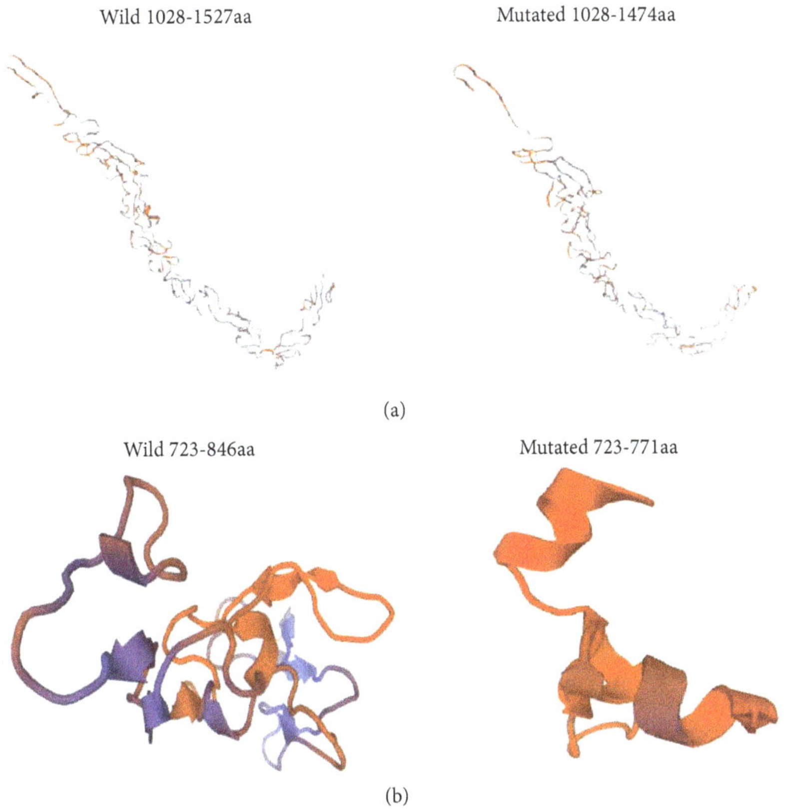

FIGURE 4: Predicted results for the frameshift variants by Swiss-Model. (a) Predicted 3D model of the variant (c.4282 delC.) in family 1. (b) Predicted 3D model of the variant (c.2192 delC.) in family 3.

carriers [19–21]. Alternatively, Wang et al. likewise observed an association with truncating mutations in MFs with cardiovascular defects [22]. For a severe ventricular phenotype, like LV dilatation, it has been reported to be related with the disorder of microfibril assembly by in-frame deletions or non-missense mutations [23, 24]. Likewise, the present study also identified similar results as in previous studies in 3 MFS pedigrees with frameshift mutations, thereby contributing to the extension of the known mutational spectrum of frameshift, for understanding the genotype–phenotype correlations in cardiovascular involvement.

In the current study, 3 novel frameshift variants, c.4282delC (p.Arg1428AlafsX47), c.7_8insTC (p.Arg3LeufsX16), and c.2192 delC (p.Pro731LeufsX41), appear to have a significant association with cardiovascular involvement and mild skeletal manifestations. Additionally, evidence from 3 pedigrees also showed varying degrees of cardiovascular phenotype owing to their own unique HI mutation. The mutation c.4282delC (p.Arg1428AlafsX47) in family 1 affects protein function by the destroyed protein structure and functions for binding of calcium and causes severe manifestations involving rapid but early onsets of aortic dissection, minor findings in both ocular and skeletal systems. Thus, c.4282delC might serve as a potential target for future research on patients with rapidly progressive aortic dissection. Another case of family 2, the proband and his young brother, though with a predominant dilated aortic sinus and LV enlargement (latent LV dysfunction, valvular insufficiency, and subsequent worse prognosis after surgery (Figure 4), have a survival time free of dissection until now. Interestingly, it is known that another mutation located on exon 2 in FBN1 results in a frameshift and premature termination of translation, and the proband likewise had major findings in the cardiovascular system, involvement of the skeletal system, and minor findings in the ocular system [21]. c.2192 delC (p.Pro731LeufsX41) mutation in patient II: 2 from family 3, which leads to a cb EGF-like domain destruction as well, just displayed a typical marfanoid dilation of aortic sinus at present, which should be monitored, since both the proband and her daughter are young, and the clinical phenotypes are not only variable but time-dependent.

Altogether, this study added 3 novel mutations to the existing spectrum of FBN1 mutations and demonstrated different frameshift genotype impacts on cardiovascular-phenotype severity in patients with MFS, which might facilitate understanding on potential phenotype–genotype associations and refine susceptible populations of deleterious cardiovascular events such as greater risk for early aortic dissection or markedly poor heart function even after prophylactic surgery. Therefore, extensive clinical and basic researches on the FBN1 mutation effects on fibrillin-1 protein are still warranted for precision medicine of cardiovascular risk classification, preventing higher-risk adverse outcomes and also avoiding unnecessary interventions in diverse patients with MFS, so as to achieve an increasingly tailored patient-specific management eventually.

Conflicts of Interest

The authors declare that they have no conflict of interest.

Authors' Contributions

Yueli Wang and Xiaoyan Li contributed equally to this work.

Acknowledgments

The study was supported by the National Natural Science Foundation of China with Grants 81501486 and 81400846 and the National Key R&D Program of China (2016YFC 0903000).

References

[1] R. Franken, G. Teixido-Tura, M. Brion et al., "Relationship between fibrillin-1 genotype and severity of cardiovascular involvement in Marfan syndrome," *Heart*, vol. 103, no. 22, pp. 1795–1799, 2017.

[2] A. Verstraeten, M. Alaerts, L. van Laer, and B. Loeys, "Marfan syndrome and related disorders: 25 years of gene discovery," *Human Mutation*, vol. 37, no. 6, pp. 524–531, 2016.

[3] G. Collod-Béroud, S. le Bourdelles, L. Ades et al., "Update of the UMD-*FBN1* mutation database and creation of an *FBN1* polymorphism database," *Human Mutation*, vol. 22, no. 3, pp. 199–208, 2003.

[4] P. N. Robinson and M. Godfrey, "The molecular genetics of Marfan syndrome and related microfibrillopathies," *Journal of Medical Genetics*, vol. 37, no. 1, pp. 9–25, 2000.

[5] H. C. Dietz, "Potential phenotype-genotype correlation in Marfan syndrome," *Circulation: Cardiovascular Genetics*, vol. 8, no. 2, pp. 256–260, 2015.

[6] L. M. Baudhuin, K. E. Kotzer, and S. A. Lagerstedt, "Increased frequency of *FBN1* truncating and splicing variants in Marfan syndrome patients with aortic events," *Genetics in Medicine*, vol. 17, no. 3, pp. 177–187, 2015.

[7] B. J. Landis, G. R. Veldtman, and S. M. Ware, "Genotype–phenotype correlations in Marfan syndrome," *Heart*, vol. 103, no. 22, pp. 1750–1752, 2017.

[8] L. Faivre, G. Collod-Beroud, B. L. Loeys et al., "Effect of mutation type and location on clinical outcome in 1,013 probands with Marfan syndrome or related phenotypes and *FBN1* mutations: an international study," *The American Journal of Human Genetics*, vol. 81, no. 3, pp. 454–466, 2007.

[9] R. Franken, T. J. Heesterbeek, V. de Waard et al., "Diagnosis and genetics of Marfan syndrome," *Expert Opinion on Orphan Drugs*, vol. 2, no. 10, pp. 1049–1062, 2014.

[10] B. L. Loeys, H. C. Dietz, A. C. Braverman et al., "The revised Ghent nosology for the Marfan syndrome," *Journal of Medical Genetics*, vol. 47, no. 7, pp. 476–485, 2010.

[11] J. M. Schwarz, C. Rödelsperger, M. Schuelke, and D. Seelow, "MutationTaster evaluates disease-causing potential of sequence alterations," *Nature Methods*, vol. 7, no. 8, pp. 575-576, 2010.

[12] R. Erbel, V. Aboyans, C. Boileau et al., "2014 ESC guidelines on the diagnosis and treatment of aortic diseases: document covering acute and chronic aortic diseases of the thoracic and abdominal aorta of the adult. The Task Force for the Diagnosis and Treatment of Aortic Diseases of the European Society of Cardiology (ESC)," *European Heart Journal*, vol. 35, no. 41, pp. 2873–2926, 2014.

[13] H. Sakai, R. Visser, S. Ikegawa et al., "Comprehensive genetic analysis of relevant four genes in 49 patients with Marfan syndrome or Marfan-related phenotypes," *American Journal of Medical Genetics Part A*, vol. 140, no. 16, pp. 1719–1725, 2006.

[14] "UniProtKB," June 2017, http://www.uniprot.org/uniprot/P35555.

[15] "MutationTaster," June 2017, http://www.mutationtaster.org/.

[16] M. N. Singh and R. V. Lacro, "Recent clinical drug trials evidence in Marfan syndrome and clinical implications," *Canadian Journal of Cardiology*, vol. 32, no. 1, pp. 66–77, 2016.

[17] T. Aoyama, U. Francke, C. Gasner, and H. Furthmayr, "Fibrillin abnormalities and prognosis in Marfan syndrome and related disorders," *American Journal of Medical Genetics*, vol. 58, no. 2, pp. 169–176, 1995.

[18] J. Wang, Y. Yan, J. Chen et al., "Novel *FBN1* mutations are responsible for cardiovascular manifestations of Marfan syndrome," *Molecular Biology Reports*, vol. 43, no. 11, pp. 1227–1232, 2016.

[19] R. Franken, M. Groenink, V. de Waard et al., "Genotype impacts survival in Marfan syndrome," *European Heart Journal*, vol. 37, no. 43, pp. 3285–3290, 2016.

[20] L. Y. Sakai, D. R. Keene, M. Renard, and J. de Backer, "*FBN1*: the disease-causing gene for Marfan syndrome and other genetic disorders," *Gene*, vol. 591, no. 1, pp. 279–291, 2016.

[21] Y. Wang, S. Chen, R. Wang et al., "Postmortem diagnosis of Marfan syndrome in a case of sudden death due to aortic rupture: detection of a novel *FBN1* frameshift mutation," *Forensic Science International*, vol. 261, pp. e1–e4, 2016.

[22] W. J. Wang, P. Han, J. Zheng et al., "Exon 47 skipping of *fibrillin-1* leads preferentially to cardiovascular defects in patients with thoracic aortic aneurysms and dissections," *Journal of Molecular Medicine*, vol. 91, no. 1, pp. 37–47, 2013.

[23] J. J. J. Aalberts, J. P. van Tintelen, L. J. Meijboom et al., "Relation between genotype and left-ventricular dilatation in patients with Marfan syndrome," *Gene*, vol. 534, no. 1, pp. 40–43, 2014.

[24] F. Loeper, J. Oosterhof, M. van den Dorpel et al., "Ventricular-vascular coupling in Marfan and non-Marfan aortopathies," *Journal of the American Heart Association*, vol. 5, no. 11, article e003705, 2016.

2

Mutations in Homocysteine Metabolism Genes Increase Keratin *N*-Homocysteinylation and Damage in Mice

Kamila Borowczyk,[1,2] **Jacek Wróblewski**,[3,4] **Joanna Suliburska**,[5] **Noriyuki Akahoshi**,[6] **Isao Ishii**,[6] **and Hieronim Jakubowski**[1,4]

[1]*Department of Microbiology, Biochemistry and Molecular Genetics, Rutgers-New Jersey Medical School, International Center for Public Health, Newark, NJ 07103, USA*
[2]*Department of Environmental Chemistry, Faculty of Chemistry, University of Łódź, 90-236 Łódź, Poland*
[3]*Institute of Bioorganic Chemistry, 71-704 Poznań, Poland*
[4]*Department of Biochemistry and Biotechnology, Poznań University of Life Sciences, 60-632 Poznań, Poland*
[5]*Institute of Human Nutrition and Dietetics, Poznań University of Life Sciences, 60-632 Poznań, Poland*
[6]*Department of Health Chemistry, Showa Pharmaceutical University, Tokyo 194-8543, Japan*

Correspondence should be addressed to Hieronim Jakubowski; jakubows@njms.rutgers.edu

Academic Editor: Michael Nonnemacher

Genetic or nutritional deficiencies in homocysteine (Hcy) metabolism increase Hcy-thiolactone, which causes protein damage by forming isopetide bonds with lysine residues, generating *N*-Hcy-protein. In the present work, we studied the prevalence and genetic determinants of keratin damage caused by homocysteinylation. We found that in mammals and birds, 35 to 98% of Hcy was bound to hair keratin via amide or isopeptide bond (Hcy-keratin), while 2 to 65% was *S*-Hcy-keratin. A major fraction of hair Hcy-keratin (56% to 93%), significantly higher in birds than in mammals, was sodium dodecyl sulfate-insoluble. Genetic hyperhomocysteinemia significantly increased *N*-Hcy-keratin levels in the mouse pelage. *N*-Hcy-keratin was elevated 3.5-, 6.3-, and 11.7-fold in hair from $Mthfr^{-/-}$, $Cse^{-/-}$, or $Cbs^{-/-}$ mice, respectively. The accumulation of *N*-Hcy in hair keratin led to a progressive reduction of *N*-Hcy-keratin solubility in sodium dodecyl sulfate, from 0.39 ± 0.04 in wild-type mice to 0.19 ± 0.03, 0.14 ± 0.01, and 0.07 ± 0.03 in $Mthfr^{-/-}$, $Cse^{-/-}$, or $Cbs^{-/-}$animals, respectively. *N*-Hcy-keratin accelerated aggregation of unmodified keratin in $Cbs^{-/-}$ mouse hair. Keratin methionine, copper, and iron levels in mouse hair were not affected by hyperhomocysteinemia. These findings provide evidence that pelage keratin is *N*-homocysteinylated *in vivo* in mammals and birds, and that this process causes keratin damage, manifested by a reduced solubility.

1. Introduction

Homocysteine (Hcy) is an important intermediate in folate and one-carbon metabolism. The only known source of Hcy in our body is the essential dietary protein amino acid methionine (Met). Hcy levels are regulated by vitamin B_{12}- and folate-dependent remethylation to Met, catalyzed by Met synthase (with methyltetrahydrofolate cofactor provided by the MTHFR enzyme) and betaine-Hcy methyltransferase, as well as by vitamin B_6-dependent transsulfuration to cysteine, the first step of which is catalyzed by cystathionine β-synthase (CBS). Genetic or nutritional deficiencies in folate/one-carbon metabolism lead to the hyperhomocysteinemia (HHcy) and are known to cause abnormalities in many organs, including the cardiovascular system and the brain [1].

Hcy is also metabolized to the thioester Hcy-thiolactone in an error-editing reaction in protein biosynthesis when Hcy is erroneously selected in place of Met by methionyl-tRNA synthetase [2]. Because Hcy-thiolactone is chemically reactive, it modifies ε-amino groups of protein lysine residues, which generates *N*-homocysteinylated protein (*N*-Hcy-protein) [2]. *N*-Homocysteinylation is an emerging posttranslational protein modification [3] that impairs or alters the protein's structure/function, causes protein damage *in vitro* [2, 4] and *in vivo* [5–7], generates amyloid-like structures [2, 8, 9], and induces proatherogenic changes in gene expression [10], an autoimmune response [2, 11], and atherothrombosis [2, 12, 13].

Although homocysteinylation can cause protein damage, the prevalence and genetic determinants of Hcy-damaged proteins *in vivo* are not well known. For this reason, the present work has been undertaken to study hair keratin from different species of mammals and birds, as well as mouse models of genetic HHcy ($Cbs^{-/-}$, $Cse^{-/-}$, and $Mthfr^{-/-}$ mice).

2. Materials and Methods

2.1. Reagents. *L*-Hcy-thiolactone·HCl, *D,L*-Hcy, horse spleen ferritin, dithiothreitol (DTT), *o*-phthaldialdehyde (OPA), sodium dodecyl sulfate (SDS), HCl, and trichloroacetic acid were purchased from Sigma-Aldrich. Suprapure nitric acid (65%) and perchloric acid (60%) were from Merck. Reagents were prepared in Milli-Q purified water.

2.2. Animal Hair. The hair samples have been collected from live domestic animals and from wild animals encountered in nature as a fresh road kill. Random feathers dropped by wild birds were collected in nature. Woolly mammoth hair was purchased from Educational Innovations.

Cystathionine *β*-synthase-deficient ($Cbs^{-/-}$) [14] and cystathionine *γ*-lyase-deficient ($Cse^{-/-}$) [15] mice on C57BL/6J background, methyltetrahydrofolate reductase-deficient ($Mthfr^{-/-}$) mice on a mixed BALB/cAnNCrlBR [16], and the corresponding wild-type control animals, 9- to 12-week-old, were used as hair donors. Hair samples were sheared off from the mouse sides.

2.3. Sample Preparation for Hair S-Hcy-Keratin and Hcy-Keratin Assays. Samples were prepared by a modification of previously described procedures as described below. The interassay and intra-assay variabilities for the quantification of various forms of Hcy were 7.3 and 11.5% [17].

2.4. S-Hcy-Keratin. Hair or feathers (2 mg) were treated in a 0.5 ml Eppendorf polyethylene tube with 50 mM Na_2HPO_4, 20 mM NaOH, 25mM DTT, and 1% SDS (200 *μ*l) for 1 h at 65°C. The extracts were collected by centrifugation, hair were extracted again with a fresh solution, and the extracts were combined (400 *μ*l). The hair pellets were saved for quantification of SDS-insoluble *N*-Hcy-keratin.

A 40 *μ*l portion of each extract was treated with DTT (4 *μ*l, 0.25M) and HCl (4 *μ*l, 12 N) at 100°C for 30 min to convert the liberated Hcy to Hcy-thiolactone, which was then quantified by HPLC [17]. Authentic samples of Hcy were similarly processed as standards for *S*-Hcy assays. This procedure liberates >95% Hcy present in hair as *S*-Hcy-keratin.

2.5. SDS-Soluble and SDS-Insoluble Hcy-Keratin. For SDS-soluble Hcy-keratin quantification, each SDS extract (300 *μ*l) was supplemented with 34 *μ*l 100% trichloroacetic acid and the precipitated keratin pellets were collected by centrifugation. SDS-soluble and SDS-insoluble pellets were transferred to 1 ml Wheaton Gold Band ampoules and hydrolyzed with 6N HCl and 50 mM DTT (110 *μ*l, 120°C, 1 h) to liberate Hcy. The hydrolysates were lyophilized on Labconco CentriVap (40 min, 70°C), dissolved in 10 *μ*l water, purified by two-dimensional TLC, and analyzed by HPLC. Horse spleen ferritin, containing 0.49 mol Hcy/mol protein, was processed in parallel as a standard. In these procedures, *N*-Hcy is quantitatively converted to Hcy-thiolactone, which is then assayed by HPLC.

2.6. HPLC, Detection, and Quantification. Quantification of Hcy-thiolactone generated from Hcy-keratin or *S*-Hcy-keratin was carried out as previously described [17] using a Beckman-Coulter System Gold Nouveau HPLC instrumentation with a manual injector (7725i Rheodyne, with 0.1 ml loop) and a Jasco 1520 fluorescence detector, controlled by a Gold Nouveau chromatography workstation software for Windows. Samples were injected onto a cation-exchange polysulfoethyl aspartamide column (35 × 2 mm, 5 *μ*m, 300 Å) (PolyLC Inc.), eluted isocratically with 30 mM NaCl and 10 mM sodium phosphate buffer (pH 6.6) at a flow rate 0.6 ml/min. The effluent was mixed in a three-way tee with 2.5 mM OPA and 0.25 M NaOH, delivered at a flow rate 0.3 ml/min, the mixture passed through a reaction coil (Teflon tubing, 0.3 mm I.D. × 3 m), and the fluorescence at 480 nm was recorded (excitation 370 nm). Hcy-thiolactone eluted at 3 min in a 4 min run.

2.7. Met-Keratin Assays. The Met HPLC assay [18] has been used with the following modification. Briefly, mouse pelage hair (~10 mg) were hydrolyzed in 1 ml Wheaton Gold Band ampoules with 6N HCl (100 *μ*l, 120°C, 1 h). The hydrolysates were dried out, dissolved in 50 *μ*l 0.2 M sodium phosphate buffer (pH 7.4), reduced with 2 *μ*l 0.25 M TCEP for 10 min, and supplemented with 10 *μ*l 0.5 M NAC. A Hewlett-Packard (Waldbronn, Germany) 1100 Series System, controlled by the HP ChemStation software, containing quaternary pump, autosampler, temperature control, vacuum degasser, and 1260 Series FL detector was used. Samples were analyzed using a reversed phase PRP-1 column (150 × 4.6 mm, 5 *μ*m; Hamilton, Energy Way, Reno, NV, USA) eluted at a flow rate 1 ml/min, 25°C, with 0.01 M OPA and 0.1 M NaOH (A) and acetonitrile (B) as follows: 0–8 min, 14–25% (B); 8–12 min, 25% (B); 12–14 min, 25–14% (B). The effluent was monitored by fluorescence (exc. 348 nm, em. 438 nm) from 0 to 7.2 min. Met was identified by coelution (at 6.2 min) with an authentic standard.

2.8. Total SDS-Soluble Keratin. Total SDS-soluble keratin was prepared by extracting mouse hair (2 mg) twice with 50 mM Na_2HPO_4, 20 mM NaOH, 25 mM DTT, and 1% SDS (200 *μ*l, 65°C, 1 h each). The extracts were collected by centrifugation and combined (400 *μ*l). A 20 *μ*l portion of each extract was analyzed by SDS-PAGE on 10% gels. Keratin bands were visualized by staining with Coomassie Blue and quantified by densitometry.

2.9. Copper and Iron Assays. Mouse hair samples were mineralized with 65% nitric acid (Merck) in a microwave oven (Mars 5 Digestion Microwave System, CEM Corporation). Iron and copper were quantified by flame atomic absorption

Table 1: Hcy- and *S*-Hcy-keratin levels and solubility in animal hair.

Animal (*n*)	Hcy-keratin (pmol/mg hair)	*S*-Hcy-keratin (pmol/mg hair)	Hcy-keratin/ (Hcy-keratin + *S*-Hcy-keratin)	SDS-insoluble Hcy-keratin fraction
Cat (5)	278 ± 157	422 ± 214	0.60 ± 0.16	0.70 ± 0.05
Cow (5)	556 ± 552	112 ± 54	0.70 ± 0.28	0.87 ± 0.11
Chipmunk (3)	200 ± 193	395 ± 189	0.56 ± 0.04	0.73 ± 0.01
Dog (3)	430 ± 80	386 ± 456	0.62 ± 0.29	0.69 ± 0.08
Goat (2)	177 ± 127	217 ± 16	0.42 ± 17	0.80± 0.10
Guinea pig (1)	56	253	0.18	0.67
Horse, tail hair base (10)	189 ± 230	48 ± 40	0.30 ± 0.25	0.92 ± 0.05
Horse, tail hair tip 60 cm (10)	1733 ± 882	31 ± 5	0.96 ± 0.01	1.00 ± 0.00
Mammoth (3)	688 ± 72	49 ± 20	0.93 ± 0.03	0.67 ± 0.20
Opossum (2)	484 ± 147			
Pig (2)	210 ± 206	131 ± 117	0.60 ± 0.04	0.67 ± 0.12
Rabbit (3)	146 ± 75	265 ± 59	0.35 ± 0.16	0.76 ± 0.03
Raccoon (3)	147 ± 58	212 ± 50	0.40 ± 0.04	0.66 ± 0.04
Sheep (3)	310 ± 15	271 ± 36	0.53 ± 0.02	0.67 ± 0.19
Squirrel (3)	352 ± 433			
Woodchuck (3)	733 ± 483	531 ± 474	0.57 ± 0.16	0.75 ± 0.09
Robin (4)	203 ± 134	76	0.80	0.92
Raven (1)	246	62	0.80	0.93
Turkey vulture (1)	125	72	0.63	0.81
Pigeon (1)	716	87	0.89	0.91
Seagull (1)	493	114	0.81	0.78

Table 2: Comparison of Hcy- and *S*-Hcy-keratin levels and solubility in mammal and bird pelage.

Animals (number of species, *n*)	Hcy-keratin (pmol/mg hair)	*S*-Hcy-keratin (pmol/mg hair)	Hcy-keratin/ (Hcy-keratin + *S*-Hcy-keratin)	SDS-insoluble Hcy-keratin fraction
Mammals (12)	277 ± 202	247 ± 146	0.47 ± 0.15	0.75 ± 0.12
Birds (5)	556 ± 552	112 ± 54	0.79 ± 0.10	0.87 ± 0.07
t-test	0.518	0.037	0.0003	0.019

spectrometry using a Zeiss AAS-3 spectrometer with deuterium background correction as previously described [19]. The accuracy of the assay was 94% for iron and 102% for copper as verified by certified reference materials (Human Hair NCS DC73347a, LGC Standards).

2.10. Statistics. The results are reported as mean ± standard deviation. Comparisons between two groups are analyzed by using two-sided Student's *t*-test. Relationships of Hcy-keratin solubility vs. levels were fitted to logarithmic equations and analyzed by linear regression. The level of statistical significance was set to $P < 0.05$.

3. Results

3.1. Hcy-Keratin and S-Hcy-Keratin Content in Animal Hair. Our previous work has shown that Hcy, bound by a disulfide or an amide bond, is a component of human hair keratin [18]. To determine whether the presence of Hcy in hair keratin is a universal feature conserved across the animal world, we analyzed Hcy content in pelages of mammals and birds. We found that both Hcy-keratin and *S*-Hcy-keratin were present both in mammal and bird pelages at levels ranging from 56 to 733 pmol/mg hair (Table 1). The higher content of Hcy-keratin and lower content of *S*-Hcy-keratin at the tip of the horse tail hair than at its base are similar to those observed previously in the long human head hair [18].

While an average Hcy-keratin level was similar in mammals and birds (277 ± 202 vs. 357 ± 244 pmol/mg hair, $P = 0.518$), significantly more *S*-Hcy-keratin was present in mammals than in birds (247 ± 146 vs. 82 ± 20 pmol/mg hair, $P = 0.037$), so that in birds, Hcy-keratin represented a significantly higher fraction of total keratin Hcy content than in mammals (0.79 ± 0.10 vs. 0.47 ± 0.15, $P = 0.0003$) (Table 2).

To determine whether the variation in keratin homocysteinylation across the mammalian and bird species can alter the protein structure, we measured the solubility of Hcy-keratin in sodium dodecyl sulfate (SDS) and examined how the solubility is affected by the level of homocysteinylation. We found that SDS-insoluble Hcy-keratin

Table 3: Hcy-keratin and S-Hcy-keratin levels and the solubility of hair Hcy-keratin in HHcy and control wild-type mice.

Genotype (n)	Hcy-keratin (pmol/mg hair)	S-Hcy-keratin (pmol/mg hair)	Hcy-keratin/ (Hcy-keratin + S-Hcy-keratin)	SDS-soluble Hcy-keratin fraction
$Cbs^{-/-}$ (4)	$1014 \pm 86^{*}$	$1937 \pm 336^{*}$	0.35 ± 0.03	$0.07 \pm 0.03^{*}$
$Cse^{-/-}$ (5)	$546 \pm 46^{*}$	$682 \pm 144^{*}$	0.55 ± 0.04	$0.14 \pm 0.01^{*}$
$Mthfr^{-/-}$ (7)	$302 \pm 71^{\#}$	$248 \pm 121^{\#}$	0.57 ± 0.17	$0.19 \pm 0.03^{\#}$
Wild type + Met (6)	97 ± 11	$303 \pm 58^{*}$	0.24 ± 0.02	0.31 ± 0.07
Wild type (28)	87 ± 10	127 ± 10	0.41 ± 0.21	0.36 ± 0.09

*,#Significantly different from wild type; $^{*}P < 0.001$, $^{\#}P < 0.05$.

varied from 78 to 93% in birds and from 56 to 92% in mammals (Table 1). Overall, birds had significantly more SDS-insoluble Hcy-keratin than mammals (0.87 ± 0.07 vs. 0.75 ± 0.12, $P = 0.019$; Table 2).

3.2. Increased Keratin Homocysteinylation and Reduced Solubility in Hyperhomocysteinemia. HHcy is known to increase the generation of Hcy-thiolactone, which modifies protein lysine residues by *N*-homocysteinylation. To examine how *N*-homocysteinylation of keratin affects its structure *in vivo*, we used mouse models of HHcy. We found that genetic HHcy significantly increased the accumulation of Hcy-keratin and *S*-Hcy-keratin in mouse pelage. In hair from $Cbs^{-/-}$, $Cse^{-/-}$, or $Mthfr^{-/-}$ mice, Hcy-keratin was elevated 11.7-, 6.3-, and 3.5-fold, respectively; *S*-Hcy-keratin was elevated 15-, 5.4-, or 2.0-fold, respectively (Table 3). We also found that dietary HHcy induced in wild-type mice by providing 1% Met in drinking water, significantly elevated *S*-Hcy-keratin levels (127 ± 10 vs. 303 ± 58 pmol/mg hair, $P < 0.001$) and had no significant effect on Hcy-keratin levels (87 ± 10 vs. 97 ± 11 pmol/mg hair) (Table 3).

The accumulation of Hcy in keratin led to a significant reduction of its solubility in SDS, from 0.39 ± 0.04 in wild-type mice to 0.19 ± 0.03, 0.14 ± 0.01, and 0.07 ± 0.03 in $Mthfr^{-/-}$, $Cse^{-/-}$, and $Cbs^{-/-}$animals, respectively (Table 3). There was a strong correlation between the solubility of *N*-Hcy-keratin in SDS and *N*-Hcy-keratin levels (Figure 1).

SDS-PAGE analysis showed that similar amounts of SDS-soluble total keratin could be extracted from hair of $Cbs^{-/-}$ mice and their $Cbs^{+/+}$ littermates (Table 4). However, after a 6-month long storage of the hair at 25°C, significantly less of SDS-soluble total keratin was extracted from hair of $Cbs^{-/-}$ mice than from hair of their $Cbs^{+/+}$ littermates (1.51 ± 0.14 vs. 2.22 ± 0.05, $P = 4.2 \times 10^{-6}$) (Figure 2, Table 4).

3.3. Hair Copper, Iron, and Met-Keratin Are Not Affected by Hyperhomocysteinemia. In a growing human hair, Hcy-keratin levels increase along the hair shaft as a result of copper and iron-dependent demethylation of methionine residues of keratin [18]. To determine whether this process could contribute to the accumulation of Hcy-keratin in HHcy, we quantified copper and iron levels in hair as well as Met content of hair keratin in $Cse^{-/-}$ and $Cse^{+/+}$ mice. We found that there were no differences in copper (4.83 ± 0.70 vs. $5.37 \pm 1.43\,\mu$g/g hair, $P = 0.296$) and iron (34.6 ± 10.7 vs. $40.1 \pm 16.0\,\mu$g/g hair, $P = 0.373$) levels between $Cse^{-/-}$ and $Cse^{+/+}$ mice (Table 5). We also found that Met-keratin

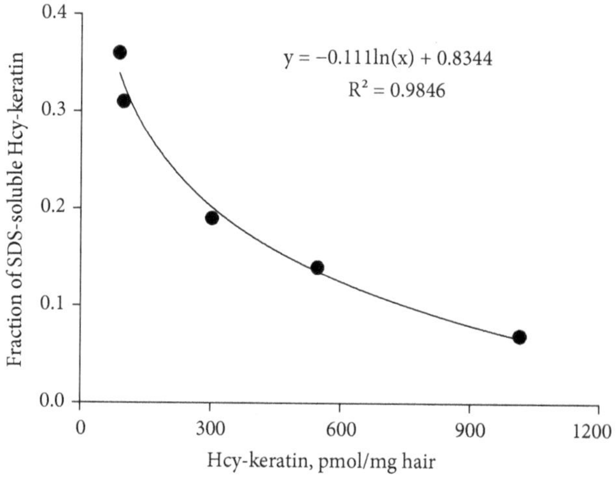

Figure 1: A relationship between the solubility of Hcy-keratin and the extent of keratin homocysteinylation. Mean values of Hcy-keratin solubility are plotted against mean Hcy-keratin levels (data from Table 3).

levels were not affected by the inactivation of the *Cse* gene (41.0 ± 10.2 in $Cse^{-/-}$ vs. 38.8 ± 6.4 nmol/mg hair in $Cse^{+/+}$ mice, $P = 0.571$) (Table 5).

4. Discussion

The present work shows that Hcy-keratin is highly prevalent in a variety of mammal and bird species. Although animal pelages contain both *S*-Hcy-keratin and Hcy-keratin, mammals have significantly more *S*-Hcy-keratin and less of the Hcy-keratin fraction than birds. At the same time, birds have significantly more SDS-insoluble Hcy-keratin than mammals, possibly suggesting higher keratin damage in birds than in mammals. We also show that the *Cbs*, *Cse*, and *Mthfr* genes control the extent of keratin *N*-homocysteinylation and damage in mice.

In hair, two mechanisms can account for the presence of Hcy linked by an amide bond to keratin: (i) *N*-homocysteinylation of keratin lysine residues by Hcy-thiolactone and (ii) copper- and iron-dependent demethylation of keratin methionine residues [18]. Although we have not directly shown that Hcy is attached to a lysine residue in keratin, we have eliminated the 2nd possibility by showing that copper, iron, and keratin methionine levels are not affected by HHcy in the $Cse^{-/-}$ mice (Table 5). In conjunction with previous findings that HHcy increases Hcy-thiolactone synthesis and

TABLE 4: Effect of mouse *Cbs* genotype on the solubility of total hair keratin.

Genotype (n)	Total SDS-soluble keratin, relative values (n)	
	Hair stored at −80°C	Hair stored at 25°C, 6 months
$Cbs^{-/-}$ (5–8)	1.17 ± 0.11 (8)	1.51 ± 0.14 (5)
$Cbs^{+/+}$ (5–8)	1.18 ± 0.11 (8)	2.22 ± 0.05 (5)
P value	0.842	4.2×10^{-6}

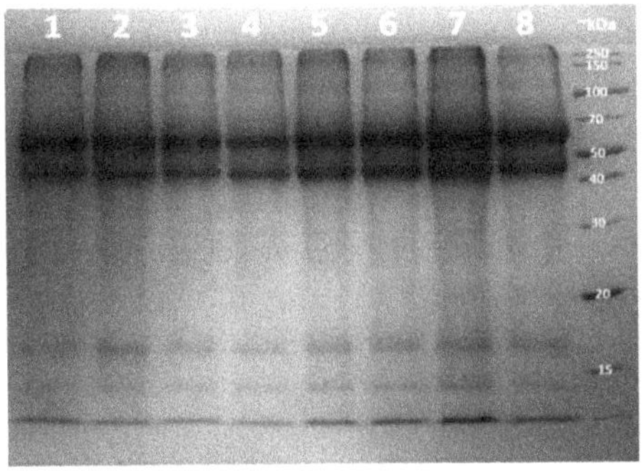

FIGURE 2: SDS-PAGE analysis of hair keratin from $Cbs^{-/-}$ and $Cbs^{+/+}$ mice. Keratin was extracted from mouse hair (stored at 25°C for 6 months) with SDS/DTT solution [18], and the extracts were analyzed on 10% SDS-PAGE gels. Lanes 1–4, keratin from $Cbs^{-/-}$ mice; lanes 5–8, keratin from their $Cbs^{+/+}$ littermates. Quantification by densitometry is shown in Table 4. The last lane on the right shows molecular weight markers.

TABLE 5: Levels of copper, iron, and Met-keratin in $Cse^{-/-}$ and $Cse^{+/+}$ mouse pelage hair.

Genotype (n)	Copper (μg/g hair)	Iron (μg/g hair)	Met-keratin (nmol/mg hair)
$Cse^{-/-}$ (10)	4.83 ± 0.70	34.6 ± 10.7	41.0 ± 10.2
$Cse^{+/+}$ (12)	5.37 ± 1.43	40.1 ± 16.0	38.8 ± 6.4
P value	0.296	0.373	0.571

protein *N*-homocysteinylation in mice and humans [2], our present data strongly suggest that the increased hair Hcy-keratin levels in HHcy mice are due to keratin *N*-homocysteinylation.

The higher content of Hcy-keratin and lower content of *S*-Hcy-keratin at the tip of the horse tail hair than at its base are reminiscent of those observed in the long human head hair [18]. Elevated Hcy-keratin at the tip of the horse tail hair relative to its base is most likely caused by demethylation of methionine residues in hair keratin, as we have previously shown for the human hair.

Previous work shows that *N*-homocysteinylation alters protein structure/function and leads to the generation of insoluble protein aggregates [20] with amyloid-like properties [2, 8]. The mechanism underlying these structural changes involves free Hcy thiol in *N*-Hcy-protein, which is prone to one-electron redox reactions that generate radicals and/or radial ions [21]. These radicals promote the formation of disulfide bonds leading to protein multimerization [20] and generate other reaction products [4]. Thiyl radicals produced from *N*-Hcy-protein undergo hydrogen atom transfer reactions, which generate C^{α}-centered radicals, well-known precursors of protein carbonyls [4, 21], which can account for an increased susceptibility of *N*-Hcy-proteins to oxidative damage [2]. These processes can also account for the findings of the present work showing that increasing *N*-Hcy-keratin levels in mouse hair leads to a progressive loss of its solubility.

To substantiate a conclusion that *N*-homocysteinylation causes keratin damage, we used mouse models of HHcy in which *N*-homocysteinylation is induced by the inactivation of the *Cse* (the present work, Table 3), *Cbs*, and *Mthfr* genes [2]. These mice have 3- to12-fold higher *N*-Hcy-keratin levels than the wild-type animals. Our findings that the solubility of *N*-Hcy-keratin in SDS is inversely correlated with the *N*-Hcy-keratin content in these mice suggests that the increasing extent of *N*-homocysteinylation causes progressive keratin damage. This damage is also reflected by significantly reduced solubility of total keratin from hair of $Cbs^{-/-}$ mice compared to their $Cbs^{+/+}$ littermates (Table 4).

The $Cbs^{-/-}$ mice have sparse hair with a significantly smaller diameter, compared to the $Cbs^{+/+}$ animals [14, 22]. It is likely that this morphological hair defect could be caused, at least in part, by the physicochemical impairments of keratin structure caused by greatly increased *N*-homocysteinylation in $Cbs^{-/-}$ mice, which is reflected by its reduced solubility in SDS (Tables 3 and 4).

Assuming that 1 mg hair contains about 20 nmol keratin, the 32% loss of solubility of total keratin (Table 4) corresponds to 6.4 nmol/mg hair more of SDS-insoluble keratin in $Cbs^{-/-}$ mice relative to their $Cbs^{+/+}$ littermates. This value is 6-fold greater than the *N*-Hcy-keratin content in $Cbs^{-/-}$ mouse hair (1014 ± 86 pmol/mg hair, Table 3), suggesting that the aggregated *N*-Hcy-keratin (Table 3) serves as a "seed" that causes aggregation and a loss of solubility of unmodified keratin. These findings strongly suggest that the mechanism of protein aggregation induced by *N*-homocysteinylation demonstrated *in vitro* for the aggregation of albumin in the presence of a small amount of *N*-Hcy-albumin "seed" [8] also occurs *in vivo* in the mouse hair.

Protein damage can also be induced by *S*-thiolation, as recently shown for human serum albumin from hyperlipidemia patients, which carries Hcy and Cys bound via a disulfide bond to albumin Cys residues normally engaged in intrachain disulfide bonds [23]. It cannot be excluded that keratin damage can also be caused by the disruption of keratin disulfide bonds by *S*-homocysteinylation. However, this has not been studied in the present work, because *S*-Hcy-keratin could not be separated from Hcy-keratin.

About three dozens of individual Hcy-proteins have been identified in humans and animals, including albumin, fibrinogen [2], collagen [7], dynein [5], actin and E-cadherin [6], and major urinary protein [17]. The present study adds mammalian and bird keratins to the list of Hcy-proteins identified *in vivo*.

In conclusion, our findings show that hair keratin in animals and birds is a target for homocysteinylation *in vivo* and that this process causes keratin damage, manifested by the reduction of its solubility.

Abbreviations

DTT: Dithiothreitol
Cbs: Cystathionine β-synthase
Cse: Cystathionine γ-lyase
Mthfr: Methyltetrahydrofolate reductase
Cys: Cysteine
Hcy: Homocysteine
HHcy: Hyperhomocysteinemia
HPLC: High-performance liquid chromatography
Hcy-protein: Hcy linked to a protein via an amide (peptide or isopeptide) bond
Met: Methionine
NAC: *N*-acetyl cysteine
N-Hcy-protein: Hcy linked to a protein via an isopeptide bond
OPA: *o*-Phthaldialdehyde
S-Hcy-protein: Hcy linked to a protein via a disulfide bond
SDS: Sodium dodecyl sulfate
TCEP: Tris(2-carboxyethyl)phosphine.

Conflicts of Interest

The authors declare that they have no conflict of interest.

Authors' Contributions

KB performed *S*-Hcy-keratin, *N*-Hcy-keratin, and Met-keratin assays; JW carried out the SDS-PAGE analyses of total mouse hair keratin; JS quantified copper and iron contents in mouse hair; NA and II provided hair samples from the $Cbs^{-/-}$ and $Cse^{-/-}$ mice; HJ designed the study, supervised the experiments, analyzed the data, drafted the paper, and had the primary responsibility for the final content. All the authors reviewed the results and approved the final version of the manuscript.

Acknowledgments

This study was supported in part by grants 2012/07/B/NZ7/01178, 2013/09/B/NZ5/02794, 2013/11/B/NZ1/00091, 2013/09/D/ST5/03909, 2014/15/N/NZ5/01646, and 2016/23/B/NZ5/00573 from the National Science Center, Poland.

References

[1] S. H. Mudd, H. L. Levy, and J. P. Kraus, "Disorders of transsulfuration," in *The Metabolic and Molecular Bases of Inherited Disease*, C. R. Scriver, A. L. Beaudet, W. S. Sly, and D. Valle, Eds., vol. 2, pp. 2007–2056, Mc Graw-Hill, New York, 2001.

[2] H. Jakubowski, *Homocysteine in Protein Structure/Function and Human Disease-Chemical Biology of Homocysteine-Containing Proteins*, Springer, Wien, 2013.

[3] H. Jakubowski, "Homocysteine modification in protein structure/function and human disease," *Physiological Reviews*, 2018, accepted for publication.

[4] M. Sibrian-Vazquez, J. O. Escobedo, S. Lim, G. K. Samoei, and R. M. Strongin, "Homocystamides promote free-radical and oxidative damage to proteins," *Proceedings of the National Academy of Sciences of the United States of America*, vol. 107, no. 2, pp. 551–554, 2010.

[5] N. Akchiche, C. Bossenmeyer-Pourie, R. Kerek et al., "Homocysteinylation of neuronal proteins contributes to folate deficiency-associated alterations of differentiation, vesicular transport, and plasticity in hippocampal neuronal cells," *The FASEB Journal*, vol. 26, no. 10, pp. 3980–3992, 2012.

[6] C. Bossenmeyer-Pourie, G. Pourie, V. Koziel et al., "Early methyl donor deficiency produces severe gastritis in mothers and offspring through N-homocysteinylation of cytoskeleton proteins, cellular stress, and inflammation," *The FASEB Journal*, vol. 27, no. 6, pp. 2185–2197, 2013.

[7] J. Perla-Kajan, O. Utyro, M. Rusek, A. Malinowska, E. Sitkiewicz, and H. Jakubowski, "*N*-Homocysteinylation impairs collagen cross-linking in cystathionine β-synthase-deficient mice: a novel mechanism of connective tissue abnormalities," *The FASEB Journal*, vol. 30, no. 11, pp. 3810–3821, 2016.

[8] P. Paoli, F. Sbrana, B. Tiribilli et al., "Protein N-homocysteinylation induces the formation of toxic amyloid-like protofibrils," *Journal of Molecular Biology*, vol. 400, no. 4, pp. 889–907, 2010.

[9] G. S. Sharma, T. Kumar, T. A. Dar, and L. R. Singh, "Protein N-homocysteinylation: from cellular toxicity to neurodegeneration," *Biochimica et Biophysica Acta*, vol. 1850, no. 11, pp. 2239–2245, 2015.

[10] D. Gurda, L. Handschuh, W. Kotkowiak, and H. Jakubowski, "Homocysteine thiolactone and N-homocysteinylated protein induce pro-atherogenic changes in gene expression in human vascular endothelial cells," *Amino Acids*, vol. 47, no. 7, pp. 1319–1339, 2015.

[11] K. J. Denny, C. F. Kelly, V. Kumar et al., "Autoantibodies against homocysteinylated protein in a mouse model of folate deficiency-induced neural tube defects," *Birth Defects Research. Part A, Clinical and Molecular Teratology*, vol. 106, no. 3, pp. 201–207, 2016.

[12] Y. C. Chung, A. Kruyer, Y. Yao et al., "Hyperhomocysteinemia exacerbates Alzheimer's disease pathology by way of the β-amyloid fibrinogen interaction," *Journal of Thrombosis and Haemostasis*, vol. 14, no. 7, pp. 1442–1452, 2016.

[13] D. L. Sauls, E. Lockhart, M. E. Warren, A. Lenkowski, S. E. Wilhelm, and M. Hoffman, "Modification of fibrinogen by homocysteine thiolactone increases resistance to fibrinolysis: a potential mechanism of the thrombotic tendency in hyperhomocysteinemia," *Biochemistry*, vol. 45, no. 8, pp. 2480–2487, 2006.

[14] N. Akahoshi, C. Kobayashi, Y. Ishizaki et al., "Genetic background conversion ameliorates semi-lethality and permits behavioral analyses in cystathionine beta-synthase-deficient mice, an animal model for hyperhomocysteinemia," *Human Molecular Genetics*, vol. 17, no. 13, pp. 1994–2005, 2008.

[15] I. Ishii, N. Akahoshi, H. Yamada, S. Nakano, T. Izumi, and M. Suematsu, "Cystathionine gamma-lyase-deficient mice require dietary cysteine to protect against acute lethal myopathy and oxidative injury," *The Journal of Biological Chemistry*, vol. 285, no. 34, pp. 26358–26368, 2010.

[16] Z. Chen, A. C. Karaplis, S. L. Ackerman et al., "Mice deficient in methylenetetrahydrofolate reductase exhibit hyperhomocysteinemia and decreased methylation capacity, with neuropathology and aortic lipid deposition," *Human Molecular Genetics*, vol. 10, no. 5, pp. 433–443, 2001.

[17] H. Jakubowski, "Quantification of urinary S- and N-homocysteinylated protein and homocysteine-thiolactone in mice," *Analytical Biochemistry*, vol. 508, pp. 118–123, 2016.

[18] K. Borowczyk, J. Suliburska, and H. Jakubowski, "Demethylation of methionine and keratin damage in human hair," *Amino Acids*, vol. 50, no. 5, pp. 537–546, 2018.

[19] J. Suliburska, "A comparison of levels of select minerals in scalp hair samples with estimated dietary intakes of these minerals in women of reproductive age," *Biological Trace Element Research*, vol. 144, no. 1–3, pp. 77–85, 2011.

[20] H. Jakubowski, "Protein homocysteinylation: possible mechanism underlying pathological consequences of elevated homocysteine levels," *The FASEB Journal*, vol. 13, no. 15, pp. 2277–2283, 1999.

[21] C. Schoneich, "Sulfur radical-induced redox modifications in proteins: analysis and mechanistic aspects," *Antioxidants & Redox Signaling*, vol. 26, no. 8, pp. 388–405, 2017.

[22] K. Robert, N. Maurin, A. Ledru, J. Delabar, and N. Janel, "Hyperkeratosis in cystathionine beta synthase-deficient mice: an animal model of hyperhomocysteinemia," *The Anatomical Record. Part A, Discoveries in Molecular, Cellular, and Evolutionary Biology*, vol. 280A, no. 2, pp. 1072–1076, 2004.

[23] F. Nakashima, T. Shibata, K. Kamiya et al., "Structural and functional insights into S-thiolation of human serum albumins," *Scientific Reports*, vol. 8, no. 1, p. 932, 2018.

3

Prediction of Blood Lipid Phenotypes using Obesity-Related Genetic Polymorphisms and Lifestyle Data in Subjects with Excessive Body Weight

Omar Ramos-Lopez,[1] Jose I. Riezu-Boj,[1,2] Fermin I. Milagro,[1,3] Marta Cuervo,[1,2,3] Leticia Goni,[1] and J. A. Martinez[1,2,3,4]

[1]*Department of Nutrition, Food Science and Physiology, and Center for Nutrition Research, University of Navarra, Pamplona, Spain*
[2]*Navarra Institute for Health Research (IdiSNA), Pamplona, Spain*
[3]*CIBERobn, Fisiopatología de la Obesidad y la Nutrición, Carlos III Health Institute, Madrid, Spain*
[4]*Madrid Institute of Advanced Studies (IMDEA Food), Madrid, Spain*

Correspondence should be addressed to J. A. Martinez; jalfmtz@unav.es

Academic Editor: Sang Hong Lee

Background and Aim. Individual lipid phenotypes including circulating total cholesterol (TC), low-density lipoprotein cholesterol (LDL-c), high-density lipoprotein cholesterol (HDL-c), and triglycerides (TG) determinations are influenced by gene-environment interactions. The aim of this study was to predict blood lipid level (TC, LDL-c, HDL-c, and TG) variability using genetic and lifestyle data in subjects with excessive body weight-for-height. *Methods.* This cross-sectional study enrolled 304 unrelated overweight/obese adults of self-reported European ancestry. A total of 95 single nucleotide polymorphisms (SNPs) related to obesity and weight loss were analyzed by a targeted next-generation sequencing system. Relevant genotypes of each SNP were coded as 0 (nonrisk) and 1 (risk). Four genetic risk scores (GRS) for each lipid phenotype were calculated by adding the risk genotypes. Information concerning lifestyle (diet, physical activity, alcohol drinking, and smoking) was obtained using validated questionnaires. Total body fat (TFAT) and visceral fat (VFAT) were determined by dual-energy X-ray absorptiometry. *Results.* Overall, 45 obesity-related genetic variants were associated with some of the studied blood lipids. In addition to conventional factors (age, sex, dietary intakes, and alcohol consumption), the calculated GRS significantly contributed to explain their corresponding plasma lipid trait. Thus, HDL-c, TG, TC, and LDL-c serum concentrations were predicted by approximately 28% (optimism-corrected adj. $R^2 = 0.28$), 25% (optimism-corrected adj. $R^2 = 0.25$), 24% (optimism-corrected adj. $R^2 = 0.24$), and 21% (optimism-corrected adj. R^2=0.21), respectively. Interestingly, GRS were the greatest contributors to TC (squared partial correlation (PC^2) = 0.18) and LDL-c ($PC^2 = 0.18$) features. Likewise, VFAT and GRS had a higher impact on HDL-c ($PC^2 = 0.09$ and $PC^2 = 0.06$, respectively) and TG levels ($PC^2 = 0.20$ and $PC^2 = 0.07$, respectively) than the rest of variables. *Conclusions.* Besides known lifestyle influences, some obesity-related genetic variants could help to predict blood lipid phenotypes.

1. Introduction

Triglycerides, cholesterol, and related lipoproteins are major constituents of the lipid fraction of the human body, playing essential physiological roles such as cell membrane stability, energy storage, hormone and bile acid syntheses, dietary fat absorption and assembling, stress response, cell signaling, and calcium metabolism [1, 2]. However, abnormalities in lipid metabolism may lead to the onset and development of several metabolic disorders, including cardiovascular disease features [3]. In this context, elevated plasma levels of total cholesterol (TC), low-density lipoprotein cholesterol (LDL-c), and triglycerides (TG) have been associated with the risk of coronary heart disease, whereas high concentrations of high-density lipoprotein cholesterol (HDL-c) may exert a protective effect [4].

Growing scientific evidence suggests that gene-environment interactions may influence plasma lipid phenotypes [5]. Lifestyle factors such as diet, physical activity, alcohol drinking, and smoking have been recognized as important determinants of the blood lipid profile [5]. Moreover, genome-wide association studies (GWAS) and gene-candidate analyses have identified a number of common genetic variants associated with diverse lipid traits [6]. Also, specific genetic risk scores (GRS) including multiple gene loci have accounted for dyslipidemia susceptibilities and predisposition to related health risks in some populations [7, 8].

Furthermore, differences in cholesterol and TG outcomes according to genotypes of single nucleotide polymorphisms (SNPs) in response to dietary interventions have been reported [9–11]. Nevertheless, most available studies mainly include SNPs in genes directly implicated in lipid metabolism (uptake, transport, and signaling) [12, 13], whereas those related to body weight regulation and obesity remain less explored. Together, these insights reveal a genetic component implicated in lipid homeostasis that may partially explain the variability in circulating lipids among individuals. In addition, this knowledge can help to specifically establish personalized nutritional guidelines that complement the general recommendations to the prevention and precision management of dyslipidemia [14, 15]. Hence, the aim of this research was to predict blood lipid profiles using genetic and environmental data in subjects with excessive body weight-for-height.

2. Materials and Methods

2.1. Subjects. This cross-sectional study enrolled 304 unrelated (nonconsanguineous) Spanish adults of self-reported European ancestry, who presented overweight (BMI: 25–29.9 kg/m^2) and obesity (BMI: 30–40 kg/m^2). Subjects were recruited at the Center for Nutrition Research of the University of Navarra in the city of Pamplona, Navarra, Spain. Major exclusion criteria included a history of diabetes mellitus, cardiovascular disease and hypertension, pregnant or lactating women, and current use of lipid-lowering drugs. Patients with diagnosed primary hyperlipidemia were also excluded. This investigation followed the ethical principles for medical research in humans from the 2013 Helsinki Declaration [16]. Moreover, the research protocol was properly approved by the Research Ethics Committee of the University of Navarra (ref. 132/2015). A written informed consent from each participant was obtained before the inclusion in the study.

2.2. Anthropometry and Blood Pressure. Anthropometric measurements such as height (cm), body weight (kg), and waist circumference (WC, cm) were collected at the fasting state by trained nutritionists following validated procedures [17]. BMI was calculated as the ratio between weight and squared height (kg/m^2). Total body fat (TFAT, kg) and visceral fat (VF, kg) were quantified by dual-energy X-ray absorptiometry according to instructions provided by the supplier (Lunar Prodigy, software version 6.0, Madison, WI, USA). Systolic blood pressure (SBP, mmHg) and diastolic blood pressure (DBP, mmHg) were measured with an automated sphygmomanometer according to standardized criteria as described by the World Health Organization and the International Society of Hypertension [18].

2.3. Biochemical Measurements. Blood samples were drawn by venipuncture after an overnight fast. Biochemical measurements including glucose (mg/dl), total cholesterol (TC, mg/dl), high-density lipoprotein cholesterol (HDL-c, mg/dl), and triglycerides (TG, mg/dl) were determined in an automatic analyzer (Pentra C200, HORIBA Medical) using appropriate kits provided by the company. Low-density lipoprotein cholesterol (LDL-c, mg/dl) was calculated with the Friedewald formula [19]. The following cutoffs for the Spanish population were used to the diagnosis of dyslipidemia: hypercholesterolemia (TC $\geq$ 200 mg/dl), high LDL-c (LDL-c $\geq$ 130 mg/dl), hypoalphalipoproteinemia (HDL-c < 40 mg/dl in men and < 50 mg/dl in women), and hypertriglyceridemia (TG $\geq$ 150 mg/dl), as reported elsewhere [20].

2.4. Lifestyle Factors. A validated semiquantitative food frequency questionnaire was used to evaluate habitual consumption (daily, weekly, monthly, or never) of 137 foods during the previous year [21]. Energy and nutrient intakes were further calculated with an ad hoc computer program based on the standard Spanish food composition tables [22].

The physical activity level was estimated using a validated questionnaire [23]. The volume of activity was expressed in metabolic equivalents (METs), as described elsewhere [24].

Current smoking and drinking habits were evaluated through valid medical questionnaires. Alcohol consumption higher than 40 g of ethanol/d in men and 20 g of ethanol/d in women was considered clinically significant [25].

2.5. SNP Selection and Genotyping. A total of 95 genetic variants related to obesity and weight loss as well as interactions with dietary prescriptions were analyzed after an exhaustive bibliographic review following PRISMA criteria [14, 15, 26, 27], whose genomic characteristics are presented (Supplementary Table 1).

Buccal samples were collected with a liquid-based kit (ORAcollect-DNA, OCR-100, DNA Genotek Inc., Ottawa, Canada). Subsequently, genomic DNA was isolated using the Maxwell® 16 Buccal Swab LEV DNA Purification Kit in the Maxwell® 16 Instrument (Promega Corp., Madison, WI, USA) according to the manufacturer's protocol. A customized panel of primers to amplify the regions containing the selected SNPs was designed using the "online" application of Thermo Fisher AmpliSeq Designer (https://www.ampliseq.com). Overall, the amplicon average size was 185 bp. The amplicon library for massive sequencing was constructed with the custom-designed panel and the Ion AmpliSeq™ Library Kit 2.0 (Thermo Fisher Scientific Inc., Waltham, MA, USA) according to the manufacturer's protocol.

Genotyping was performed by targeted next-generation sequencing in the Ion Torrent PGM™ equipment (Thermo Fisher Scientific Inc., Waltham, MA, USA), as described elsewhere [9, 10]. Raw data were processed in the Ion Torrent Suite™ Server version 5.0.4 (Thermo Fisher Scientific Inc.,

Waltham, MA, USA) using the *Homo sapiens* (HG19) as the reference genome for the alignment. A custom-designed Bed file was used to locate the SNPs of interest. Genetic variants were identified with the Torrent Variant Caller 5.0 (Thermo Fisher Scientific Inc., Waltham, MA, USA) with a minimum coverage value of 25 sequences [28]. Hardy-Weinberg equilibrium (HWE) was estimated with the Convert (version 1.31) and the Arlequin software (version 3.0). Furthermore, the analysis of molecular variance (AMOVA) test using the 95 SNPs was performed in the Arlequin software in order to corroborate the homogeneity of the sample.

2.6. GRS Calculation. Once the 95 SNPs were genotyped, four individual GRS were calculated for each lipid trait (TC, LDL-c, HDL-c, and TG) according to the following steps. First, in order to avoid bias and overfitting in the preselection of SNPs [29], ANOVA tests were run to discard those clearly not associated ($P > 0.25$) with some of the four blood lipid phenotypes. The genotypes of the rest of SNPs ($n = 74$) were differentially coded as 0 (nonrisk) and 1 (risk) based on the observed average values of each lipid between the three genotypes using post hoc tests (Bonferroni or Dunnett's T3). A risk genotype was defined as the one that was associated with increased concentrations of fasting TC, LDL-c, and TG or decreased HDL-c levels. Genotypes with similar effects were grouped in a single category. In a third step, Student's t-tests were further applied to assess statistical differences between the categorized genotype groups (risk vs. nonrisk). Then SNPs showing at least a marginal statistical trend ($P < 0.10$) were selected ($n = 54$) to design each specific GRS, excluding those with a low prevalence (<10%) in either genotype category (risk and nonrisk) to avoid model instability ($n = 9$). From the remaining 45 SNPs, four different GRS (GRS_TC, GRS_LDL-c, GRS_HDL-c, and GRS_TG) were constructed by adding the risk genotypes of the corresponding SNPs for each study lipid trait (Supplementary Tables 2a–2d). Analyses were performed using the four GRS as continuous and categorical variables.

2.7. Statistical Analyses. Continuous variables were expressed as means ± standard deviations, while dichotomous variables were presented as numbers and percentages. Normality of study variables was screened by the Kolmogorov-Smirnov test. All principal variables including TC, LDL-c, HDL-c, and TG were normally distributed ($P > 0.05$).

In addition to genetic variants, other conventional predictors of blood lipid levels were evaluated including age, sex, BMI (kg/m^2), adiposity markers (TFAT and VFAT), physical activity (METs), total energy (kcal), and macronutrient intakes (% E) as well as smoking and drinking habits. Relevant interactions between genetic and lifestyle factors were calculated with simple linear regression tests. Statistical differences in blood lipids by predictor categories were assessed by Student's t-tests.

The prediction of the variability in all blood lipid levels was performed using multiple linear regression models. For this purpose, three statistical approaches were used: least-angle regression (LARS) [30], best subset regression procedure (BSRP) [31], and bootstrapping stepwise method (BSM) [32]. In order to select the most robust model, all candidate predictive models were corrected for optimism and overfitting following Harrell's bootstrapping algorithm [33]. This method is based on using bootstrapped datasets to internally validate the linear regression models as well as to repeatedly quantify the degree of overfitting in the model-building process. Moreover, squared partial correlations (PC^2) were used to estimate the individual contribution of each predictor to the blood lipid variability.

Statistical analyses were performed in the statistical program STATA 12 (StataCorp LLC, College Station, TX, USA; http://www.stata.com). A Venn diagram was constructed online (http://bioinfogp.cnb.csic.es/tools/venny/) in order to show common and uncommon SNPs associated with each of the studied blood lipids. Figure plots concerning comparisons of blood lipid levels between predictor categories were created using the GraphPad Prism® software, version 6.0C (La Jolla, CA, USA). Statistical significance was based on a P value lower than 0.05.

3. Results

The anthropometric, biochemical, and nutritional characteristics of the study population are reported (Table 1). Overall, 70% ($n = 212$) of subjects were women. According to the BMI classification criteria of the World Health Organization, 38% of individuals were overweight ($n = 114$), and 62% ($n = 190$) presented obesity. The average values of TC and LDL-c were above the reference limits. The frequencies of hypercholesterolemia, high LDL-c, and low HDL-c (also known as hypoalphalipoproteinemia) were 65% ($n = 199$), 59% ($n = 179$), and 23% ($n = 69$), respectively, whereas 15% of the study population had hypertriglyceridemia ($n = 45$). The nutritional pattern of the study population was characterized by a high consumption of energy derived from fat (40.4%) and a concomitant low intake of carbohydrates (40.7%) with respect to general nutritional recommendations for the Spanish population. The frequencies of smoking and drinking habits were 21.9 and 13.5%, respectively (Table 1).

A total of 45 obesity-related genetic variants were associated with some of the studied blood lipid levels (Supplementary Tables 2a–2d). Of these, 2 SNPs were common among all lipids: rs1685325 (*UCP3*) and rs894160 (*PLIN1*). On the other hand, 22 SNPs were exclusively related to a specific lipid—4 for TC: rs569805 (*ABCB11*), rs494874 (*ABCB11*), rs1801260 (*CLOCK*), and rs6013029 (*CTNNBL1*); 3 for LDL-c: rs7799039 (*LEP*), rs7498665 (*SH2B1*), and rs7359397 (*SH2B1*); 9 for HDL-c: rs2815752 (*NEGR1*), rs2943641 (*IRS1*), rs2419621 (*ACSL5*), rs6265 (*BDNF*), rs11030104 (*BDNF*), rs4769873 (*ALOX5AP*), rs9939609 (*FTO*), rs6567160 (*MC4R*), and rs2287019 (*QPCTL*); and 6 for TG: rs324420 (*FAAH*), rs2959272 (*PPARG*), rs1386835 (*PPARG*), rs709158 (*PPARG*), rs1175540 (*PPARG*), rs1800544 (*ADRA2A*) (Figure 1). The distribution of genotypes of most obesity-predisposing SNPs was concordant with the Hardy-Weinberg equilibrium principle, except for rs1386835 (*PPARG*), rs17782313 (*MC4R*), rs2287019 (*QPCTL*), and rs3813929 (*HTR2C*), as shown in Supplementary Table 1.

Table 1: Anthropometric, biochemical, and nutritional characteristics of the study population ($n = 304$).

Variable	Average values
Age (years)	45.8 ± 10.5
Sex (F/M)	212/92
Anthropometrics and clinical data	
Weight (kg)	87.7 ± 13.0
BMI (kg/m^2)	31.6 ± 3.5
WC (cm)	102 ± 11
TFAT (kg)	36.9 ± 7.6
VFAT (kg)	1.48 ± 0.90
SBP (mmHg)	128 ± 17
DBP (mmHg)	79 ± 11
Biochemical profile	
Glucose (mg/dl)	96.6 ± 14.1
TC (mg/dl)	216 ± 38
HDL-c (mg/dl)	55.3 ± 12.9
LDL-c (mg/dl)	140 ± 34
TG (mg/dl)	104 ± 56
Dietary intake/day	
Energy (Kcal)	2970 ± 934
Carbohydrates (% E)	40.7 ± 6.8
Protein (% E)	17.0 ± 2.9
Fat (% E)	40.4 ± 5.8
Lifestyle	
Smokers (%)	21.9
Drinkers (%)	13.5
METs	23.8 ± 20.0

Variables are expressed as means ± standard deviations. BMI: body mass index; WC: waist circumference; TFAT: total body fat; VFAT: visceral fat; SBP: systolic blood pressure; DBP: diastolic blood pressure; TC: total cholesterol; HDL-c: high-density lipoprotein cholesterol; LDL-c: low-density lipoprotein cholesterol; TG: triglycerides; METs: metabolic equivalents.

AMOVA analyses revealed no significant differentiation within the sample ($P > 0.05$).

The performance of the three multiple linear regression models predicting blood lipid profiles are reported (Supplementary Tables 3a–3d). After optimism correction, the best model explaining TC, LDL-c, and HDL-c serum concentrations was then obtained using the BSRP approach, whereas TG levels were better predicted by the BSM method (Table 2). Of note, all models included the calculated GRS in addition to conventional factors such as age, sex, dietary intakes, and alcohol consumption. The highest number of predictors was found for HDL-c, whereas LDL-c was only influenced by the GRS_LDL-c and age. No statistically significant interactions between the 4 GRS and lifestyle variables were found. Overall, HDL-c, TG, TC, and LDL-c variabilities were explained in approximately 28% (optimism-corrected adj. $R^2 = 0.28$), 25% (optimism-corrected adj. $R^2 = 0.25$), 24% (optimism-corrected adj. $R^2 = 0.24$), and 21% (optimism-corrected adj. $R^2 = 0.21$), respectively (Table 2).

Moreover, estimations regarding the individual contribution of each independent predictor to blood lipid levels using PC^2 are presented (Table 2). Interestingly, GRS_TC and GRS_LDL-c were the greatest contributors to TC and LDL-c features, respectively, with about 18% for both lipids (PC$^2 = 0.18$). Likewise, VFAT and the respective GRS (GRS_HDL-c and GRS_TG) had a higher impact on HDL-c, with 9% (PC$^2 = 0.09$) and 6% (PC$^2 = 0.06$), respectively, as well as on TG concentrations, with 20% (PC$^2 = 0.20$) and 7% (PC$^2 = 0.07$), respectively. Additionally, comparisons of average blood lipid levels by predictor clusters based on median values are plotted (Figure 2). Greater differences in TC and LDL-c values were found by GRS_TC and GRS_LDL-c categorized by the median number of risk genotypes. Meanwhile, VFAT and the corresponding GRS categories (GRS_HDL-c and GRS_TG) accounted for higher variances in HDL-c and TG, respectively, as compared to other factors including energy intake, alcohol consumption, cholesterol intake, and TFAT (Figure 2).

4. Discussion

In the last years, multiple genetic variants have been found to be associated with specific phenotypes and metabolic disorders, including dyslipidemia [6]. In the current investigation, 45 obesity-related SNPs were associated with circulating lipids (TC, LDL-c, HDL-c, and TG). Of note, some of such associations are reported for the first time, except for rs1799883 (*FABP2*) [34], rs660339 (*UCP2*) and rs659366 (*UCP2*) [35], rs1052700 (*PLIN1*) [36], rs17782313 (*MC4R*) [37], rs7799039 (*LEP*) [38], rs2943641 (*IRS1*) [39], rs9939609 (*FTO*) [40], and rs324420 (*FAAH*) [41]. Interestingly, about 50% of SNPs were related to one specific circulating lipid, whereas only rs1685325 (*UCP3*) and rs894160 (*PLIN1*) were common among all lipids. This finding is consistent with previous studies illustrating the number of loci influencing blood lipid phenotypes using genome-wide and customized genotyping approaches [6, 42]. In contrast to UCP1, it has been postulated that UCP3 regulates cellular lipid metabolism by exporting those fatty acids that cannot be oxidized from the mitochondrial matrix to prevent their deleterious accumulation [43]. Meanwhile, PLIN1 is an adipocyte-specific lipid-coated protein involved in the regulation of lipolysis by regulating lipase interactions [44]. Also, PLIN1 promotes the efficient lipid droplet formation in adipocytes [45].

The magnitude of associations between individual gene variants and metabolic traits is generally modest. Therefore, effect size estimations based on the combination of multiple loci into a GRS are a common method to improve the predictive value of simple SNPs [46, 47]. In this study, GRS adding risk genotypes were major predictors of their respective plasma lipid in all performed linear regression models, mainly for TC and LDL-c blood concentrations (both 18%) and followed by TG (7%) and HDL-c (6%). Lower effects were reported for different GRS constructed from published meta-analyses of individuals of European ancestry, explaining 7%, 6%, 4%, and 3% of the total variance in HDL-c, TC, LDL-c, and TG, respectively [48]. Also, the combination of

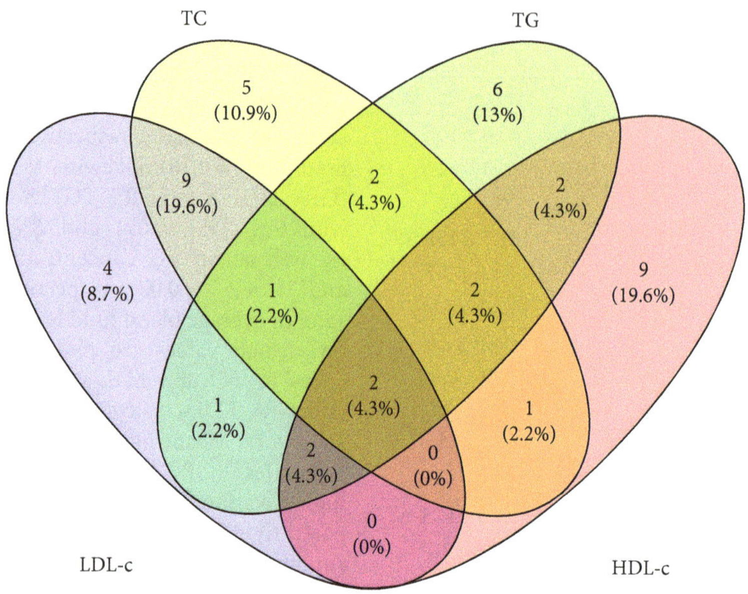

Figure 1: Venn diagram showing the number of SNPs associated with blood lipid levels. TC: total cholesterol; TG: triglycerides; LDL-c: low-density lipoprotein cholesterol; HDL-c: high-density lipoprotein cholesterol.

Table 2: Best multiple linear regression models explaining blood lipid levels as dependent variables.

Predictors	TC		LDL-c		HDL-c		TG	
	β	PC^2	β	PC^2	β	PC^2	β	PC^2
Age (years)	0.80 ± 0.19	0.06	0.58 ± 0.18	0.04	0.22 ± 0.07	0.04		
Sex					4.51 ± 1.98	0.02		
Energy intake (100 kcal)	0.38 ± 0.21	0.01			0.18 ± 0.12	0.009		
Protein intake (%)					0.93 ± 0.30	0.04		
Cholesterol intake (mg)					−0.01 ± 0.004	0.02		
Alcohol					5.83 ± 2.05	0.03	−19.21 ± 9.33	0.02
TFAT (kg)							−1.14 ± 0.42	0.03
VFAT (kg)					−5.22 ± 1.05	0.09	30.95 ± 3.92	0.20
GRS_TC	6.55 ± 0.83	0.18						
GRS_LDL-c			6.79 ± 0.87	0.18				
GRS_HDL-c					−1.12 ± 0.27	0.06		
GRS_TG							4.20 ± 0.97	0.07
Constant	93.40 ± 12.99		45.74 ± 11.43		42.72 ± 7.71		61.58 ± 17.31	
R^2	0.2578		0.2217		0.3394		0.2828	
Adj. R^2	0.2501		0.2160		0.3192		0.2715	
Optimism correction coefficient for R^2	0.0112		0.0083		0.0373		0.0211	
Optimism correction coefficient for adj. R^2	0.0113		0.0084		0.0384		0.0214	
Optimism-corrected R^2	0.2466		0.2134		0.3021		0.2617	
Optimism-corrected adj. R^2	0.2388		0.2076		0.2808		0.2501	

Data are expressed as β values ± standard errors. The best models for each lipid phenotype were TC (BSRP, AIC/AICC); LDL-c (BSRP, BIC); HDL (BSRP, AICC); TG (BSM). BSRP: best subset regression procedure; AIC: akaike information criterion; AICC: corrected akaike information criterion; BIC: bayesian information criterion; BSM: bootstrapping stepwise method; PC^2: squared partial correlation; TC: total cholesterol; LDL-c: low-density lipoprotein cholesterol; HDL-c: high-density lipoprotein cholesterol; TG: triglycerides; TFAT: total body fat; VFAT: visceral fat; GRS_TC: genetic risk score for total cholesterol; GRS_LDL-c: genetic risk score for low-density lipoprotein cholesterol; GRS_HDL-c: genetic risk score for high-density lipoprotein cholesterol; GRS_TG: genetic risk score for triglycerides.

GWAS-identified or well-established lipid-related genetic loci into a weighted GRS explained no more than 11% of the blood lipid oscillations in major ethnic groups living in the United States, with no evidence of interactions between GRS and ethnicity [49]. In a cross-sectional study, 4 weighted GRS of lipid-associated SNPs accounted for 8% (TC), 7%

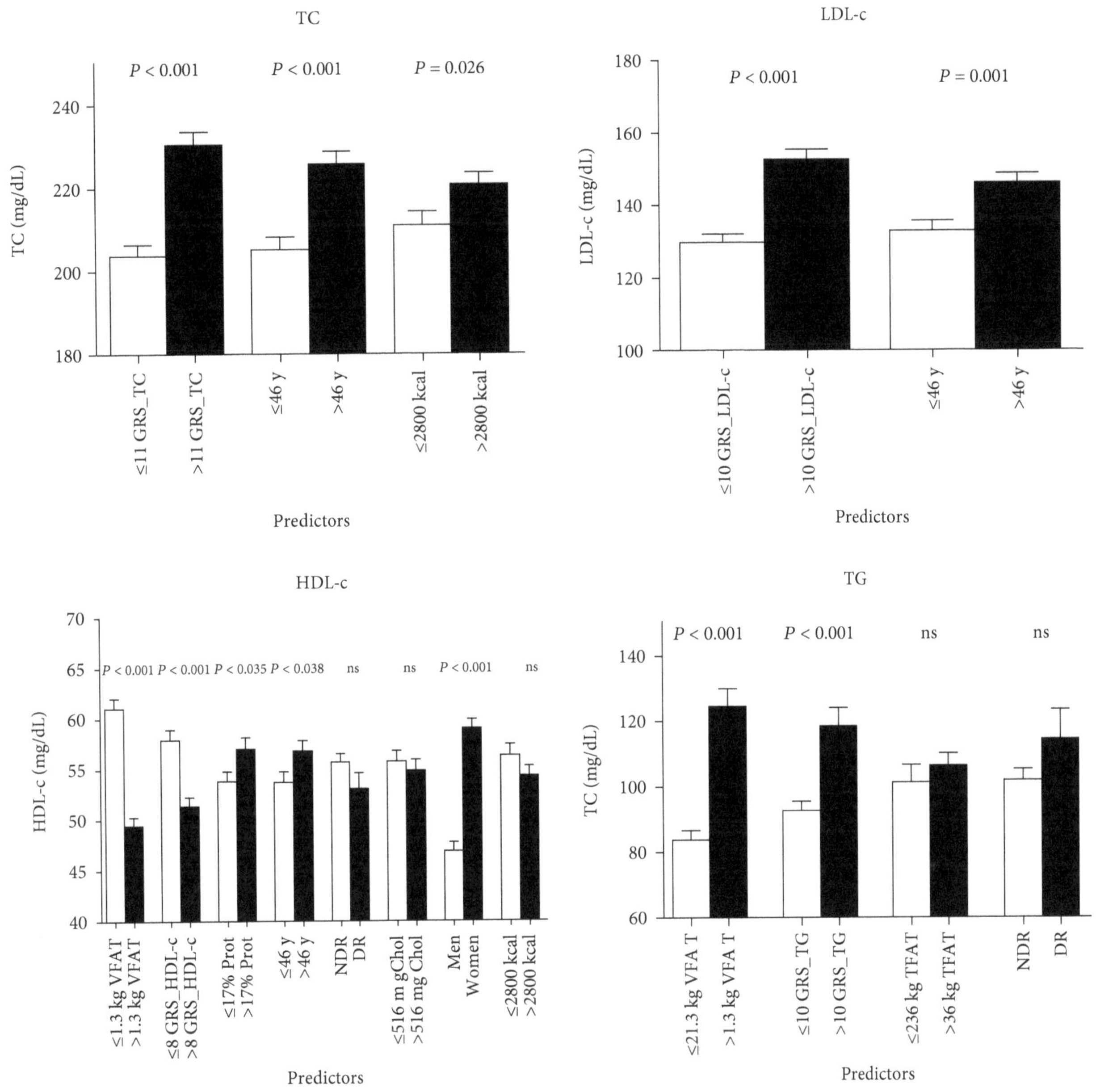

FIGURE 2: Comparisons of average blood lipids levels between predictor categories based on the median values. Data are expressed as means ± standard errors and sorted in descending order by the effect sizes. TC: total cholesterol; LDL-c: low-density lipoprotein cholesterol; HDL-c: high-density lipoprotein cholesterol; TG: triglycerides; GRS_TC: genetic risk score for total cholesterol; GRS_LDL-c: genetic risk score for low-density lipoprotein cholesterol; GRS_HDL-c: genetic risk score for high-density lipoprotein cholesterol; GRS_TG: genetic risk score for triglycerides; TFAT: total body fat; VFAT: visceral fat; DR: drinkers; NDR: nondrinkers.

(HDL-c), 6% (LDL-c), and 5% (TG) of the total variance in two Danish cohorts [50]. Furthermore, the highest quartile (more than 8 risk alleles) of a calculated GRS from obesity-predisposing variants was significantly associated with lower HDL-c levels compared to the lowest GRS quartile (lower than 4 risk alleles) in women with type 2 diabetes mellitus [51].

To date, most available studies analyzing the association of GRS with dyslipidemia and cardiovascular risk use an additive model of allele risk codification (0, 1, 2) across a number of genetic variants [48–51]. In this investigation, no additive effects in any included SNP were detected, so GRS were constructed according to different genotype categories. Interestingly, heterozygous genotypes of some SNPs were associated with the most favorable blood lipid phenotype compared to both homozygous groups, including rs8192678 (*PPARGC1A*), rs1052700 (*PLIN1*), rs894160 (*PLIN1*), rs7799039 (*LEP*), rs6567160 (*MC4R*), rs3813929 (*HTR2C*), rs11091046 (*AGTR2*), rs1386835 (*PPARG*), and rs1805081 (*NPC1*). This finding, known as heterozygote advantage, is a genetic condition in which heterozygous individuals for a locus have greater biological efficacy than the homozygous ones for the same locus [52]. Indeed, quantitative genetics theory predicts that this phenomenon, related to individual genetic diversity, should influence the variation in genetic predisposition to metabolic risks that show dominance variance. Therefore, it has been suggested that heterozygosity must be considered in genetic epidemiological studies concerning common disease traits [53].

Excessive adiposity is generally accompanied by unfavorable blood lipid patterns, which may depend upon regional fat distribution [54]. Here, VFAT has been associated with high TG levels but negatively correlated with HDL-c levels. Instead, TFAT increases tended to diminish circulating TG. In agreement with our findings, visceral adiposity has been shown to have a detrimental effect on plasma lipids, even after adjusting for abdominal subcutaneous adipose tissue [55]. For example, central fat accumulation showed a stronger association with metabolic risks than total fat mass in normal-weight Chinese adults [56].

Besides the genetic background, modifiable environmental factors may also influence serum lipids and related cardiovascular risk [5]. In this research, protein intake and alcohol drinking were positively associated with circulating HDL-c but negatively correlated with dietary cholesterol. Consistently, higher HDL-c concentrations have been reported in people consuming high-protein diets, accounting for a lower risk of developing cardiometabolic disease [57, 58]. Also, most available randomized-controlled trials have reported modest but significant increases in serum HDL-c concentrations after cholesterol supplementation with eggs [59]. Additionally, favorable lipid outcomes (higher levels of HDL-c) have been linked to moderate ethanol consumption, providing indirect evidence for a protective effect of alcohol on cardiovascular risk [60, 61].

The main strengths of this investigation include the analysis of the genetic influence on blood lipids using GRS from obesity-related SNPs instead of conventional lipid-protein genes as well as the use of different multiple linear regression tests to evaluate the contribution of genetic and lifestyle factors to plasma lipid profiles. Although SNPs were located on obesity-related genes, some of the genes also have a direct role in lipid metabolism including *PPARG*, *FABP2*, *PLIN1*, *NPC1*, *ACSL5*, and *FAAH*, suggesting relationships between genetics, adiposity, and plasma lipid profiles. Also, the results found in this research are unlikely to be confounded by population stratification since the studied sample was ethnically homogeneous (Spanish individuals of European ancestry) as revealed by AMOVA analyses. As for drawbacks, our findings may be not generalizable to other ethnic groups and populations, especially those who are exposed to different gene-environmental interactions. Moreover, this study enrolled subjects with excessive body weight-for-height; thus, further research is needed concerning the analysis of lean individuals. In addition, type I and type II errors cannot be completely ruled out, especially those related to the selection of SNPs to be introduced into the GRS. However, as previously reviewed [62], genomic profile risk scoring analyses can tolerate, at balance, some of these biases due to the use of less stringent P value thresholds compared to association studies of single variants. Likewise, although all linear regression models were internally validated by the bootstrapping method, it is not likely that the overfitting problem is totally ruled out. Also, because our findings were not assessed in an independent validation data set, replications in external populations may be required in a further study. Another way of validation could consist in splitting the original data set into two subsets, separating a discovery sample (training) and a target sample (testing), but given the relatively low sample analyzed, the statistical power of the study concerning main outcomes would be lowered. Furthermore, the role of new SNPs associated with excessive adiposity and accompanying metabolic alterations through a GRS approach needs to be explored. As a final point, while several gene-gene or gene-environment interactions in relation to lipid traits have been reported [5], no relevant relationships were found in this study.

In conclusion, our results suggest that multiple obesity-related genetic variants are important predictors of blood lipid phenotypes, in addition to environmental influences in subjects with excessive body weight-for-height. Together, these insights may contribute to design and implement precision lifestyle strategies to the control of lipid disorders.

Conflicts of Interest

The authors declare that they have no conflicts of interest concerning this investigation.

Authors' Contributions

Omar Ramos-Lopez and Jose I. Riezu-Boj contributed equally to this research.

Acknowledgments

Authors thank technical and laboratory assistance from Laura Olazaran, Iosune Zubieta (dietitians), and Ana Lorente (technician). The tractor role from CINFA concerning the genetic tools is also appreciated. This research was supported by grants from the Government of Navarra (PT024 OBEKIT), CIBERobn (CB12/03/30002 to JAM), and NUTRIGENIO (AGL2013-45554-R). ORL was supported by a 2-year postdoctoral grant from National Council of Science and Technology, Mexico (CONACyT, CVU 444175) in cooperation with the PhD Program in Molecular Biology in Medicine, University of Guadalajara, Mexico (CONACyT, PNPC 000091), and the Center for Nutrition Research, University of Navarra, Spain (LE/97).

Supplementary Materials

Supplementary Table 1: genomic characteristics of the 95 obesity-predisposing SNPs. Supplementary Table 2a: list of SNPs associated with circulating total cholesterol levels and related genotype codifications. Supplementary Table 2b: list of SNPs associated with circulating LDL-c levels and related genotype codifications. Supplementary Table 2c: list of SNPs associated with circulating HDL-c levels and related genotype codifications. Supplementary Table 2d: list of SNPs associated with circulating triglyceride levels and related genotype codifications. Supplementary Table 3a: multiple linear regression models explaining total cholesterol levels as the dependent variable. Supplementary Table 3b: multiple linear regression models explaining LDL-c levels as the dependent variable. Supplementary Table 3c: multiple linear regression models explaining HDL-c levels as the dependent variable. Supplementary Table 3d: multiple linear regression models explaining triglyceride levels as the dependent variable. *(Supplementary Materials)*

References

[1] G. AbouRjaili, N. Shtaynberg, R. Wetz, T. Costantino, and G. S. Abela, "Current concepts in triglyceride metabolism, pathophysiology, and treatment," *Metabolism*, vol. 59, no. 8, pp. 1210–1220, 2010.

[2] T. L. Errico, X. Chen, J. M. Martin Campos, J. Julve, J. C. Escolà-Gil, and F. Blanco-Vaca, "Basic mechanisms: structure, function and metabolism of plasma lipoproteins," *Clinic and Research in Arteriosclerosis*, vol. 25, no. 2, pp. 98–103, 2013.

[3] L. D. Colantonio, V. Bittner, K. Reynolds et al., "Association of serum lipids and coronary heart disease in contemporary observational studies," *Circulation*, vol. 133, no. 3, pp. 256–264, 2016.

[4] M. F. Piepoli, A. W. Hoes, S. Agewall et al., "2016 European guidelines on cardiovascular disease prevention in clinical practice," *Revista Española de Cardiología*, vol. 69, no. 10, p. 939, 2016.

[5] C. B. Cole, M. Nikpay, and R. McPherson, "Gene-environment interaction in dyslipidemia," *Current Opinion in Lipidology*, vol. 26, no. 2, pp. 133–138, 2015.

[6] F. W. Asselbergs, R. C. Lovering, and F. Drenos, "Progress in genetic association studies of plasma lipids," *Current Opinion in Lipidology*, vol. 24, no. 2, pp. 123–128, 2013.

[7] P. L. Lutsey, L. J. Rasmussen-Torvik, J. S. Pankow et al., "Relation of lipid gene scores to longitudinal trends in lipid levels and incidence of abnormal lipid levels among individuals of European ancestry: the Atherosclerosis Risk in Communities (ARIC) study," *Circulation. Cardiovascular Genetics*, vol. 5, no. 1, pp. 73–80, 2012.

[8] A. Isaacs, S. M. Willems, D. Bos et al., "Risk scores of common genetic variants for lipid levels influence atherosclerosis and incident coronary heart disease," *Arteriosclerosis, Thrombosis, and Vascular Biology*, vol. 33, no. 9, pp. 2233–2239, 2013.

[9] O. Ramos-Lopez, J. I. Riezu-Boj, F. I. Milagro, L. Goni, M. Cuervo, and J. A. Martinez, "Association of the Gly482Ser PPARGC1A gene variant with different cholesterol outcomes in response to two energy-restricted diets in subjects with excessive weight," *Nutrition*, vol. 47, pp. 83–89, 2018.

[10] O. Ramos-Lopez, J. I. Riezu-Boj, F. I. Milagro, L. Goni, M. Cuervo, and J. A. Martinez, "Differential lipid metabolism outcomes associated with ADRB2 gene polymorphisms in response to two dietary interventions in overweight/obese subjects," *Nutrition, Metabolism, and Cardiovascular Diseases*, vol. 28, no. 2, pp. 165–172, 2018.

[11] L. Goni, D. Sun, Y. Heianza et al., "Macronutrient-specific effect of the MTNR1B genotype on lipid levels in response to 2 year weight-loss diets," *Journal of Lipid Research*, vol. 59, no. 1, pp. 155–161, 2018.

[12] A. Moleres, F. I. Milagro, A. Marcos et al., "Common variants in genes related to lipid and energy metabolism are associated with weight loss after an intervention in overweight/obese adolescents," *Nutrición Hospitalaria*, vol. 30, no. 1, pp. 75–83, 2014.

[13] J. Hallmann, S. Kolossa, K. Gedrich et al., "Predicting fatty acid profiles in blood based on food intake and the FADS1 rs174546 SNP," *Molecular Nutrition & Food Research*, vol. 59, no. 12, pp. 2565–2573, 2015.

[14] L. Goni, M. Cuervo, F. I. Milagro, and J. A. Martínez, "Future perspectives of personalized weight loss interventions based on nutrigenetic, epigenetic, and metagenomic data," *The Journal of Nutrition*, vol. 146, no. 4, pp. 905S–912S, 2016.

[15] O. Ramos-Lopez, F. I. Milagro, H. Allayee et al., "Guide for current nutrigenetic, nutrigenomic, and nutriepigenetic approaches for precision nutrition involving the prevention and management of chronic diseases associated with obesity," *Journal of Nutrigenetics and Nutrigenomics*, vol. 10, no. 1-2, pp. 43–62, 2017.

[16] World Medical Association, "World Medical Association Declaration of Helsinki: ethical principles for medical research involving human subjects," *Journal of the American Medical Association*, vol. 310, no. 20, pp. 2191–2194, 2013.

[17] P. Lopez-Legarrea, R. de la Iglesia, I. Abete et al., "Short-term role of the dietary total antioxidant capacity in two hypocaloric regimes on obese with metabolic syndrome symptoms: the RESMENA randomized controlled trial," *Nutrition & Metabolism*, vol. 10, no. 1, p. 22, 2013.

[18] J. A. Whitworth and J. Chalmers, "World health organisation-international society of hypertension (WHO/ISH) hypertension guidelines," *Clinical and Experimental Hypertension*, vol. 26, no. 7&8, pp. 747–752, 2004.

[19] A. J. Tremblay, H. Morrissette, J. M. Gagné, J. Bergeron, C. Gagné, and P. Couture, "Validation of the Friedewald formula for the determination of low-density lipoprotein cholesterol compared with beta-quantification in a large population," *Clinical Biochemistry*, vol. 37, no. 9, pp. 785–790, 2004.

[20] P. Guallar-Castillón, M. Gil-Montero, L. M. León-Muñoz et al., "Magnitude and management of hypercholesterolemia in the adult population of Spain, 2008-2010: the ENRICA study," *Revista Española de Cardiología*, vol. 65, no. 6, pp. 551–558, 2012.

[21] C. de la Fuente-Arrillaga, Z. Vázquez Ruiz, M. Bes-Rastrollo, L. Sampson, and M. A. Martinez-González, "Reproducibility of an FFQ validated in Spain," *Public Health Nutrition*, vol. 13, no. 09, pp. 1364–1372, 2010.

[22] O. Moreiras, A. Carbajal, L. Cabrera, and C. Cuadrado, *Tablas de composición de alimentos. Guía de prácticas*, Pirámide, Madrid, 16th edition, 2013.

[23] M. A. Martínez-González, C. López-Fontana, J. J. Varo, A. Sánchez-Villegas, and J. A. Martinez, "Validation of the Spanish version of the physical activity questionnaire used in the Nurses' Health Study and the Health Professionals' Follow-up Study," *Public Health Nutrition*, vol. 8, no. 7, pp. 920–927, 2005.

[24] F. J. Basterra-Gortari, M. Bes-Rastrollo, M. Pardo-Fernández, L. Forga, J. A. Martinez, and M. A. Martínez-González, "Changes in weight and physical activity over two years in Spanish alumni," *Medicine and Science in Sports and Exercise*, vol. 41, no. 3, pp. 516–522, 2009.

[25] O. Ramos-Lopez, E. Martinez-Lopez, S. Roman, N. A. Fierro, and A. Panduro, "Genetic, metabolic and environmental factors involved in the development of liver cirrhosis in Mexico," *World Journal of Gastroenterology*, vol. 21, no. 41, pp. 11552–11566, 2015.

[26] L. Goni, F. I. Milagro, M. Cuervo, and J. A. Martínez, "Single-nucleotide polymorphisms and DNA methylation markers associated with central obesity and regulation of body weight," *Nutrition Reviews*, vol. 72, no. 11, pp. 673–690, 2014.

[27] Y. Heianza and L. Qi, "Gene-diet interaction and precision nutrition in obesity," *International Journal of Molecular Sciences*, vol. 18, no. 4, 2017.

[28] F. Guo, Y. Zhou, H. Song et al., "Next generation sequencing of SNPs using the HID-Ion AmpliSeq™ Identity Panel on the Ion Torrent PGM™ platform," *Forensic Science International. Genetics*, vol. 25, pp. 73–84, 2016.

[29] D. W. Hosmer and S. Lemeshow, *Applied Logistic Regression. Wiley Series in Probability and Statistics*, John Wiley & Sons, Inc, USA, 2nd edition, 2000.

[30] B. Efron, T. Hastie, I. Johnstone, and R. Tibshirani, "Least angle regression," *The Annals of Statistics*, vol. 32, no. 2, pp. 407–499, 2004.

[31] C. Lindsey and S. Sheather, "Variable selection in linear regression," *Stata Journal*, vol. 10, no. 4, pp. 650–669, 2010.

[32] P. C. Austin and J. V. Tu, "Bootstrap methods for developing predictive models," *The American Statistician*, vol. 58, no. 2, pp. 131–137, 2004.

[33] F. E. Harrell, K. L. Lee, and D. B. Mark, "Multivariable prognostic models: issues in developing models, evaluating assumptions and adequacy, and measuring and reducing errors," *Statistics in Medicine*, vol. 15, no. 4, pp. 361–387, 1996.

[34] B. E. Ríos-González, B. Ibarra-Cortés, G. Ramírez-López, J. Sánchez-Corona, and M. T. Magaña-Torres, "Association of polymorphisms of genes involved in lipid metabolism with blood pressure and lipid values in Mexican hypertensive individuals," *Disease Markers*, vol. 2014, Article ID 150358, 9 pages, 2014.

[35] S. Oktavianthi, H. Trimarsanto, C. A. Febinia et al., "Uncoupling protein 2 gene polymorphisms are associated with obesity," *Cardiovascular Diabetology*, vol. 11, no. 1, p. 41, 2012.

[36] T. Kawai, M. C. Y. Ng, M. G. Hayes et al., "Variation in the perilipin gene (PLIN) affects glucose and lipid metabolism in non-Hispanic white women with and without polycystic ovary syndrome," *Diabetes Research and Clinical Practice*, vol. 86, no. 3, pp. 186–192, 2009.

[37] L. Tao, Z. Zhang, Z. Chen et al., "A common variant near the melanocortin 4 receptor is associated with low-density lipoprotein cholesterol and total cholesterol in the Chinese Han population," *Molecular Biology Reports*, vol. 39, no. 6, pp. 6487–6493, 2012.

[38] V. Manriquez, J. Aviles, L. Salazar et al., "Polymorphisms in genes involved in the leptin-melanocortin pathway are associated with obesity-related cardiometabolic alterations in a southern Chilean population," *Molecular Diagnosis & Therapy*, vol. 22, no. 1, pp. 101–113, 2018.

[39] R. Sharma, S. Prudente, F. Andreozzi et al., "The type 2 diabetes and insulin-resistance locus near IRS1 is a determinant of HDL cholesterol and triglycerides levels among diabetic subjects," *Atherosclerosis*, vol. 216, no. 1, pp. 157–160, 2011.

[40] M. S. Khella, N. M. Hamdy, A. I. Amin, and H. O. El-Mesallamy, "The (FTO) gene polymorphism is associated with metabolic syndrome risk in Egyptian females: a case-control study," *BMC Medical Genetics*, vol. 18, no. 1, p. 101, 2017.

[41] Y. Zhang, G. E. Sonnenberg, T. M. Baye et al., "Obesity-related dyslipidemia associated with FAAH, independent of insulin response, in multigenerational families of Northern European descent," *Pharmacogenomics*, vol. 10, no. 12, pp. 1929–1939, 2009.

[42] Global Lipids Genetics Consortium, C. J. Willer, E. M. Schmidt et al., "Discovery and refinement of loci associated with lipid levels," *Nature Genetics*, vol. 45, no. 11, pp. 1274–1283, 2013.

[43] P. Schrauwen, J. Hoeks, and M. Hesselink, "Putative function and physiological relevance of the mitochondrial uncoupling protein-3: involvement in fatty acid metabolism?," *Progress in Lipid Research*, vol. 45, no. 1, pp. 17–41, 2006.

[44] J. G. Granneman, H. P. H. Moore, R. Krishnamoorthy, and M. Rathod, "Perilipin controls lipolysis by regulating the interactions of AB-hydrolase containing 5 (Abhd5) and adipose triglyceride lipase (Atgl)," *The Journal of Biological Chemistry*, vol. 284, no. 50, pp. 34538–34544, 2009.

[45] Z. Sun, J. Gong, H. Wu et al., "Perilipin1 promotes unilocular lipid droplet formation through the activation of Fsp27 in adipocytes," *Nature Communications*, vol. 4, no. 1, p. 1594, 2013.

[46] R. Moonesinghe, T. Liu, and M. J. Khoury, "Evaluation of the discriminative accuracy of genomic profiling in the prediction of common complex diseases," *European Journal of Human Genetics*, vol. 18, no. 4, pp. 485–489, 2010.

[47] L. Goni, M. Cuervo, F. I. Milagro, and J. A. Martínez, "A genetic risk tool for obesity predisposition assessment and personalized nutrition implementation based on macronutrient intake," *Genes & Nutrition*, vol. 10, no. 1, p. 445, 2015.

[48] P. T. Williams, "Quantile-specific penetrance of genes affecting lipoproteins, adiposity and height," *PLoS One*, vol. 7, no. 1, article e28764, 2012.

[49] M.-H. Chang, R. M. Ned, Y. Hong et al., "Racial/ethnic variation in the association of lipid-related genetic variants with blood lipids in the US adult population," *Circulation. Cardiovascular Genetics*, vol. 4, no. 5, pp. 523–533, 2011.

[50] J. M. Justesen, K. H. Allin, C. H. Sandholt et al., "Interactions of lipid genetic risk scores with estimates of metabolic health in a Danish population," *Circulation. Cardiovascular Genetics*, vol. 8, no. 3, pp. 465–472, 2015.

[51] M. He, M. C. Cornelis, P. W. Franks, C. Zhang, F. B. Hu, and L. Qi, "Obesity genotype score and cardiovascular risk in women with type 2 diabetes mellitus," *Arteriosclerosis, Thrombosis, and Vascular Biology*, vol. 30, no. 2, pp. 327–332, 2010.

[52] D. Sellis, B. J. Callahan, D. A. Petrov, and P. W. Messer, "Heterozygote advantage as a natural consequence of adaptation in diploids," *Proceedings of the National Academy of Sciences of the United States of America*, vol. 108, no. 51, pp. 20666–20671, 2011.

[53] H. Campbell, A. D. Carothers, I. Rudan et al., "Effects of genome-wide heterozygosity on a range of biomedically relevant human quantitative traits," *Human Molecular Genetics*, vol. 16, no. 2, pp. 233–241, 2007.

[54] S. Kachur, C. J. Lavie, A. de Schutter, R. V. Milani, and H. O. Ventura, *Minerva Medica*, vol. 108, no. 3, pp. 212–228, 2017.

[55] T. B. Nguyen-Duy, M. Z. Nichaman, T. S. Church, S. N. Blair, and R. Ross, "Visceral fat and liver fat are independent predictors of metabolic risk factors in men," *American Journal of Physiology. Endocrinology and Metabolism*, vol. 284, no. 6, pp. E1065–E1071, 2003.

[56] X. Fu, F. Zhu, X. Zhao, X. Ma, and S. Zhu, "Central fat accumulation associated with metabolic risks beyond total fat in normal BMI Chinese adults," *Annals of Nutrition & Metabolism*, vol. 64, no. 2, pp. 93–100, 2014.

[57] P. Mirmiran, M. Hajifaraji, Z. Bahadoran, F. Sarvghadi, and F. Azizi, "Dietary protein intake is associated with favorable cardiometabolic risk factors in adults: Tehran Lipid and Glucose Study," *Nutrition Research*, vol. 32, no. 3, pp. 169–176, 2012.

[58] S. M. Pasiakos, H. R. Lieberman, and V. L. Fulgoni, "Higher-protein diets are associated with higher HDL cholesterol and lower BMI and waist circumference in US adults," *The Journal of Nutrition*, vol. 145, no. 3, pp. 605–614, 2015.

[59] J. D. Griffin and A. H. Lichtenstein, "Dietary cholesterol and plasma lipoprotein profiles: randomized-controlled trials," *Current Nutrition Reports*, vol. 2, no. 4, pp. 274–282, 2013.

[60] S. E. Brien, P. E. Ronksley, B. J. Turner, K. J. Mukamal, and W. A. Ghali, "Effect of alcohol consumption on biological markers associated with risk of coronary heart disease: systematic review and meta-analysis of interventional studies," *BMJ*, vol. 342, p. d636, 2011.

[61] S. Huang, J. Li, G. C. Shearer et al., "Longitudinal study of alcohol consumption and HDL concentrations: a community-based study," *The American Journal of Clinical Nutrition*, vol. 105, no. 4, pp. 905–912, 2017.

[62] N. R. Wray, S. H. Lee, D. Mehta, A. A. E. Vinkhuyzen, F. Dudbridge, and C. M. Middeldorp, "Research review: polygenic methods and their application to psychiatric traits," *Journal of Child Psychology and Psychiatry*, vol. 55, no. 10, pp. 1068–1087, 2014.

4

The miRNA Pull Out Assay as a Method to Validate the miR-28-5p Targets Identified in other Tumor Contexts in Prostate Cancer

Milena Rizzo,[1,2] Gabriele Berti,[1,3] Francesco Russo,[4,5] Monica Evangelista,[1] Marco Pellegrini,[4] and Giuseppe Rainaldi[1,2]

[1]*Non-coding RNA Laboratory, Institute of Clinical Physiology (IFC), CNR, Via Moruzzi 1, 56124 Pisa, Italy*
[2]*Tuscan Tumor Institute (ITT), Via Alderotti 26/N, 50139 Firenze, Italy*
[3]*Instute of Life Science, Scuola Superiore Sant'Anna, Piazza Martiri delle Libertà 33, 56127 Pisa, Italy*
[4]*Laboratory of Integrative Systems Medicine (LISM), Institute of Informatics and Telematics (IIT) and Institute of Clinical Physiology (IFC), CNR, Via Moruzzi 1, 56124 Pisa, Italy*
[5]*Novo Nordisk Foundation Center for Protein Research, Faculty of Health and Medical Sciences, University of Copenhagen, Blegdamsvej 3b, 2200 Copenhagen, Denmark*

Correspondence should be addressed to Milena Rizzo; milena.rizzo@ifc.cnr.it

Academic Editor: Davide Barbagallo

miR-28-5p is an intragenic miRNA which is underexpressed in several tumor types showing a tumor suppressor (TS) activity. Routinely, the known miR-28-5p targets are validated in specific tumor contexts but it is unclear whether these targets are also being regulated in other tumor types. To this end, we adopted the miRNA pull out assay to capture the miR-28-5p targets in DU-145 prostate cancer (PCa) cells. Firstly, we demonstrated that miR-28-5p acts as a TS-miRNA in PCa, affecting cell proliferation, survival, and apoptosis. Secondly, we evaluated the enrichment of the 10 validated miR-28-5p targets in the pull out sample. We showed that E2F6, TEX-261, MAPK1, MPL, N4BP1, and RAP1B but not BAG1, OTUB1, MAD2L1, and p21 were significantly enriched, suggesting that not all the miR-28-5p targets are regulated by this miRNA in PCa. We then verified whether the miR-28-5p-interacting targets were regulated by this miRNA. We selected E2F6, the most enriched target in the pull out sample, and demonstrated that miR-28-5p downregulated E2F6 at the protein level suggesting that our approach was effective. In general terms, these findings support the miRNA pull out assay as a useful method to identify context-specific miRNA targets.

1. Introduction

It is well known that the deregulation of miRNA expression is one of the causes or contributory causes of cancer development. miRNAs may act as tumor suppressors (TS), oncogenes, or both depending on the tumor context [1]. miR-28-5p is an intragenic miRNA downregulated in several tumor types, such as hepatocellular carcinoma [2], renal cell carcinoma [3], natural killer/T-cell lymphoma [4], B-cell lymphoma [5], colorectal cancer (CRC) [6], and CRC liver metastasis [7, 8], although in some cases, an increased expression level of miR-28-5p has been observed (ovarian, esophageal, and cervical cancer) [9–11]. Most of the papers regarding the role of miR-28-5p in tumors suggested a prevalent tumor suppressor activity of this miRNA *in vitro* [2, 3, 5, 6, 12]. Very recently, it has been demonstrated that the miR-28-5p reexpression in xenograft models of Burkitt (BL) and diffuse large B-cell lymphoma (DLBCL) as well as in a BL murine model blocked tumor growth, opening the way to miR-28-5p-based replacement therapy as a novel therapeutic strategy for these diseases [13].

The molecular targets through which miR-28-5p exerts its anti- or proproliferative role are only partially known. For example, miR-28-5p reduced cell growth and migration in hepatocellular carcinoma [2] and in CRC [6] cells inhibiting the expression of IGF-1, CCND1, and HOXB3 genes. In addition, miR-28-5p acted as a TS-miRNA in renal cell carcinoma by directly repressing the expression of RAP1B [3] and in B-cell lymphoma by directly inhibiting BAG1 expression, a gene involved in the MAP-kinase pathway regulation [5].

To date, there are no data either on the role of miR-28-5p in prostate cancer (PCa) or on the targets regulated by this miRNA in PCa cells. In this work, we evaluated whether the miR-28-5p targets validated in other types of tumors were regulated in PCa cells using the miRNA pull out assay, a technique that allows the isolation of all the targets of a given miRNA in specific biological contexts. We demonstrated that miR-28-5p exerted TS activity in PCa cells and that not all validated miR-28-5p targets are regulated by this miRNA in the PCa context.

2. Materials and Methods

2.1. Cell and Culture Conditions. DU-145 and A-549 cells were grown in RPMI Medium 1640 (EuroClone) whereas PC-3 cells were grown in HAM's medium (EuroClone) and MCF-7 cells in DMEM low glucose (EuroClone). 10% FBS (fetal bovine serum, EuroClone), 1% penicillin/streptomycin (2 mM, EuroClone), and 1% L-glutamine (2 mM, Sigma-Aldrich) were added to the medium. The cells were incubated at 37°C in a humidified atmosphere containing 5% CO_2.

2.2. Transfection. Transient transfections of double-stranded miRNA mimics (miR-28a-5p) or negative control (miR-NC) (GenePharma) in DU-145 cells were carried out using Lipofectamine 2000 (Thermo Fisher): 1.5×10^5 cells were seeded in P30 dishes and after 48 hours, cells were transfected with miRNA mimic using 10 μl of Lipofectamine according to the protocol provided by the manufacturer. The suspension of the transfected cells was used for cellular and molecular assays.

2.3. Cell Proliferation. 1×10^5 cells were seeded in a series of 30 mm diameter dishes and grown for 96 hours. At 24-hour intervals, cells were collected and counted.

2.4. Cell Cycle Analysis. Cell cycle analysis was performed as follows: 5×10^5 cells were fixed with 95% cold ethanol and labelled with 300 μl of 50 μl/ml propidium iodide (Sigma-Aldrich) solution. After overnight incubation at 4°C, the cell cycle analysis was performed with Accuri™ C6 flow cytometer (BD Biosciences). Using specific software supplied with the instrument, the percentage of cells in each phase of the cell cycle was determined considering the parameters SSC-H/FL2-A.

2.5. Survival Assay. Survival was measured as follows: cells were collected and seeded at cell density of 200 cells/60 mm diameter culture dish to allow colony formation. After 10–12 days, dishes were stained with 0.1% CV and the ratio (number of colonies/number of seeded cells) was used to calculate the fraction of surviving cells.

2.6. Apoptosis Assay. Apoptosis was measured as follows: 1×10^6 cells were suspended in 300 μl Binding Buffer 1X and left at room temperature for 15 minutes. Thereafter, cell labelling was done according to the kit Annexin V-FITC. Cells were then passed through flow cytometer BD Accuri C6 (BD Biosciences) and analyzed using FL3-H/FL1-H parameters.

2.7. miRNA Pull Out Assay. The miRNA pull out assay was performed as described in Rizzo et al. [14]. Briefly, DU-145 cells were transfected using Lipofectamine 2000 (Thermo Fisher) with 60 nM of either miR-28a-5p duplex (ds-miR-28a$_{CT}$) or a mix of 3′ biotin-tagged miR-28a-5p 8tU (nucleotide 8 was a thiouridine) and miR-28a-5p 18tU duplexes (ds-miR-28a$_{BIO}$). The day after transfection, cells were irradiated with UV (365 nm, 2 J/cm^2) using Bio-Link crosslinking (BLX) (Ambrose Lourmat) and total RNA extracted adding TRIzol reagent (Thermo Fisher) directly to adherent cells and following the instructions provided by the manufacturer. 15 μg of RNA was incubated for 4 hours at 4°C with 100 μl of streptavidin-conjugated beads (Streptavidin Sepharose High Performance, GE Healthcare), and the RNA complexed with the beads recovered using the Trizol protocol. We performed three biological replicates obtaining three miR-28$_{CT}$ (miR-28 control) and three miR-28$_{BIO}$ (miR-28) pull out samples.

2.8. Quantification of miRNAs and mRNAs (qRT-PCR). Total RNA was extracted from 1×10^6 cells using the miRNeasy mini kit following the manufacturer's recommendations. 1 μg of total RNA was retrotranscribed using either the miScript II RT kit (Qiagen) or the QuantiTect Reverse Transcription Kit (Qiagen) for the miRNA or the mRNA quantification, respectively. The reverse transcription was performed following the manufacturer's instructions. miRNAs and the mRNAs were quantified with Rotor-Gene Q 2plex (Qiagen), using the miScript SYBR Green PCR Kit (Qiagen) and the SsoAdvanced™ SYBR® Green Supermix (Bio-Rad), respectively, according to the manufacturer protocols. The relative quantification was performed using the Rotor-Gene Q software, normalizing to the internal controls (U6, SNORD55, and SNORD110 for miRNA and GAPDH, ACTB, and HPRT for mRNA). The relative miR-28a-5p expression level in tumor cell lines was evaluated with respect to the normal cell RNA (FirstChoice Human Total RNA, Ambion). All reactions were performed in triplicate, and the results are the mean of three biological replicates.

2.9. Western Blot Analysis. Proteins were extracted from cell pellets using lysis buffer (1 M Tris HCl pH 8, Triton X-100 1%, and Na deoxycholate 0.25%) with the addition of PMSF 1 mM. The proteins were then quantified colorimetrically using the BioRad protein Assay Reagent (Bio-Rad). Absorbance was measured at 595 nm with ChroMate microplate reader (Awareness Technology). The proteins were separated on polyacrylamide gels SDS-PAGE (10%, gel precast Mini-PROTEAN® TGX Stain-Free™, Bio-Rad) and transferred to 0.2 μm nitrocellulose membranes by electro blotting using the Trans-Blot Turbo Blotting System (Bio-Rad). The resulting blots were blocked with 5% nonfat dry milk solution in TBST. Anti-GAPDH (Cell Signaling) (1 : 20000), anti-E2F6 (Santa Cruz Biotechnology) (1 : 500), and PARP-1 (Santa

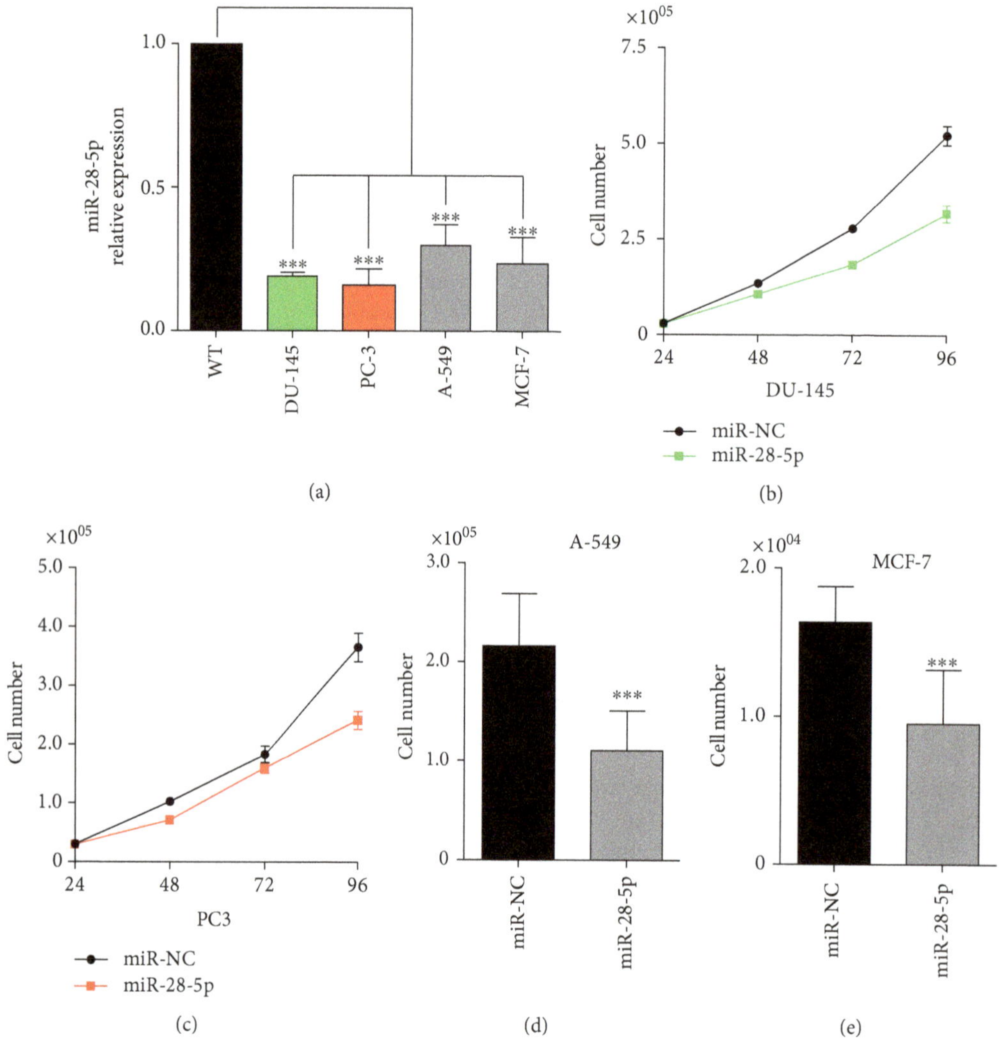

FIGURE 1: miR-28-5p expression and effect on tumor cells. (a) Analysis of the miR-28-5p expression level with qRT-PCR in prostate (PC-3 and DU-145), lung (A-549), and breast (MCF-7) cancer cell lines compared to the normal cell RNA. Cell proliferation of DU-145 (b), PC-3 (c), A-549 (d), and MCF-7 (e) cells at different time points or at 96 hours after the miR-28-5p reexpression. $^{***}P < 0.001$, unpaired t-test.

Cruz Biotechnology) (1 : 500) primary antibodies were used. Incubation was performed overnight at 4°C, and bands were revealed after incubation with the recommended secondary antibody coupled to peroxidase using ECL (GE Healthcare). Scanned images were quantified using ImageJ software and normalized to GAPDH.

2.10. Statistical Analyses. Results are expressed as mean SD of at least three independent experiments, and data are analyzed using Student's t-test ($^{*}P < 0.05$, $^{**}P < 0.01$, $^{***}P < 0.001$, and $^{****}P < 0.0001$).

3. Results

3.1. miR-28-5p Exerted a TS Activity in PCa Cells. To investigate the role of miR-28-5p in PCa, we first evaluated its expression in two PCa cell lines (DU-145 and PC-3) compared to normal cells (Figure 1(a)). We also evaluated the expression of miR-28-5p in lung (A-549) and breast (MCF-7) cancer cells, demonstrating that miR-28-5p was markedly downregulated in all the analyzed cancer cell lines.

In order to test whether miR-28-5p behaves as a TS in PCa cells, we first measured cell proliferation of DU-145 and PC-3 cells after miR-28-5p reexpression. Data showed a significant inhibition of cell proliferation of both PCa cell lines (Figures 1(b) and 1(c)). Similar results were obtained when the miR-28-5p was transfected in breast and colon cancer cell lines (Figures 1(d) and 1(e)).

To further investigate the biological effects of the miR-28-5p in PCa, we checked the colony-forming ability (CFA) and the cell cycle after miRNA reexpression in DU-145 cells. Data showed that miR-28-5p reexpression resulted in both a significant reduction of CFA (Figure 2(a)) and a slight but significant increase of cells in G1 phase (Figure 2(b)), suggesting that the proliferation was negatively affected. Finally, we demonstrated that the miR-28-5p reexpression increased apoptosis in DU-145 cells (Figures 2(c) and 2(d)). Overall data suggested that the miR-28-5p acts as a TS-miRNA in

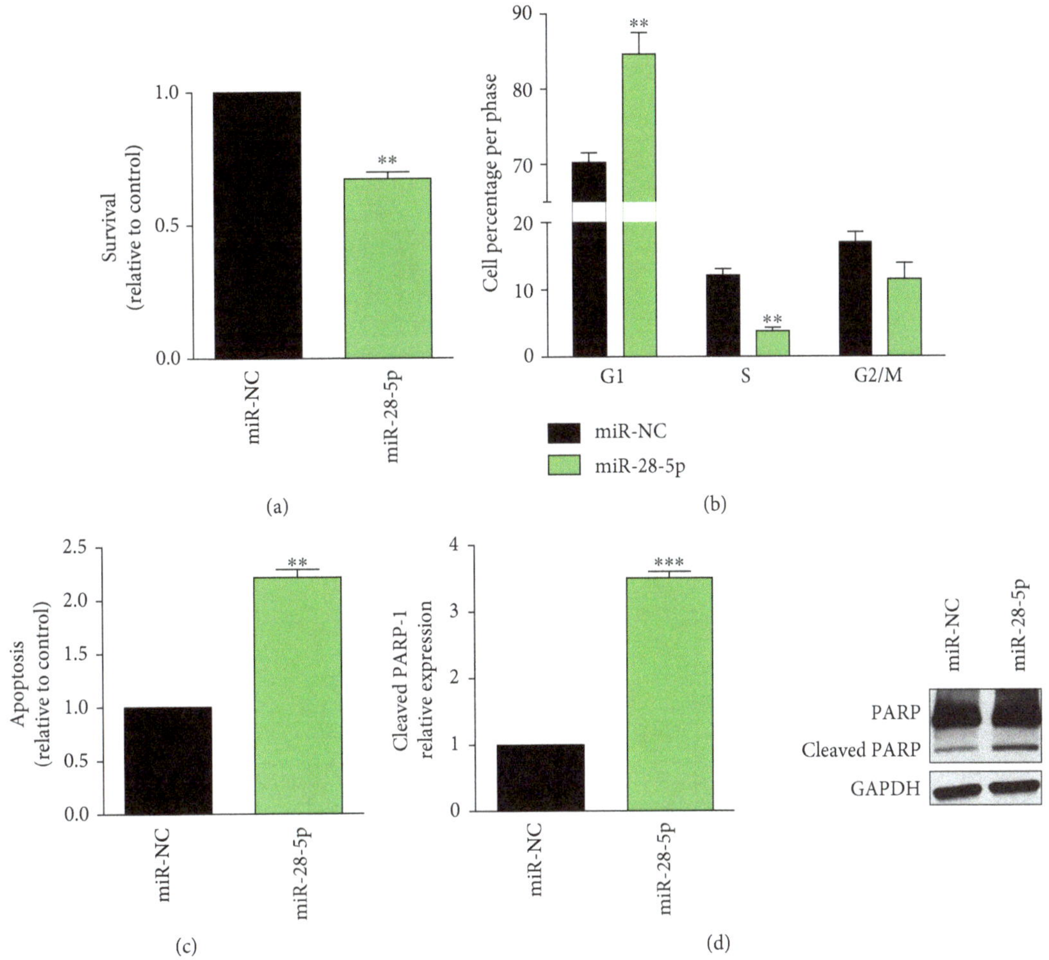

Figure 2: Effects of miR-28-5p reexpression on DU-145 cells. Cell survival (a) and cell cycle (b) in DU-145 cells after miR-28-5p reexpression. Apoptosis analysis measured with both annexin assay (c) and western blot of PARP-1 and cleaved PARP-1 (d) in miR-28-5p-transfected DU-145 cells. $^{**}P < 0.01$ and $^{***}P < 0.001$, unpaired t-test.

PCa, probably by regulating key pathways involved not only in tumor cell proliferation but also in tumor cell survival.

3.2. Some Validated miR-28-5p Targets Interacted with miR-28-5p in PCa. To investigate which targets were regulated by miR-28-5p in PCa, we transfected this miRNA in DU-145 cells and the miRNA pull out assay was performed [14]. This technique allowed the capture and the isolation of the miR-28-5p/target complexes using a biotinylated version of miR-28-5p. We considered the miR-28-5p targets deposited in miRTarBase (http://mirtarbase.mbc.nctu.edu.tw), in particular, the ones validated with the luciferase reporter assay (Table 1).

Using qRT-PCR, we checked the enrichment of these targets in the pool of miR-28-5p-captured targets (miR-28 pull out sample) and found that RAP1B, N4BP1, MPL, MAPK1, TEX-261, and E2F6 were enriched by more than 2-fold in the miR-28-5p pull out sample (Figure 3(a)). These results suggested that not all the miR-28-5p-validated targets interact and, as a consequence, may be regulated by miR-28-5p in PCa cells.

To verify whether the enrichment of the selected targets in the miR-28-5p pull out sample was indicative of the miRNA regulatory function, we selected the most enriched one, that is, E2F6, and determined its expression after the miR-28-5p reexpression in DU-145 cells. We demonstrated

Table 1: miR-28-5p targets validated with gene reporter assay according to miRTarBase.

miR-28-5p target	Tumor type	Reference
p21	Choriocarcinoma cells	[15]
MPL	Myeloproliferative neoplasms	[16]
N4BP1	Myeloproliferative neoplasms/ ovarian cancer	[9]
OTUB1	Myeloproliferative neoplasms	[16]
TEX-261	Myeloproliferative neoplasms	[16]
MAPK1	Myeloproliferative neoplasms	[16]
E2F6	Myeloproliferative neoplasms	[16]
MAD2L1	B-cell lymphomas	[5]
BAG1	B-cell lymphomas	[5]
RAP1B	B-cell lymphomas/renal cell carcinoma	[3, 5]

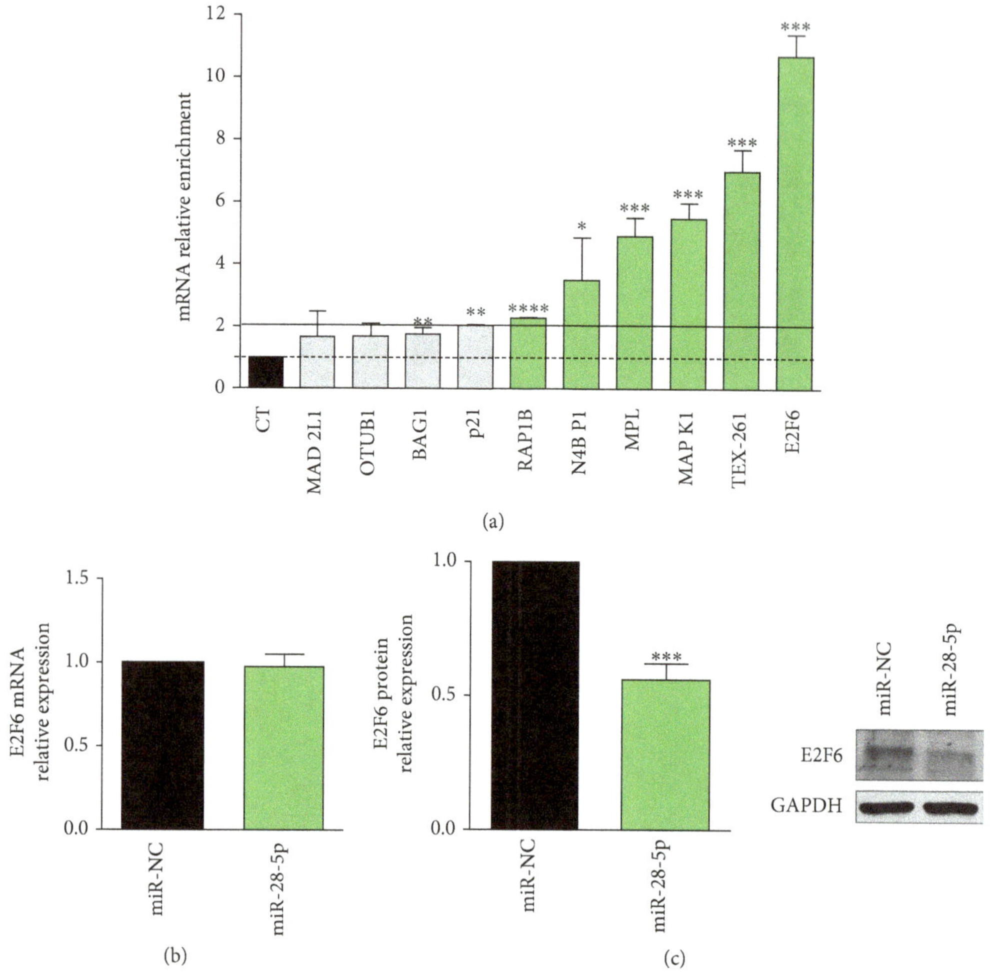

Figure 3: miR-28-5p target validation. (a) miR-28-5p-validated target enrichment quantified by qRT-PCR in miR-28-5p pull out sample compared to miR-28-5p control pull out sample (CT). E2F6 mRNA (b) and protein (c) quantification in DU-145 cells transfected with miR-28-5p or miR-NC. $^{*}P < 0.05$, $^{**}P < 0.01$, $^{***}P < 0.001$, and $^{****}P < 0.0001$ unpaired t-test.

that E2F6 was inhibited by miR-28-5p reexpression only at the protein level (Figures 3(b) and 3(c)), indicating that E2F6 was regulated by miR-28-5p in PCa.

4. Discussion

miRNAs are key inhibitors of gene expression that play a pivotal role in tumor development and progression affecting genes and pathways involved in all the hallmarks of cancer [17, 18]. By regulating oncogenes or TS genes, they can act, respectively, as a TS or oncogenes, although it is known that the role of a specific miRNA in cancer is not absolute but strongly related to the tumor context [1]. Therefore, it is not surprising that miR-28-5p may act as an oncogene (e.g., [9]) or as a TS (e.g., [13]), even though in most tumors, it showed an antiproliferative effect. It is possible that, depending on the expression level of the targets or on the effect of regulators which interfere with miRISC binding or function, not all the targets of a particular miRNA can be bound and regulated by the miRNA in a specific context [19, 20].

In this work, performing the miR-28-5p pull out assay, we have explored whether the known miR-28-5p targets, deposited in miRTarBase and validated with at least the luciferase reporter assay, interacted and were regulated by miR-28-5p in PCa. Under this strategy was that, among all miR-28-5p-validated targets, the ones that interact with miR-28-5p in PCa have higher chance to be regulated by the miRNA in this tumor. We first evaluated the miR-28-5p role in PCa, and we demonstrated for the first time that this miRNA is underexpressed in these cells and that its reexpression inhibited cell proliferation and survival. These data led us to conclude that miR-28-5p acted as a TS-miRNA in PCa. As it has been demonstrated that miR-28-5p negatively regulated genes involved in tumor cell growth in lung [12] as well as in colon rectal [6] cancer cells, we checked the expression level and the effects of miR-28-5p in lung and colon cancer cell lines as a positive control.

Using the miRNA pull out assay, we found that not all the validated miR-28-5p targets were enriched in the miR-28 pull out sample. Among the enriched targets that are more strongly associated with cancer (i.e., RAP1B, N4BP1, MAPK1, and E2F6), almost all are protumoral. Indeed, both MAPK1 and RAP1B, a Ras-related small GTP-binding protein that acts as GTPase in several signaling cascades, are proproliferative proteins involved as oncogenes in the development and progression of several tumor types (e.g., [21, 22]). In particular, it has been demonstrated that miR-28-5p suppressed cell proliferation and migration by directly inhibiting RAP1B in renal cell carcinoma [22]. These observations are consistent with a possible regulation by miR-28-5p, suggesting the utility of our approach to identify context-specific miRNA targets. In the same way, it has been reported that when PCa evolves from a benign to more aggressive stage, it becomes resistant to apoptosis due to the increased expression of antiapoptotic proteins [23] such as E2F6 [24]. Indeed, we demonstrated that E2F6, the most enriched target in the miR-28-5p pull out sample, was regulated by miR-28-5p at the posttranscriptional level. We can speculate that the target enrichment level in the miRNA pull out sample might facilitate the identification of the targets affected by the regulation of the miRNA. In addition, we also showed that the miR-28-5p reexpression induced apoptosis in DU-145 cells. Given that DU-145 cells (androgen independent PCa cell line) are an advanced PCa *in vitro* model, miR-28-5p reexpression may be taken into consideration as a novel therapeutic approach for PCa at this stage. Furthermore, we demonstrated that E2F6 is regulated by miR-28-5p in DU-145 cells; thus, it is conceivable that E2F6 could be one of the mediators of the apoptosis induced by miR-28-5p reexpression.

In conclusion, in this work, we demonstrated that the capture of the targets that interact with a given miRNA in a specific tumor is a suitable approach to identify the subset of targets that have a higher probability of being regulated by that miRNA in the context under evaluation. In the future, the identification of all the miR-28-5p targets (miR-28-5p targetome) could help to decipher the genes and pathways affected by the regulation of this miRNA in PCa.

Conflicts of Interest

The authors declare that there are no conflicts of interest.

Acknowledgments

The authors would like to thank Dr. Marcella Simili for critical reading and Dr. Mike Minks for the revision of the manuscript. This work was supported by the Istituto Toscano Tumori (Grant 2010, Giuseppe Rainaldi; Grant 2013, Milena Rizzo).

References

[1] H. Ling, L. Girnita, O. Buda, and G. A. Calin, "Non-coding RNAs: the cancer genome dark matter that matters!," *Clinical Chemistry and Laboratory Medicine*, vol. 55, no. 5, pp. 705–714, 2017.

[2] X. Shi and F. Teng, "Down-regulated miR-28-5p in human hepatocellular carcinoma correlated with tumor proliferation and migration by targeting insulin-like growth factor-1 (IGF-1)," *Molecular and Cellular Biochemistry*, vol. 408, no. 1-2, pp. 283–293, 2015.

[3] C. Wang, C. Wu, Q. Yang et al., "miR-28-5p acts as a tumor suppressor in renal cell carcinoma for multiple antitumor effects by targeting RAP1B," *Oncotarget*, vol. 7, no. 45, pp. 73888–73902, 2016.

[4] S. B. Ng, J. Yan, G. Huang et al., "Dysregulated microRNAs affect pathways and targets of biologic relevance in nasal-type natural killer/T-cell lymphoma," *Blood*, vol. 118, no. 18, pp. 4919–4929, 2011.

[5] C. Schneider, M. Setty, A. B. Holmes et al., "MicroRNA 28 controls cell proliferation and is down-regulated in B-cell lymphomas," *Proceedings of the National Academy of Sciences of the United States of America*, vol. 111, no. 22, pp. 8185–8190, 2014.

[6] M. I. Almeida, M. S. Nicoloso, L. Zeng et al., "Strand-specific miR-28-5p and miR-28-3p have distinct effects in colorectal cancer cells," *Gastroenterology*, vol. 142, no. 4, pp. 886–896.e9, 2012.

[7] J. M. Sayagués, L. A. Corchete, M. L. Gutiérrez et al., "Genomic characterization of liver metastases from colorectal cancer patients," *Oncotarget*, vol. 7, no. 45, pp. 72908–72922, 2016.

[8] P. Vychytilova-Faltejskova, M. Pesta, L. Radova et al., "Genome-wide microRNA expression profiling in primary tumors and matched liver metastasis of patients with colorectal cancer," *Cancer Genomics Proteomics*, vol. 13, no. 4, pp. 311–316, 2016.

[9] J. Xu, N. Jiang, H. Shi, S. Zhao, S. Yao, and H. Shen, "miR-28-5p promotes the development and progression of ovarian cancer through inhibition of N4BP1," *International Journal of Oncology*, vol. 50, 2017.

[10] Z. Huang, L. Zhang, D. Zhu et al., "A novel serum microRNA signature to screen esophageal squamous cell carcinoma," *Cancer Medicine*, vol. 6, no. 1, pp. 109–119, 2017.

[11] S. M. Wilting, P. J. Snijders, W. Verlaat et al., "Altered microRNA expression associated with chromosomal changes contributes to cervical carcinogenesis," *Oncogene*, vol. 32, no. 1, pp. 106–116, 2013.

[12] Y. C. Choi, S. Yoon, Y. Byun et al., "MicroRNA library screening identifies growth-suppressive microRNAs that regulate genes involved in cell cycle progression and apoptosis," *Experimental Cell Research*, vol. 339, no. 2, pp. 320–332, 2015.

[13] N. Bartolomé-Izquierdo, V. G. de Yébenes, A. F. Álvarez-Prado et al., "miR-28 regulates the germinal center reaction and blocks tumor growth in preclinical models of non-Hodgkin lymphoma," *Blood*, vol. 129, no. 17, pp. 2408–2419, 2017.

[14] M. Rizzo, G. Berti, F. Russo, S. Fazio, M. Evangelista, and G. Rainaldi, "Discovering the miR-26a-5p targetome in prostate cancer cells," *Journal of Cancer*, vol. 8, no. 14, pp. 2729–2739, 2017.

[15] S. Wu, S. Huang, J. Ding et al., "Multiple microRNAs modulate p21Cip1/Waf1 expression by directly targeting its 3′ untranslated region," *Oncogene*, vol. 29, no. 15, pp. 2302–2308, 2010.

[16] M. Girardot, C. Pecquet, S. Boukour et al., "miR-28 is a thrombopoietin receptor targeting microRNA detected in a fraction of myeloproliferative neoplasm patient platelets," *Blood*, vol. 116, no. 3, pp. 437–445, 2010.

[17] V. G. Manasa and S. Kannan, "Impact of microRNA dynamics on cancer hallmarks: an oral cancer scenario," *Tumour Biology*, vol. 39, no. 3, 2017.

[18] R. Rupaimoole and F. J. Slack, "MicroRNA therapeutics: towards a new era for the management of cancer and other diseases," *Nature Reviews Drug Discovery*, vol. 16, no. 3, pp. 203–222, 2017.

[19] A. E. Pasquinelli, "MicroRNAs and their targets: recognition, regulation and an emerging reciprocal relationship," *Nature Reviews Genetics*, vol. 13, no. 4, pp. 271–282, 2012.

[20] N. Cloonan, "Re-thinking miRNA-mRNA interactions: intertwining issues confound target discovery," *BioEssays*, vol. 37, no. 4, pp. 379–388, 2015.

[21] W. Li, J. Liang, Z. Zhang et al., "MicroRNA-329-3p targets MAPK1 to suppress cell proliferation, migration and invasion in cervical cancer," *Oncology Reports*, vol. 37, no. 5, pp. 2743–2750, 2017.

[22] Y. Li, Y. Liu, F. Shi, L. Cheng, and J. She, "Knockdown of Rap1b enhances apoptosis and autophagy in gastric cancer cells via the PI3K/Akt/mTOR pathway," *Oncology Research*, vol. 24, no. 5, pp. 287–293, 2016.

[23] K. L. Mahon, S. M. Henshall, R. L. Sutherland, and L. G. Horvath, "Pathways of chemotherapy resistance in castration-resistant prostate cancer," *Endocrine-Related Cancer*, vol. 18, no. 4, pp. R103–R123, 2011.

[24] N. Bhatnagar, X. Li, S. K. Padi, Q. Zhang, M. S. Tang, and B. Guo, "Downregulation of miR-205 and miR-31 confers resistance to chemotherapy-induced apoptosis in prostate cancer cells," *Cell Death & Disease*, vol. 1, article e105, 2010.

5

Insights on Genomic and Molecular Alterations in Multiple Myeloma and their Incorporation towards Risk-Adapted Treatment Strategy

Taiga Nishihori[1] and Kenneth Shain[2]

[1]*Department of Blood and Marrow Transplantation, H. Lee Moffitt Cancer Center and Research Institute, Tampa, FL, USA*
[2]*Department of Malignant Hematology, H. Lee Moffitt Cancer Center and Research Institute, Tampa, FL, USA*

Correspondence should be addressed to Taiga Nishihori; taiga.nishihori@moffitt.org

Academic Editor: Lam C. Tsoi

Although recent advances in novel treatment approaches and therapeutics have shifted the treatment landscape of multiple myeloma, it remains an incurable plasma cell malignancy. Growing knowledge of the genome and expressed genomic information characterizing the biologic behavior of multiple myeloma continues to accumulate. However, translation and incorporation of vast molecular understanding of complex tumor biology to deliver personalized and precision treatment to cure multiple myeloma have not been successful to date. Our review focuses on current evidence and understanding of myeloma biology with characterization in the context of genomic and molecular alterations. We also discuss future clinical application of the genomic and molecular knowledge, and more translational research is needed to benefit our myeloma patients.

1. Introduction

Multiple myeloma remains an all but incurable terminally differentiated B cell malignancy. We have long understood that multiple myeloma evolves from a premalignant condition of terminally differentiated B cells. Many, if not all, myeloma patients are considered to have harbored a premalignant monoclonal gammopathy of undetermined significance (MGUS) that progresses to smoldering or asymptomatic multiple myeloma and ultimately to symptomatic multiple myeloma requiring therapeutic intervention [1]. This "symptomatic" threshold is characterized as the worsening of myeloma associated multiorgan dysfunction and may be associated with altered genetic profiles [2–4]. Typical symptomatic presentation of multiple myeloma includes hypercalcemia, renal failure, anemia, and bone lesions (i.e., lytic lesions) and/or fractures, collectively known as "CRAB." The recent explosion of available antimyeloma agents targeting particular pathways or antigens including various proteasome inhibitors (such as bortezomib, carfilzomib, and ixazomib), immunomodulatory agents, pan-histone deacetylase inhibitor, and anti-CS1 and anti-CD38 monoclonal antibodies, as well as widespread adaptation of high-dose chemotherapy followed by autologous hematopoietic cell transplantation (HCT) has led to growing optimism for the treatment of myeloma and indeed has shown overall improvement in therapeutic outlook and survivals [1, 5]. Despite these remarkable achievements, long term control of myeloma can only be achieved in minority of patients in part due to inevitable development of clonal evolution and therapy resistance stemming from a number of factors including influences on the genome including the stress of therapy and interactions of myeloma cells and bone marrow microenvironment in the context of environment mediated drug resistance (EMDR) [6, 7]. Clonal evolution results in genotypically dynamic and, therefore, phenotypically dynamic myeloma cell populations with varying sensitivities to therapy. As such, there is a critical need to be able to characterize the molecular aspects of disease both at diagnosis (prognostic) and throughout the course disease to best allocate care and understand the best use of available antimyeloma agents to control this heterogeneous disease.

The last decade has observed rapid growth of knowledge pertaining to genomic and molecular characterization of multiple myeloma which was made possible due to technological advancements. Myeloma characterization has evolved from traditional microscopic examination and conventional flow cytometry to genomic analysis using methods for gene expression profiling (GEP), proteomics, and beyond. Most clinical risk stratification relies on standard karyotypic analysis and fluorescent in situ hybridization (FISH) which comprise the bulk of high-risk definition but emerging data exist to incorporate GEP data into prognostication of myeloma. In this review, we focus on the review of current knowledge with regard to risk stratification and incorporation of GEP and further genomic and molecular information to future myeloma treatment strategies.

2. Known Genetic Alterations Associated with High-Risk Myeloma

For a decade we have developed different molecular tools to characterize a classic set of myeloma-related genetic alterations involving duplication of odd numbered chromosomes, translocations of immunoglobulin heavy chain (IgH) enhancers and oncogenes, and deletions of tumor suppressors that aid us in prognostic risk stratification. The anomalies aid in the parsing of patients into low- and high-risk stages/classifications. Complete and comprehensive characterization of cytogenetic changes in multiple myeloma has been reviewed elsewhere and it is beyond the scope of this brief review [8]. The dichotomy in risk dictates that standard (low-risk) risk is defined by the presence of hyperdiploidy (multiple copies of odd numbered chromosomes) and translocations involving IgH enhancers with D-type cyclins (t(6;14) and t(11;14)). Here, we focus on a number of representative chromosomal alterations commonly seen in high-risk multiple myeloma, a disease state that remains to be the bane of myeloma therapy. The t(4;14) (p16;q32) has been shown to be associated with poor survival [9, 10]. This translocation results in increased expression of fibroblast growth factor receptor 3 (FGFR3) and multiple myeloma SET (MMSET) domain [11, 12]. Translocations involving *maf* genes have been found in minority of myeloma and these derive from IgH rearrangements with a locus in chromosome 16, most commonly t(14;16)(q32;q23). The t(14;16) or other 16q abnormalities may occur in conjunction with chromosome 13 deletion, another common genetic alteration in multiple myeloma portending increased risk [9]. One study showed deletion of 13q in 85% of myeloma patients with t(4;14) and 92% of t(14;16) suggesting strong correlations between deletion 13q and 14q32 rearrangements [13]. These chromosomal changes involving 4 and 16 are thought to be primary high-risk genetic events [14]. Further, studies have also suggested that loss of chromosome 16 and/or increased expression of the *FOPNL* gene at 16p13 may be linked to poorer outcomes in myeloma [15].

Additionally, monoallelic deletion of 17p13, the locus of tumor suppressor gene *p53*, confers aggressive clinical course, extramedullary disease, and inferior survival [10, 14]. Majority of chromosome 1 abnormalities involve rearrangements located in the pericentromeric regions and chromosome 1 abnormalities have been relatively recently added to major negative prognostic factor in myeloma including transplant-eligible patients [16–18]. Both 1q gain (or 1q21 amplification) and 1p loss could occur as chromosome 1 abnormalities. Gene expression signature for high-risk disease is also enriched disproportionately for genes located in chromosome 1 supporting the importance of chromosome 1 abnormalities in risk stratification [16]. Biologic and genetic underpinning of chromosome 1 abnormalities has been debated but some initial studies suggested possible association of *CKS1B* [19]. Deletion of chromosome 13 or 13q when encountered in conventional karyotyping (but not by FISH) is also considered high-risk.

These recurrent high-risk cytogenetic changes may coexist with other cytogenetic alterations and the combination of these features may modulate the risk status. For example, trisomies of chromosomes 3 and 5 in addition to known high-risk cytogenetics may lessen the overall high-risk significance and trisomy of chromosome 21 may worsen the outcomes [20, 21]. These represent specific examples; however, with our growing ability to molecularly characterize the genome and the expressed genome we anticipate that we may be able to provide even better risk stratification even when faced with coexistent complex molecular changes in a personalized context and possibly dictate therapy [22].

3. Current Risk Stratification and Clinical Staging of Multiple Myeloma

Prognostic staging of myeloma has long been limited to assessment of laboratory values (albumin and beta-2 microglobulin) as defined by the International Staging System (ISS) [23]. With a parallel risk stratification based on myelomatous molecular changes and in recognition of the importance of these factors collectively in portending risk for myeloma patients, the International Myeloma Working Group (IMWG) published their consensus statement on myeloma risk classification and reported that combination of (a) ISS stage II or III and (b) t(4;14) or deletion of 17p13 would be considered high-risk myeloma on the basis of limited overall survival (OS) of approximately 2 years [24]. Further, a similar incorporation of this molecular high-risk classification was used in establishing the Revised-ISS (R-ISS) with the addition of t(14;16) as well as elevated lactate dehydrogenase (LDH) and those with R-ISS stage III myeloma are projected to have median OS of 43 months highlighting the significant adverse impact of certain genomic alterations in myeloma over a simple unidimensional measurement of tumor burden as the disease stage [25]. IMWG also posed therapeutic questions for the high-risk group by asking the research community to examine novel therapeutic strategies such as allogeneic hematopoietic cell transplantation or immunotherapy to improve the outcomes [24].

4. Gene Expression Profiling in Myeloma

High-throughput genomic tools such as GEP have been extensively investigated in recent years to predict patient outcomes. A pioneering effort in multiple myeloma with GEP was made by the group at University of Arkansas for Medical Sciences (UAMS) where they identified an initial 70-gene signature allowing a new definition of high-risk myeloma by GEP under the treatment platform of Total Therapy II which was followed by a reduced number of 17-gene model providing prognostic information [16]. Of note, this high-risk gene profile seems to be enriched for genes located in chromosome 1. A recent logistic regression analysis of the 70-gene score using the data from UAMS Total Therapies II and III by van Laar et al. showed that the 70-gene prognostic risk score is continuously associated with increased risk of 5-year relapse and death reconfirming the importance of this gene score [26]. The Intergroupe Francophone du Myélome (IFM) also identified 15 genes that would predict poor prognosis in multiple myeloma [27]. GEP studies suggested that there are possibly up to 8 different molecular subtypes in multiple myeloma and opened our eyes to the complexity and difficulty of molecular characterization of myeloma [28, 29]. Indeed, there are several other GEP-based prognostic signatures that were developed including centrosome index [30], genes with homozygous deletion [31], human myeloma cell lines with IL-6 stimulation [32], proliferation index [33], and others [34, 35]. Interestingly, these individual GEP signatures demonstrate minimal overlap amongst them; however, this is likely expected as each GEP signature might be associated with different but important myeloma tumor biology (such as proliferation) related genes thus a minimum number of such genes would be represented in each signature. Although each GEP signature has been validated using different independent data sets, there is no universal agreement on the GEP signature(s) to be considered as a gold standard of high-risk definition as each of these GEP signature was developed in the different treatment and baseline risk context.

The gene expression microarray has become the most commonly used technology for genomic investigation in oncology which is also robust and has good interlaboratory agreement [36]. There are increasing number of clinical applications using GEP in medicine for more refined classifications of a tumor that was not previously possible or for treatment decisions; however, the latter lags behind likely due to multitude of unresolved issues. For example, there are several GEP signatures reported for multiple myeloma as above and these different signatures do not necessarily correlate with one another. Some investigators have reported their attempts to incorporate GEP information into clinical practice with the aim to provide a precision patient care in multiple myeloma. Meißner et al. built the GEP-report (GEP-R) using an open-source software developed in R using Affymetrix microarrays [37]. GEP-R integrated GEP-based (both UAMS- and IFM-scores) and conventional prognostic factors such as proliferation, ISS stage, t(4;14), and expression of prognostic target genes *(AURKA, IGF1R)* and developed one risk stratification termed HM-metascore [37]. The HM-metascore distinguished 3 different risk groups, low-risk, medium-risk versus high-risk with 6-year survival rates of 89.3%, 60.6%, and 18.6%, respectively. This system allows automated interpretation of Affymetrix U133 Plus 2.0 gene expression profiles which has been in use in some European centers for both routine clinical practice and clinical trials. The larger issue might be related to the fact that high-risk GEP signature has yet to be linked with specific treatment options. For example, Hose et al. identified varying frequencies of aurora-A and aurora-B expression in myeloma samples using Affymetrix DNA microarrays and presence of aurora-A expression was associated with inferior event-free and OS [38]; however, aurora kinase inhibitors have not been commercially available to translate this finding into clinical practice. Additionally, it is still possible that novel GEP signature may be identified or defined in the future which may be of clinical interest to predict specific novel therapeutic interventions especially for immunotherapeutic approaches and we expect that the landscape of high-risk myeloma GEP signatures would evolve over time. To date, GEP has not facilitated the rationale selection of antimyeloma agents and RNA sequencing may replace microarray-based GEP in the future. More comprehensive analysis of differential RNA splicing and isoform expression are expected to inform myeloma characterization.

5. Moving towards Risk-Adapted Therapeutic Approaches in Myeloma Using Genomic Information

Though risk stratification is commonly performed, there are currently no widely accepted consensus risk-adapted approaches for multiple myeloma treatment based on genomic information. National Comprehensive Cancer Network (NCCN) guidelines on multiple myeloma provide list of potential combination chemotherapy options but they do not suggest modifications of treatment strategy based on risk status [39]. Group of investigators at Mayo Clinic published myeloma treatment guidelines which suggest the utilization of myeloma risk stratification to generate risk-adapted therapy approaches termed mSMART (https://www.msmart.org/) [40]. mSMART version 2.0 classifies deletion of 17p, t(14;16) and t(14;20) as high-risk and t(4;14) and 1q gain as intermediate-risk myeloma. They suggest to consider carfilzomib/lenalidomide/dexamethasone induction in transplant-eligible high-risk myeloma population. Bortezomib-based maintenance is considered after autologous HCT in both intermediate- and high-risk population with consideration of carfilzomib-based maintenance in high-risk myeloma.

With more antimyeloma agents in hand in the past few years, data exist to consider differential usage of target specific therapeutic approaches which would be expected to evolve over time. Some studies suggested that bortezomib-based regimen may abrogate or reduce the negative survival impact of t(4;14) and t(14;16) or possibly deletion of 17p [41–44]. In addition, at least 2 reports suggested that incorporation of bortezomib before and after autologous HCT might overcome some of the poor prognostic significance posed by deletion of 17p13 [45, 46]. Of note, though thalidomide has fallen out of favor for maintenance therapy after the availability of lenalidomide, the Medical Research Council (MRC) IX study showed that patients with high-risk disease (including gain of 1q, deletion of 1p32, t(4;14), t(14;16), and t(14;20), and deletion of 17p) had worse OS with thalidomide maintenance [47]. Emerging evidence indicates that newer antimyeloma agents such as carfilzomib and pomalidomide may preferentially improve outcomes in high-risk myeloma patients [48–50].

While only limited data from randomized controlled studies are available to inform the development of universal consensus on risk-adapted treatment strategies, the risk stratification schema could serve as useful frameworks for the development of rational design and treatment selection of specific myeloma therapeutic clinical trials. For instance, Bone Marrow Transplant Clinical Trial Network (BMT CTN) has designed a multicenter randomized placebo-controlled trial evaluating a novel allogeneic HCT clinical trial using fludarabine, melphalan, and bortezomib conditioning and posttransplant ixazomib maintenance specifically targeting high-risk myeloma patients (https://clinicaltrials.gov/ct2/show/NCT02440464). Investigators at Johns Hopkins have also launched a phase 2 clinical trial evaluating the role of marrow infiltrating lymphocytes (MILs) in high-risk myeloma in the context of high-dose melphalan and autologous HCT (https://clinicaltrials.gov/ct2/show/NCT01858558) [51]. Though not universally accepted as a high-risk definition, some clinical trials use a commercially available GEP signature as one of the study eligibility criteria for high-risk myeloma. It is expected that many future novel immunotherapeutic approaches including chimeric antigen receptor (CAR) T cell therapy would be designed using the risk stratification to overcome poor overall outcomes in high-risk myeloma patients.

6. Growing Knowledge of Genomic and Molecular Information in Myeloma Characterization

Various DNA-based high-throughput technologies (also termed next generation sequencing (NGS)) including whole-genome sequencing (WGS) and whole-exome sequencing (WES) have been developed and have become the common genomic tools to evaluate the origin and evolution of cancer cells over the past 2 decades. These technologies have also been applied to expand and better understand the biology of multiple myeloma. Chapman et al. performed the initial study of WGS in 38 myeloma patients [52] which was followed by at least 3 other large scale NGS studies in myeloma to evaluate the mutational landscape [53, 54]. The study by Chapman et al. showed 10 statistically significant protein-coding mutations at a false discovery rate of ≤ 0.10 including *NRAS*, *KRAS*, *FAM46C*, *DIS3*, *TP53*, *CCND1*, *PNRC1*, *ALOX12B*, *HLA-A*, and *MAGED1* but frequencies were low [52]. Of note, *BRAF* mutations (G469A, K601N, and V600E) in myeloma were discovered albeit at low frequency (7 out of 161 patients = 4%) [52]. Walker et al. performed WES in over 400 myeloma patients through Myeloma XI trial and reported somewhat overlapping significant mutations in 15 genes [55]. Taken together, genes involved in mitogen-activated protein kinase (MAPK) pathway such as *NRAS*, *KRAS*, and *BRAF* (RAS pathway 43%) and those in the nuclear factor (NF)-kappa B pathway (17%) are recurrently mutated in myeloma and these are likely important drivers of myelomagenesis [55].

These studies showed wide variety of mutations with varying frequencies and the significance of individual gene mutations in relation to survival outcomes might be difficult to discern due to the complex subclonal mixtures in one patient. However, in one study, a complex subclonal structure and clusters of subclonal variants correlated with a high-risk of relapse and death [53]. The study also identified new candidate genes including truncations of *SP140*, *LTB*, and *ROBO1* and clustered missense mutations in *EGR1* [53]. In another study, presence of *TP53*, *ATM*, *ATR*, and *CCND1* mutations was associated with inferior OS in multivariate analysis [55]. It is conceivable that emerging mutational and molecular information may be incorporated into the future risk stratification and prognostication of myeloma patients.

Moving further into personalized treatment decision making, this mutational profiling may guide therapeutic choice in the future with ever increasing number of available antimyeloma agents or nonmyeloma specific but pathway specific drugs. In a study of 133 patients with relapsed myeloma treated with single-agent bortezomib, a targeted sequencing of a panel of 41 known oncogenes and tumor suppressor genes was performed [56]. Mutually exclusive mutations of *BRAF* and *RAS* were seen in closer to a half of patients and mutations of *NRAS* (but not *KRAS*) were associated with lower response rate to bortezomib [56]. On the other hand, mutations in *IRF4* which are believed to be the downstream of immunomodulatory drug target of cereblon seem to be associated with better survival [55]. These studies uncovered actionable mutations in multiple myeloma, and coupled with explosion of various pathways and targeted inhibitors there is a growing optimism to hope for mutation specific adapted treatment strategies becoming reality in the near future.

However, detailed characterization of mutational landscape in multiple myeloma also disclosed complexity in predicting and interpreting the NGS data due to higher number of coexisting clones and subclonal mutations. Clonal evolution in myeloma has been documented to involve linear evolution, differential clonal response, and branching evolution [53]. Careful delineation of subclonal population with sequential targeted sequencing would likely be required to accurately predict treatment outcomes using mutational information. Another challenge may involve coexistence of downstream activating mutations in the same pathway which may render some of the targeted therapies ineffective. Additionally, we have to acknowledge that genes are the essential component of a phenotype though the latter may also be influenced by other external or environmental factors. Epigenetic modifications such as histone methylation as well as the stresses of therapy and myeloma ecosystem (bone marrow microenvironment) could also potentially influence myeloma clinical behavior (i.e., clonal evolution) and further research is needed to investigate these possibilities. Therefore, there is an urgent need to develop comprehensive informatics to combine all genomic, epigenetic, and other molecular information to accurately prognosticate myeloma patients' survival and to rationally select target specific therapies based on genomic and epigenetic alterations.

Competing Interests

Taiga Nishihori receives research funding from Novartis and Signal Genetics, Inc. Kenneth Shain received research funding from AbbVie and Signal Genetics, Inc.

References

[1] A. Palumbo and K. Anderson, "Multiple myeloma," *New England Journal of Medicine*, vol. 364, no. 11, pp. 1046–1060, 2011.

[2] P. Broderick, D. Chubb, D. C. Johnson et al., "Common variation at 3p22.1 and 7p15.3 influences multiple myeloma risk," *Nature Genetics*, vol. 44, no. 1, pp. 58–61, 2012.

[3] L. López-Corral, N. C. Gutiérrez, M. B. Vidriales et al., "The progression from MGUS to smoldering myeloma and eventually to multiple myeloma involves a clonal expansion of genetically abnormal plasma cells," *Clinical Cancer Research*, vol. 17, no. 7, pp. 1692–1700, 2011.

[4] N. Weinhold, D. C. Johnson, A. C. Rawstron et al., "Inherited genetic susceptibility to monoclonal gammopathy of unknown significance," *Blood*, vol. 123, no. 16, pp. 2513–2517, 2014.

[5] S. K. Kumar, S. V. Rajkumar, A. Dispenzieri et al., "Improved survival in multiple myeloma and the impact of novel therapies," *Blood*, vol. 111, no. 5, pp. 2516–2520, 2008.

[6] K. H. Shain and W. S. Dalton, "Environmental-mediated drug resistance: a target for multiple myeloma therapy," *Expert Review of Hematology*, vol. 2, no. 6, pp. 649–662, 2009.

[7] K. H. Shain, W. S. Dalton, and J. Tao, "The tumor microenvironment shapes hallmarks of mature B-cell malignancies," *Oncogene*, vol. 34, no. 36, pp. 4673–4682, 2015.

[8] R. Fonseca, P. L. Bergsagel, J. Drach et al., "International myeloma working group molecular classification of multiple myeloma: spotlight review," *Leukemia*, vol. 23, no. 12, pp. 2210–2221, 2009.

[9] R. Fonseca, E. Blood, M. Rue et al., "Clinical and biologic implications of recurrent genomic aberrations in myeloma," *Blood*, vol. 101, no. 11, pp. 4569–4575, 2003.

[10] H. Avet-Loiseau, M. Attal, P. Moreau et al., "Genetic abnormalities and survival in multiple myeloma: the experience of the Intergroupe Francophone du Myélome," *Blood*, vol. 109, no. 8, pp. 3489–3495, 2007.

[11] M. Chesi, E. Nardini, R. S. C. Lim, K. D. Smith, W. Michael Kuehl, and P. L. Bergsagel, "The t(4;14) translocation in myeloma dysregulates both FGFR3 and a novel gene, MMSET, resulting in IgH/MMSET hybrid transcripts," *Blood*, vol. 92, no. 9, pp. 3025–3034, 1998.

[12] M. Chesi, E. Nardini, L. A. Brents et al., "Frequent translocation t(4;14)(p16.3;q32.3) in multiple myeloma is associated with increased expression and activating mutations of fibroblast growth factor receptor 3," *Nature Genetics*, vol. 16, no. 3, pp. 260–264, 1997.

[13] H. Avet-Loiseau, T. Facon, B. Grosbois et al., "Oncogenesis of multiple myeloma: 14q32 and 13q chromosomal abnormalities are not randomly distributed, but correlate with natural history, immunological features, and clinical presentation," *Blood*, vol. 99, no. 6, pp. 2185–2191, 2002.

[14] W. M. Kuehl and P. L. Bergsagel, "Multiple myeloma: evolving genetic events and host interactions," *Nature Reviews Cancer*, vol. 2, no. 3, pp. 175–187, 2002.

[15] E. Ziv, E. Dean, D. Hu et al., "Genome-wide association study identifies variants at 16p13 associated with survival in multiple myeloma patients," *Nature Communications*, vol. 6, article 7539, 2015.

[16] J. D. Shaughnessy Jr., F. Zhan, B. E. Burington et al., "Avalidated gene expression model of high-risk multiple myeloma is defined by deregulated expression of genes mapping to chromosome 1," *Blood*, vol. 109, no. 6, pp. 2276–2284, 2007.

[17] J. R. Sawyer, G. Tricot, S. Mattox, S. Jagannath, and B. Barlogie, "Jumping translocations of chromosome 1q in multiple myeloma: evidence for a mechanism involving decondensation of pericentromeric heterochromatin," *Blood*, vol. 91, no. 5, pp. 1732–1741, 1998.

[18] F. Bock, G. Lu, S. A. Srour et al., "Outcome of patients with multiple myeloma and CKS1B gene amplification after autologous hematopoietic stem cell transplantation," *Biology of Blood and Marrow Transplantation*, vol. 22, no. 12, pp. 2159–2164, 2016.

[19] R. Fonseca, S. A. Van Wier, W. J. Chng et al., "Prognostic value of chromosome 1q21 gain by fluorescent in situ hybridization and increase CKS1B expression in myeloma," *Leukemia*, vol. 20, no. 11, pp. 2034–2040, 2006.

[20] P. Moreau, M. Attal, F. Garban et al., "Heterogeneity of t(4;14) in multiple myeloma. Long-term follow-up of 100 cases treated with tandem transplantation in IFM99 trials," *Leukemia*, vol. 21, no. 9, pp. 2020–2024, 2007.

[21] S. Kumar, R. Fonseca, R. P. Ketterling et al., "Trisomies in multiple myeloma: impact on survival in patients with high-risk cytogenetics," *Blood*, vol. 119, no. 9, pp. 2100–2105, 2012.

[22] S. Roychowdhury, M. K. Iyer, D. R. Robinson et al., "Personalized oncology through integrative high-throughput sequencing: a pilot study," *Science Translational Medicine*, vol. 3, no. 111, Article ID 111ra121, 2011.

[23] P. R. Greipp, J. S. Miguel, B. G. M. Dune et al., "International staging system for multiple myeloma," *Journal of Clinical Oncology*, vol. 23, no. 15, pp. 3412–3420, 2005.

[24] W. J. Chng, A. Dispenzieri, C.-S. Chim et al., "IMWG consensus on risk stratification in multiple myeloma," *Leukemia*, vol. 28, no. 2, pp. 269–277, 2014.

[25] A. Palumbo, H. Avet-Loiseau, S. Oliva et al., "Revised international staging system for multiple myeloma: a report from international myeloma working group," *Journal of Clinical Oncology*, vol. 33, no. 26, pp. 2863–2869, 2015.

[26] R. van Laar, R. A. Bender, A. Zielinski, K. Leigh, B. Barlogie, and G. J. Morgan, "Use of multiple myeloma 70-gene prognostic risk score as a continuous predicitor of patient outcome," *Blood*, vol. 128, no. 22, p. 5614, 2016.

[27] O. Decaux, L. Lodé, F. Magrangeas et al., "Prediction of survival in multiple myeloma based on gene expression profiles reveals cell cycle and chromosomal instability signatures in high-risk patients and hyperdiploid signatures in low-risk patients: A Study of The Intergroupe Francophone du Myélome," *Journal of Clinical Oncology*, vol. 26, no. 29, pp. 4798–4805, 2008.

[28] F. Zhan, Y. Huang, S. Colla et al., "The molecular classification of multiple myeloma," *Blood*, vol. 108, no. 6, pp. 2020–2028, 2006.

[29] A. Broyl, D. Hose, H. Lokhorst et al., "Gene expression profiling for molecular classification of multiple myeloma in newly diagnosed patients," *Blood*, vol. 116, no. 14, pp. 2543–2553, 2010.

[30] W. J. Chng, E. Braggio, G. Mulligan et al., "The centrosome index is a powerful prognostic marker in myeloma and identifies a cohort of patients that might benefit from aurora kinase inhibition," *Blood*, vol. 111, no. 3, pp. 1603–1609, 2008.

[31] N. J. Dickens, B. A. Walker, P. E. Leone et al., "Homozygous deletion mapping in myeloma samples identifies genes and an expression signature relevant to pathogenesis and outcome," *Clinical Cancer Research*, vol. 16, no. 6, pp. 1856–1864, 2010.

[32] J. Moreaux, B. Klein, R. Bataille et al., "A high-risk signature for patients with multiple myeloma established from the molecular classification of human myeloma cell lines," *Haematologica*, vol. 96, no. 4, pp. 574–582, 2011.

[33] D. Hose, T. Rème, T. Hielscher et al., "Proliferation is a central independent prognostic factor and target for personalized and risk-adapted treatment in multiple myeloma," *Haematologica*, vol. 96, no. 1, pp. 87–95, 2011.

[34] R. Kuiper, A. Broyl, Y. De Knegt et al., "A gene expression signature for high-risk multiple myeloma," *Leukemia*, vol. 26, no. 11, pp. 2406–2413, 2012.

[35] T.-H. Chung, G. Mulligan, R. Fonseca, and W. J. Chng, "A novel measure of chromosome instability can account for prognostic difference in multiple myeloma," *PLoS ONE*, vol. 8, no. 6, Article ID e66361, 2013.

[36] L. Shi, G. Campbell, W. D. Jones et al., "The MicroArray Quality Control (MAQC)-II study of common practices for the development and validation of microarray-based predictive models," *Nature Biotechnology*, vol. 28, no. 8, pp. 827–838, 2010.

[37] T. Meißner, A. Seckinger, T. Rème et al., "Gene expression profiling in multiple myeloma—reporting of entities, risk, and targets in clinical routine," *Clinical Cancer Research*, vol. 17, no. 23, pp. 7240–7247, 2011.

[38] D. Hose, T. Rème, T. Meissner et al., "Inhibition of aurora kinases for tailored risk-adapted treatment of multiple myeloma," *Blood*, vol. 113, no. 18, pp. 4331–4340, 2009.

[39] K. C. Anderson, M. Alsina, D. Atanackovic et al., "NCCN guidelines insights: multiple myeloma, version 3.2016," *JNCCN Journal of the National Comprehensive Cancer Network*, vol. 14, no. 4, pp. 389–400, 2016.

[40] J. R. Mikhael, D. Dingli, V. Roy et al., "Management of newly diagnosed symptomatic multiple myeloma: updated Mayo Stratification of Myeloma and Risk-Adapted Therapy (mSMART) consensus guidelines 2013," *Mayo Clinic Proceedings*, vol. 88, no. 4, pp. 360–376, 2013.

[41] J. F. San Miguel, R. Schlag, N. K. Khuageva et al., "Bortezomib plus melphalan and prednisone for initial treatment of multiple myeloma," *New England Journal of Medicine*, vol. 359, no. 9, pp. 906–917, 2008.

[42] M. Cavo, P. Tacchetti, F. Patriarca et al., "Bortezomib with thalidomide plus dexamethasone compared with thalidomide plus dexamethasone as induction therapy before, and consolidation therapy after, double autologous stem-cell transplantation in newly diagnosed multiple myeloma: a randomised phase 3 study," *The Lancet*, vol. 376, no. 9758, pp. 2075–2085, 2010.

[43] P. Sonneveld, H. Goldschmidt, L. Rosiñol et al., "Bortezomib-based versus nonbortezomib-based induction treatment before autologous stem-cell transplantation in patients with previously untreated multiple myeloma: a meta-analysis of phase III randomized, controlled trials," *Journal of Clinical Oncology: Official Journal of the American Society of Clinical Oncology*, vol. 31, no. 26, pp. 3279–3287, 2013.

[44] H. Avet-Loiseau, X. Leleu, M. Roussel et al., "Bortezomib plus dexamethasone induction improves outcome of patients with t(4;14) myeloma but not outcome of patients with del(17p)," *Journal of Clinical Oncology*, vol. 28, no. 30, pp. 4630–4634, 2010.

[45] K. Neben, H. M. Lokhorst, A. Jauch et al., "Administration of bortezomib before and after autologous stem cell transplantation improves outcome in multiple myeloma patients with deletion 17p," *Blood*, vol. 119, no. 4, pp. 940–948, 2012.

[46] J. D. Shaughnessy, Y. Zhou, J. Haessler et al., "TP53 deletion is not an adverse feature in multiple myeloma treated with total therapy 3," *British Journal of Haematology*, vol. 147, no. 3, pp. 347–351, 2009.

[47] G. J. Morgan, W. M. Gregory, F. E. Davies et al., "The role of maintenance thalidomide therapy in multiple myeloma: MRC Myeloma IX results and meta-analysis," *Blood*, vol. 119, no. 1, pp. 7–15, 2012.

[48] H. Avet-Loiseau, R. Fonseca, D. Siegel et al., "Carfilzomib significantly improves the progression-free survival of high-risk patients in multiple myeloma," *Blood*, vol. 128, no. 9, pp. 1174–1180, 2016.

[49] M. A. Dimopoulos, K. C. Weise, K. W. Song et al., "Cytogenetics and long-term survival of patients with refractory or relapsed and refractory multiple myeloma treated with pomalidomide and low-dose dexamethasone," *Haematologica*, vol. 100, no. 10, pp. 1327–1333, 2015.

[50] S. Z. Usmani, Q. Zhang, K. Stratton et al., "Phase II study of pomalidomide in high-risk relapsed and refractory multiple myeloma," *Leukemia*, vol. 28, no. 12, pp. 2413–2415, 2014.

[51] K. A. Noonan, C. A. Huff, J. Davis et al., "Adoptive transfer of activated marrow-infiltrating lymphocytes induces measurable antitumor immunity in the bone marrow in multiple myeloma," *Science Translational Medicine*, vol. 7, no. 288, Article ID 288ra278, 2015.

[52] M. A. Chapman, M. S. Lawrence, J. J. Keats et al., "Initial genome sequencing and analysis of multiple myeloma," *Nature*, vol. 471, no. 7339, pp. 467–472, 2011.

[53] N. Bolli, H. Avet-Loiseau, D. C. Wedge et al., "Heterogeneity of genomic evolution and mutational profiles in multiple myeloma," *Nature Communications*, vol. 5, article 2997, 2014.

[54] J. Lohr, P. Stojanov, S. Carter et al., "Widespread genetic heterogeneity in multiple myeloma: implications for targeted therapy," *Cancer Cell*, vol. 25, no. 1, pp. 91–101, 2014.

[55] B. A. Walker, E. M. Boyle, C. P. Wardell et al., "Mutational spectrum, copy number changes, and outcome: results of a sequencing study of patients with newly diagnosed myeloma," *Journal of Clinical Oncology*, vol. 33, no. 33, pp. 3911–3920, 2015.

[56] G. Mulligan, D. I. Lichter, A. D. Bacco et al., "Mutation of NRAS but not KRAS significantly reduces myeloma sensitivity to single-agent bortezomib therapy," *Blood*, vol. 123, no. 5, pp. 632–639, 2014.

6

Protein Engineering Strategies to Expand CRISPR-Cas9 Applications

Lucas F. Ribeiro,[1] **Liliane F. C. Ribeiro**,[2] **Matheus Q. Barreto**,[2] **and Richard J. Ward**[3]

[1]*Department of Biology, Faculdade de Filosofia, Ciências e Letras de Ribeirão Preto, University of São Paulo, São Paulo, SP, Brazil*
[2]*Department of Biochemistry and Immunology, Faculdade de Medicina de Ribeirão Preto, University of São Paulo, São Paulo, SP, Brazil*
[3]*Department of Chemistry, Faculdade de Filosofia, Ciências e Letras de Ribeirão Preto, University of São Paulo, São Paulo, SP, Brazil*

Correspondence should be addressed to Lucas F. Ribeiro; lucasfribeiro@usp.br

Academic Editor: Raul A. Platero

The development of precise and modulated methods for customized manipulation of DNA is an important objective for the study and engineering of biological processes and is essential for the optimization of gene therapy, metabolic flux, and synthetic gene networks. The clustered regularly interspaced short palindromic repeat- (CRISPR-) associated protein 9 is an RNA-guided site-specific DNA-binding complex that can be reprogrammed to specifically interact with a desired DNA sequence target. CRISPR-Cas9 has been used in a wide variety of applications ranging from basic science to the clinic, such as gene therapy, gene regulation, modifying epigenomes, and imaging chromosomes. Although Cas9 has been successfully used as a precise tool in all these applications, some limitations have also been reported, for instance (i) a strict dependence on a protospacer-adjacent motif (PAM) sequence, (ii) aberrant off-target activity, (iii) the large size of Cas9 is problematic for CRISPR delivery, and (iv) lack of modulation of protein binding and endonuclease activity, which is crucial for precise spatiotemporal control of gene expression or genome editing. These obstacles hinder the use of CRISPR for disease treatment and in wider biotechnological applications. Protein-engineering approaches offer solutions to overcome the limitations of Cas9 and generate robust and efficient tools for customized DNA manipulation. Here, recent protein-engineering approaches for expanding the versatility of the *Streptococcus pyogenes* Cas9 (SpCas9) is reviewed, with an emphasis on studies that improve or develop novel protein functions through domain fusion or splitting, rational design, and directed evolution.

1. Introduction

Progress in genetic engineering, metabolic engineering, and synthetic biology will be determined by the development of versatile, user-friendly technologies for the precise and efficient manipulation of cells. The clustered regularly interspaced short palindromic repeat- (CRISPR-) associated protein 9 (CRISPR-Cas9) system is a particularly attractive tool for gene/base editing, gene regulation studies, epigenetic modulation, genome imaging, and manipulation of chromatin topology [1–7]. This technology has attracted considerable attention due to its simplicity for targeting and modifying a specific DNA sequence. Use of the Cas9 protein circumvents the requirement of other DNA-binding proteins such as zinc finger nucleases (ZFNs) and transcription activator-like effector (TALE) to be reengineered each time in order to bind different target DNA sequences. In CRISPR-Cas9, the element that specifies the DNA target is not the protein itself, but a single-guide RNA (sgRNA) molecule, which is straightforward to design and to synthesize [8]. The sgRNA-Cas9 complex binds to the DNA to create a double-stranded break. Cleavage by the sgRNA-Cas9 requires both sequence complementarity between the sgRNA (spacer sequence) and the target DNA sequence ("protospacer") as well as the presence of an appropriate protospacer-adjacent motif (PAM) sequence at the 3′ end of the protospacer sequence [9]. CRISPR-Cas9 is part of the adaptive immune system in prokaryotes, and the PAM sequence allows this immune system to distinguish between self and nonself DNA target [10]. In addition, there is significant variation in PAM specificity between Cas9 orthologs [11], which extends the applications of Cas9 to complex systems that require orthogonality or multiplexing [12].

The Cas9 enzyme is an endonuclease found in some bacteria and Archaea [13–15]. The Cas9 of *Streptococcus pyogenes* (SpCas9) is the best characterized and most widely applied for DNA sequence manipulation. The SpCas9 is comprised of 1368 amino acids and is organized in multiple domains each with a distinct function (Figure 1(a)) [16]. The HNH-like and RuvC-like domains have nuclease activity and are so named because they present sequence similarity to other endonucleases. The HNH-like domain cleaves the DNA strand that is complementary to the sgRNA (target strand), and the RuvC-like domain cleaves the noncomplementary DNA strand (nontarget strand) (Figure 1(b)) [8, 17, 18]. Alanine substitution of key residues in the RuvC domain (D10A) produces a nick in the targeting strand, while the H840A mutation of the HNH domain nicks only the nontargeting strand. The double D10A/H840A mutant eliminates nuclease activity in both domains producing a catalytically inactive, or "dead," Cas9 (dCas9) [8]. Nevertheless, dCas9 retains the ability of Cas9 to bind specifically to a target DNA sequence.

The structure of Cas9 presents two lobes, an α-helical recognition lobe (REC) and a nuclease lobe (NUC) (Figures 1(a) and 1(b)) [16]. The REC is comprised of three α-helical domains and has no structural similarity to any other known protein. The NUC lobe includes the nuclease domains RuvCs, HNH, and also the C-terminal domain (CTD). The two lobes are connected by two linker sequences, the arginine-rich bridge helix (Arg) and the disordered linker (DL). The sgRNA-DNA complex is bound at the interface between the two lobes. The CTD domain contains the PAM-interaction residues necessary for the recognition of the PAM sequence, in which arginine residues at positions 1333 and 1335 play an important role [19]. The apo Cas9 protein (without sgRNA) adopts an inactive conformation, in which the PAM-interaction region is largely disordered [20], and sgRNA binding is a key event for Cas9 activation [20]. When bound to sgRNA, the REC lobe of Cas9 undergoes a large conformational change and the Cas9-sgRNA complex is ready to probe for the target DNA sequence. Single-molecule experiments have demonstrated that the recognition of target DNA by Cas9-sgRNA occurs through three-dimensional collision [21]. The process starts by probing the correct PAM sequence; if PAM is not found, the protein rapidly dissociates from the DNA. When PAM is present, the DNA adjacent to the PAM sequence begins to denature and subsequent base pairing of the sgRNA forms a RNA:DNA hybrid [21, 22]. DNA cleavage requires perfect complementarity between the sgRNA and 10–12 nucleotides located at the $3'$ end of the 20-nt spacer sequence, and this hybrid structure is denominated as the "seed" region [8, 21, 23, 24]. Imperfect base pairing outside the seed region can be tolerated and results in off-target Cas9 activity [25]. This detailed understanding of the structural basis of Cas9 activity is the starting point for improving and expanding the functions and applications of this protein.

Despite its proven potential, CRISPR-Cas9 system has limitations that restrict its wide use for disease treatment or in other biotechnological applications. These limitations include the strict dependence of a PAM sequence, off-target DNA cleavage, and problems for CRISPR delivery posed by the large size of the protein [26–30]. Not only could protein engineering offer elegant approaches to overcome these limitations but may also be used to improve the kinetic and biophysical properties of the enzyme. Engineered Cas9 could provide robust and efficient tools for DNA manipulation tailored to specified applications. In this review, we will discuss recent protein-engineering approaches to expand Cas9 functions, emphasizing those studies that have improved or created SpCas9 functions through domain fusion or splitting, rational design, and directed evolution.

2. Engineering Cas9 by End-to-End Fusion or Protein Domain Insertion

Protein domains are evolutionarily conserved polypeptide units that generally show independent structural or functional properties. Proteins containing multiple domains represent more than two-thirds of the proteins found in prokaryotes and eukaryotes [31]. Moreover, protein domains can act as structural reservoirs to generate new protein architectures [32]. Bioengineers have used domains as building blocks to generate new proteins with novel biotechnological applications [33–38]. Multidomain proteins can be generated by either end-to-end fusion or by domain insertion. End-to-end fusion consists of creating a peptidyl linkage between the N-terminal residue of one domain and the C-terminal residue of another. Since no prior knowledge of the protein structure is necessary, end-to-end fusion is amongst the most widely used strategies in protein engineering [37], and several studies have used this approach to engineer Cas9 with the aim of manipulating its properties for different applications (Figure 2(a)) [39–42].

Alternatively, multidomain proteins can be formed by domain insertion in which one domain (insert) is spliced into another domain (acceptor) either at a specific position or by random insertion. Domain insertion can create structural coupling among the combined domains, with the emergence of new functions (Figure 2(b)) [43]. Furthermore, in contrast to end-to-end fusion in which proteins are linked by a single contact point, when two domains are fused by domain insertion, they will be linked by peptidyl bonds at two contact points, which can generate more stable structures [36, 44–46]. In the following examples, we discuss current strategies for engineering SpCas9 through end-to-end fusion and domain insertion [43].

2.1. Reducing Off-Target Events. One of the main limitations of the CRISPR-Cas9 system is the high level of off-target DNA-cleavage events [47]. In order to improve specificity, dimerization-dependent Cas9-based nucleases have been created by end-to-end fusion of dCas9 with a dimerization-dependent FokI nuclease domain (Figure 2(a)) [48, 49]. The fusion protein doubled the sequence length required for

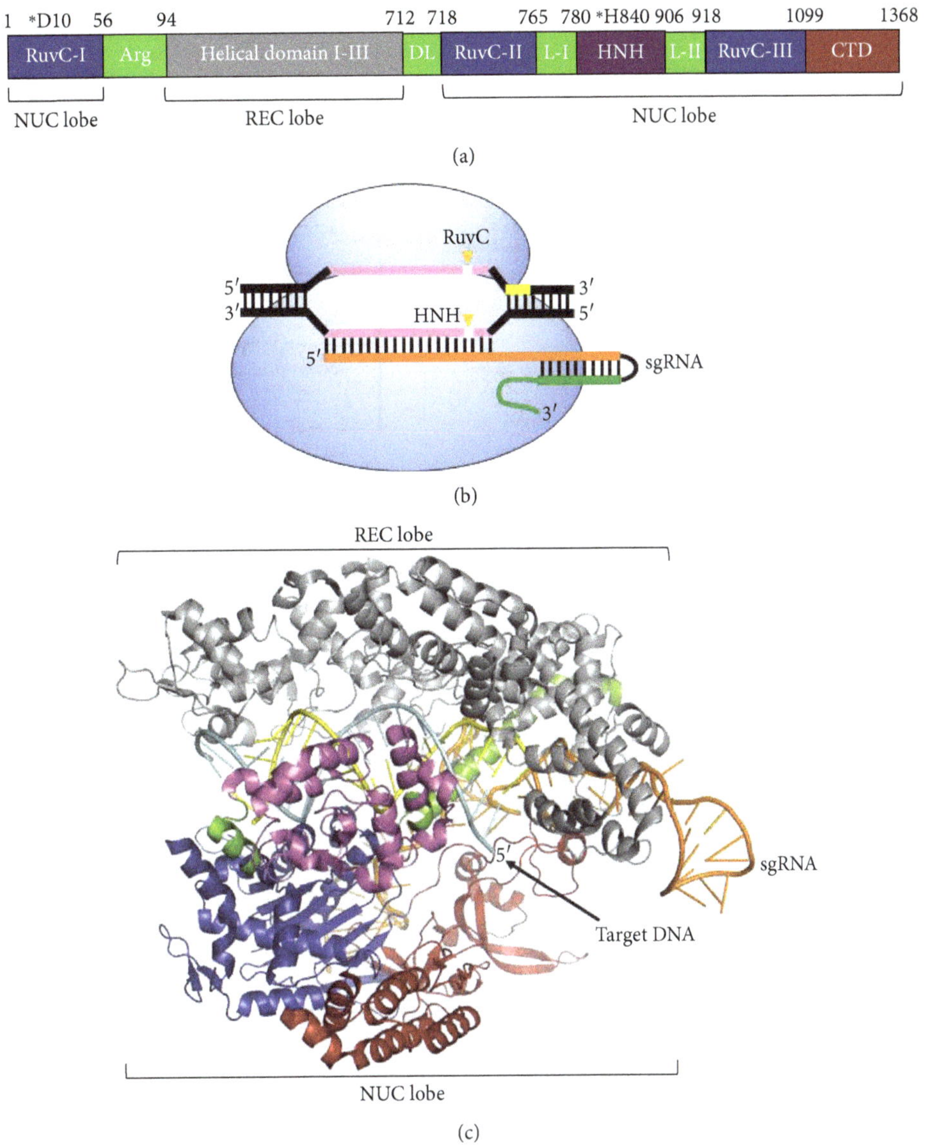

FIGURE 1: Overall organization, structure, and function of CRISPR-associated protein 9 (Cas9) from *Streptococcus pyogenes* (SpCas9). (a) Schematic representation of the domain organization of the SpCas9. Asterisks denote catalytical residues. (b) Cas9 (blue) requires a sgRNA that has a 20 bp region complementary to the target DNA. Cas9 requires two RNA components—CRISPR RNA (crRNA; orange) and transactivating RNA (tracrRNA; green). sgRNA is a chimeric RNA in which crRNA and tracrRNA are fused through a linker. PAM sequence (5′-NGG-3′) is shown in yellow and is crucial for binding and cleavage. DNA cleavage occurs in two different domains: the HNH domain that cuts the target strand and RuvC domain that cleaves the nontarget strand. (c) Cartoon representation of the crystal structure of SpCas9 (PDB 4UN3). Cas9 domains are colored according to the scheme in (a). Abbreviations: Arg: arginine-rich bridge helix; DL: disordered linker; CTD: C-terminal domain; NUC: nuclease lobe; PAM: protospacer-adjacent motif; REC: recognition lobe.

target DNA recognition and efficient cleavage, resulting in a >140-fold higher specificity in human cells as compared to wild-type dCas9 [49]. However, this strategy restricts the overall targetable sequence space because it requires two Cas9-compatible sequences to be close enough (~15 to 25 bp) in order for *Fok*I to dimerize and cleave the target DNA. In an alternative strategy to improve Cas9 DNA-binding precision [41], a DNA-binding domain (either ZFP (zinc finger proteins) or a TALE (transcription activator-like effector)) was fused to a Cas9 in which key PAM-interacting residues were mutated to reduce DNA-binding affinity. Three different Cas9-ZFP end-to-end fusion proteins were designed to bind to 12 base pair sequences, and their activities were tested at previously defined off-target sites. The approach not only dramatically decreased off-target cleavage events but the Cas9-ZFP chimeras also reduced the size of the engineered protein, which is advantageous for the viral delivery systems.

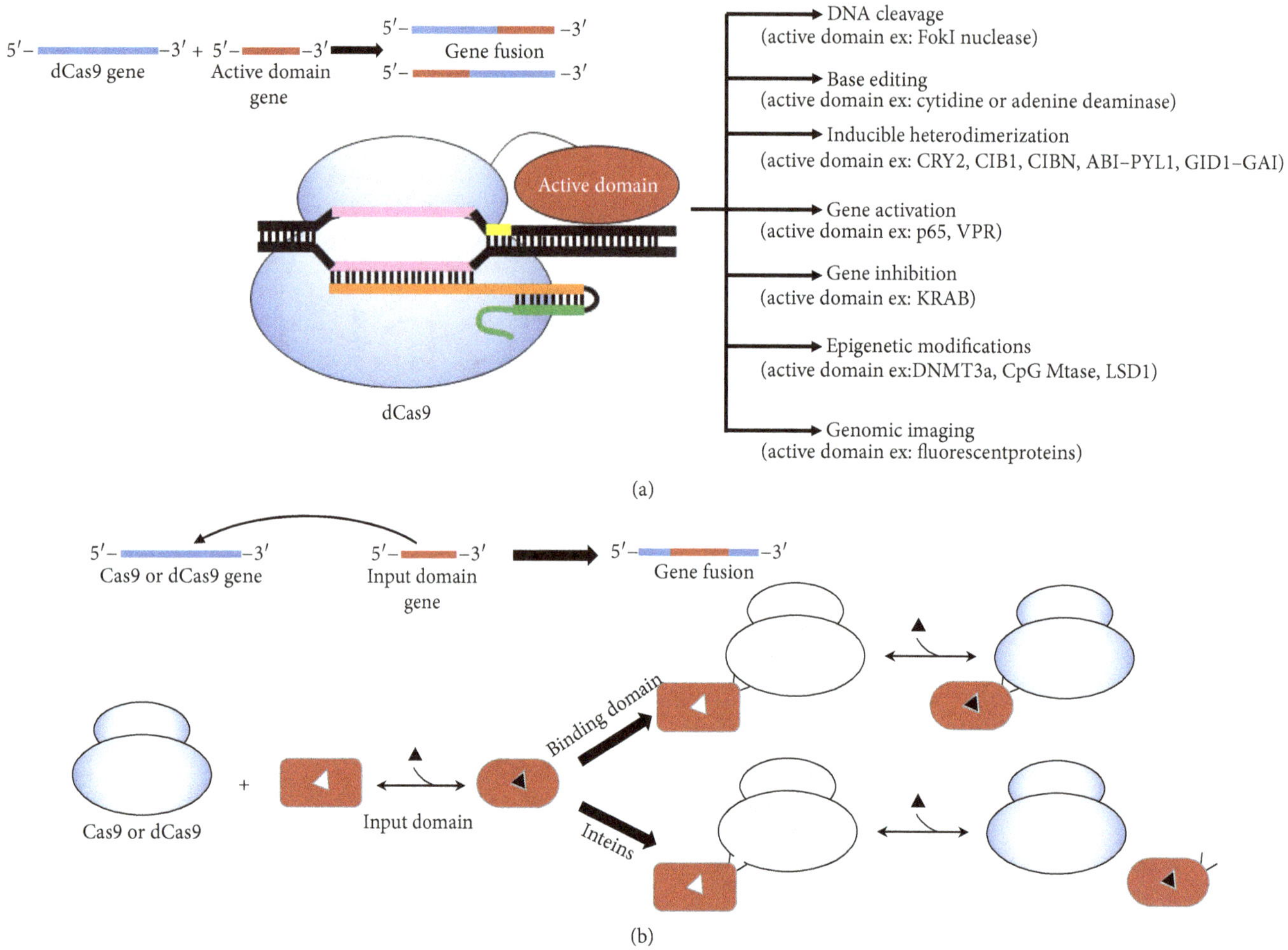

Figure 2: Schematic representations of Cas9 engineering by domain fusion. (a) End-to-end gene fusion approach. A diverse range of new technologies for DNA manipulation can be created by end-to-end fusion of Cas9 with an active domain. These fusions have been used to expand Cas9 applications. (b) Schematic depiction of Cas9 engineering by domain insertion. A switch behaviour could emerge in such a way that the Cas9 protein could be regulated by the input domain's recognition of an input signal. Either binding domains or inteins have been used to generate Cas9 with a switch response. DNA sequences are depicted as lines. A light gray color of the Cas9 indicates that the protein is inactive or less active while blue represents the active state. The signal that modulates the switch is showed as a black triangle.

In a further attempt to decrease the off-target events, the SpCas9 was fused to the structurally unstable protein domain: dihydrofolate reductase (DHFR) or the estrogen receptor (ER50) [42], which target the fusion protein for rapid proteasomal degradation [50]. Switched systems were created by fusion of DHFR or ER50 to both the N- and C-termini of SpCas9. These fusion proteins bind to the DNA target with full endonuclease activity only in the presence of trimethoprim (TMP) or 4-hydroxytamoxifen (4-HT), small molecules that stabilize DHFR and ER50, respectively. In the absence of TMP or 4-HT, fused proteins were targeted for degradation by the proteasome [42]. This approach demonstrates that limiting Cas9 action to a short and controlled period may decrease off-target activity.

2.2. Base Editing with Cas9 Fusion Chimeras. Cas9 has also been fused end-to-end with cytidine deaminase [51] and adenine deaminase [52] (Figure 2(a)) in order to develop nucleotide base editors. In both cases, the goal was to efficiently correct point mutations related to diseases, without the generation of random insertions and deletions (indels). To achieve this goal, dCas9 was used as a DNA-binding domain and fused to a deaminase. In 2016, Komor et al. created a so-called "base editor" to convert cytidines into uridine within a sequence of five nucleotides located between the protospacer and PAM [51]. A cytidine deaminase from Rat known as APOBEC1 (apolipoprotein B mRNA editing enzyme, catalytic polypeptide-like 1) was fused to the N-terminus of dCas9 using a linker in order to maintain deaminase activity. Iterative optimizations in the linker and chimera were performed, and each step of optimization produced a new "generation" of the construction. Therefore, the third generation of base editors (BE3) consisted of APOBEC1 linked to the N-terminus of dCas9 with the catalytic His840 restored and a uracil glycosylase inhibitor linked at Cas9 C-terminus (APOBEC-linker-dCas9(A840H)-UGI). This chimera showed a permanent point mutation correction of up to 75% of total cellular DNA with 1% indel formation [51]. Since the use of this chimera is limited by off-target activity, the BE3 has been improved by incorporating four point mutations (N497A, R661A, Q695A, and Q926A) to generate a high-fidelity base editor (HF-BE3) with reduced off-target activity [53]. However, efficient CRISPR-Cas9-based editors require a 5′-NGG-3′ PAM, limiting the sequence space in the genome that can be efficiently targeted. Furthermore, these base editors are only able to efficiently convert C to T within ~5 bp of the editing window, and this window could introduce undesired changes around the base target. In an attempt to address these limitations, SpCas9 was substituted for dCas9 from *Staphylococcus aureus* (SaCas9) in the BE3 to generate the APOBEC1–SaCas9n–UGI (SaBE3). In order to obtain a narrower editing window, a series of iterative optimizations were conducted resulting in base editors with altered PAM specificities (YE1-BE3, EE-BE3, YE2-BE3, or YEE-BE3) [54]. Although mutations in the cytidine deaminase enzyme could narrow the editing window, the variants were not able to discriminate among the cytidines within the window. An alternative base-editing chimera has been described that uses an engineered human APOBEC3a (eA3A) domain in the BE3 [55]. The eA3A-BE3 chimera was able to correct a mutation in a human beta-thalassemia promoter with >40-fold higher precision than BE3. Furthermore, the eA3A-BE3 had lower off-target activity as compared to BE3 [55].

Using an alternative approach, Gaudelli et al. sought to develop a base editor to convert A:T base pairs to G:C by fusing dCas9 to an adenosine deaminase [52]. However, since there is no enzyme known to deaminate adenine in DNA, the first step was to evolve *E. coli* TadA, a tRNA adenine deaminase that converts adenine to inosine. Mutations in the vicinity of the D108 residue in TadA were sufficient to introduce adenine deamination activity against DNA substrates. The TadA A106V/D108N mutant was fused to the N-terminus of the Cas9 (D10A mutant with nickase activity) [52]. After several rounds of iterative optimization, evolved fused proteins were capable of converting A:T base pairs into G:C base pairs in human cells with approximately 50% efficiency, high specificity (~99%), and low rates of indel formation (<0.1%). These studies highlight both the potential and the importance of improving the precision of base editors using CRISPR-Cas9 for use in medical therapies.

2.3. Cas9 Chimeras for Controlling Transcriptional Regulation Activated by Light. Several studies have reported the use of CRISPR-Cas9 to control gene expression [40, 56–58]. Two similar systems have been developed by fusion of CRISPR-Cas9 and a light-inducible heterodimerizing cryptochrome 2 (CRY2) together with the calcium- and integrin-binding protein 1 (CIB1) (Figure 2(a)) [40, 56]. Nihongaki et al. fused dCas9 with CIB1 and CRY2 with a transcriptional activator [56]. In this system, dCas9 is expressed and binds to a target DNA sequence guided by the gRNA. Upon blue light irradiation, CRY2 and CIB1 heterodimerize, and the transcriptional activator is recruited to the target locus, activating gene expression. After optimization, a fusion protein was generated in which the N-terminus of dCas9 was linked to a nuclear localization signal (NLS), and the C-terminus of dCas9 was linked to CIB1 with a truncated C-terminus. The activator includes three NLS for correct nuclear localization, the photolyase homology region of CRY2 (CRY2PHR), and the activator domain (p65)-NLSdCas9-trCIB1 and $(NLS)_3$-CRY2PHR-p65, and the final fusion protein demonstrated a 31-fold induction. Polstein and Gersbach created a light-activated CRISPR/Cas9 effector (LACE) system in which a different fusion protein combination was used [40]. Optimized LACE consisted of two CIB domains fused to both the N- and C-termini of Cas9 together with a CRY2 fused to a transcriptional activator. This system was used in HEK293T cells to control expression of human IL1RN and presented a 400-fold increase after 30 hours of blue light irradiation.

In addition to light activation, systems for chemical activation of dCas9 to regulate gene expression have been developed [57, 58]. A variant of Cas9 with nuclease activity modulated by the presence of the tamoxifen analogue 4-HT (Figure 3(d)) [58] was based on fusions of human estrogen receptor 2 (ERT2) domain at either the N- or C-terminus of Cas9, in which the positions of the nuclear localization signal (NLS) was varied. The result was a switch protein working as a rapidly inducible conditional genome-editing system, with high DNA-cleavage efficiency after induction with 4-HT and low background in the absence of inducer [58]. Variants of Cas9 responding to both chemical and light inducers have also been developed [57] where an activator domain (VPR) was fused to six chemical and light-inducible heterodimerization domains (Figure 2(a)): abscisic acid- (ABA-) inducible ABI–PYL1, gibberellin- (GA-) inducible GID1–GAI, rapamycin-inducible FKBP–FRB, phytochrome-based red-light-inducible PHYB–PIF, cryptochrome-based blue-light-inducible CRY2PHR–CIBN, and light-oxygen-voltage-based blue-light-inducible FKF1–GI [57]. The best results were obtained with ABA- and GA-dimerized VPR–SpdCas9-activated systems (Figure 3(f)), in which EGFP protein expression increased 165- and 94-fold, respectively, upon induction. These efficiencies were similar to those obtained for the VPR-dCas9 direct fusion protein, demonstrating the potential for the use of these chimeras as gene expression regulators. The platform developed in this work was also created to be used as AND, OR, NAND, and NOR dCas9 logic gates, offering a protein-based alternative to produce a functional output from multiple inputs [57].

2.4. Epigenetic Regulation by Cas9 Chimeras. Cas9 chimeras have also been created with the aim of controlling and manipulating epigenetic modifications (Figure 2(a)) [59–62]. Fusion of a DNA-binding domain to a cytosine DNA methyltransferase has been a common strategy in order to elucidate the effects of DNA methylation in mammalian cells, in which dCas9 was fused to a DNA methyltransferase domain named DNMT3a [59–63]. In these chimeras, the dCas9 acts as a DNA-binding domain and directs binding to a specific DNA promotor target determined by the sgRNA control. The methyltransferase then modulates DNA methylation of the target promoter resulting in the downregulation of target genes.

In a different strategy, DNMT3a was artificially split generating N- and C-terminal fragments. Subsequently, the DNMT3a C-terminal fragment was fused to the C-terminus

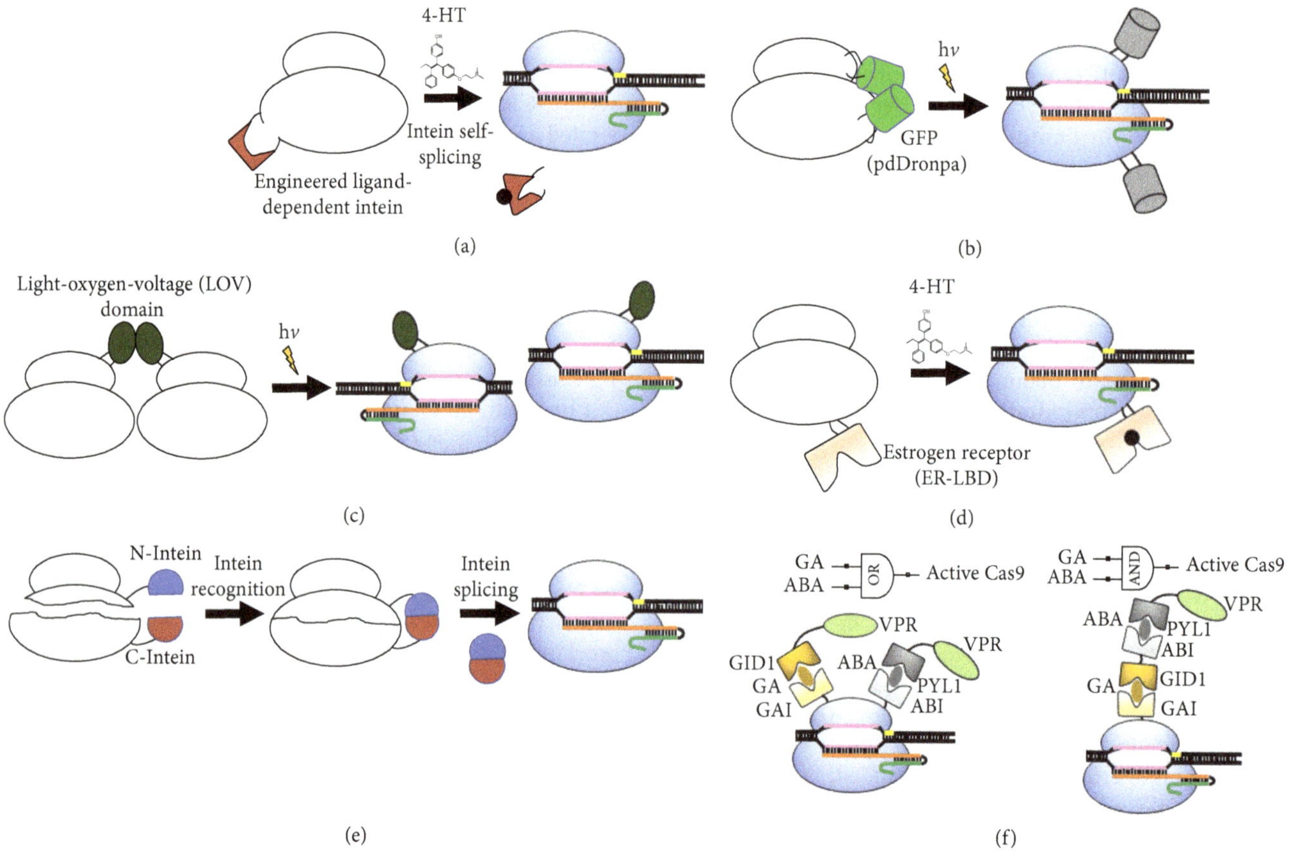

Figure 3: Selected case studies of engineering dCas9 by domain fusion. (a) Intein insertion inactivates dCas9, and in the presence of 4-HT, the inserted intein (red) undergoes self-splicing and restores the active Cas9 structure and function [70]. (b) Insertion of an engineered GFP (pdDronpa) that dimerizes in the dark and prevents DNA binding. On light illumination, pdDronpa dissociates and enables the binding of dCas9 to DNA [71]. (c) In the absence of light, the LOV domain (dark green) maintains a stable dimer that sterically blocks dCas9 function. Light induces LOV dissociation and Cas9 activation [74]. (d) Domain insertion of a ligand-binding domain (LBD) of the estrogen receptor ER leads to an allosteric activation by 4-HT [75]. Rather than domain insertion, an end-to-end fusion with ER-LBD results to nuclear translocation by 4-HT [58]. (e) A split Cas9 composed of two separate fragments is fused to intein sequences that perform self-splicing upon dimerization, leading to fully active Cas9 [83]. (f) The dCas9 fusion with multiple domains can generate a multi-input system and produce logic gates [57]. A VPR–SpdCas9 construct induces gene expression in the presence of gibberellin (GA) OR/AND abscisic acid (ABA). A light gray color of the Cas9 indicates that the protein is inactive or less active while blue represents the active state.

of dCas9 using a 15-amino-acid linker [20]. The dCas9 directed the assembly of methyltransferase fragments at the CpG site, resulting in an efficient (~70%) and predictable DNA methylation [63]. Application of Cas9 fusions for epigenetic manipulation also includes the fusion of a histone demethylase LSD1 (Lys-specific histone demethylase) to Cas9 from *Neisseria meningitides* [64]. This chimera was used to target the distal enhancer region of the endogenous transcription factor gene Oct4 in mouse embryonic stem cells, resulting in the repression of Oct4 expression and loss of pluripotency [64]. In a separate study, targeting three endogenous promoters using a chimera in which the C-terminus of a dCas9 was fused to the highly conserved acetyltransferase p300 resulted in acetylation of the histone H3 lysine 27 at the target site and transcriptional activation of target genes [65]. Finally, dCas9 was fused to two families of proteins directly involved in gene silencing through methylation, the KRAB (Krüppel-associated box) that forms a complex with two histone methyltransferases, and the DNMT [59]. This study showed that the combination of dCas9-KRAB with dCas9-DNMT improved silencing efficiency. These examples highlight the versatility of the Cas9 as a tool for silencing transcription and targeting regulatory regions.

2.5. Cas9 Chimera for Genome Imaging. With the aim of simplifying the study of spatial genome organization, fusion of fluorescent protein domains to dCas9 has been applied to the visualization of genomic loci and chromatin spatio-temporal dynamics in live cells (Figure 2(a)) [39, 66, 67]. Although the fusion of single-color protein fluorescent labels (eGFP) to SpdCas9 has been described [39, 66], multiple labels are required to differentiate interchromosomal and intrachromosomal loci within the nucleus. The design of multicolor versions of dCas9 from three bacterial orthologs, *S. pyogenes*, *N. meningitides*, and *Streptococcus thermophiles*, has been reported [67]. Each dCas9 ortholog, targeting the human telomere DNA repeat, was fused to a fluorescent protein (green fluorescent protein (GFP), red fluorescent protein (RFP), or blue fluorescent protein (BFP)). The construction NLS-dCas9-$(NLS)_{1-3}$-$(XFP)_3$, in which XFP means GFP, RFP, or BFP, permitted the estimation of the intranuclear distance between loci in different chromosomes and the linear distance between two loci in the same chromosome, allowing assessment of the DNA compaction in these regions in a live cell. The development of these tools may enable the study of the 4D nucleome and the regulation of gene expression in different eukaryotic cell types and at various stages of development and differentiation [67].

2.6. Cas9 Chimeras Created by Domain Insertion. A limitation of many Cas9-derived systems is that its activity is not directly modulated, and domain insertion has been used to create new functions in Cas9, including inducible response. Fine tuning of the Cas9 is of key importance to reduce the off-target activity and to allow a spatial-temporal control of the protein. Inducible Cas9 can be created by fusing two domains in such a way that the Cas9 activity is regulated by the recognition of an input signal by a sensor domain.

Inteins (intervening proteins) are proteins that perform protein-splicing posttranslational modifications [68] with the goal of creating a regulated Cas9; SpCas9 has been recombined with the intein 37R3-2 [69, 70]. Since inteins are self-excised from the "immature" polypeptide producing a functional mature protein, the insertion of an intein into the Cas9 may lead to its inactivation, and its excision may result in Cas9 activation. The 37R3-2 intein was engineered to perform protein splicing only in the presence of chemical 4-HT. Subsequently, the engineered 37R3-2 was inserted at 15 positions of Cas9 distributed throughout its different domains. Insertions at Ala728, Thr995, and Ser1154 did not lead to a significant inactivation of Cas9 in the absence of 4-HT, suggesting that these positions tolerate large protein insertions. Insertions at positions S219 and C574 produced low nuclease activity in the absence of 4-HT and an ~4-fold increase of this activity in the presence of the compound (Figure 3(a)). The on-target/off-target ratio was 6-folds higher on average and up to 25-folds higher when compared to the wild-type Cas9, demonstrating that the insertion of the intein into specific sites of Cas9 was able to increase the protein precision for gene editing. However, a decrease in nonhomologous end joining (NHEJ) efficiency was observed, and in addition, the activation of the Cas9 activity by the intein domain is irreversible.

2.7. Photoregulation of Cas9 Chimeras Created by Domain Insertion. The ideal tool for DNA editing would be an inducible Cas9 that could be readily modulated by the presence of a nontoxic, nonmetabolizable, inexpensive signal. A recent study explored the photomodulated dimerizing protein domains known as pdDronpa [71], a green fluorescent protein which dimerizes in the dark but dissociates to monomers upon illumination with light at 500 nm [72]. A single-chain photoswitchable Cas9 (ps-SpCas9) was engineered by insertion of the pdDronpa protein into Cas9 at two positions, and in the absence of light, the dimerization of the pdDronpa protein blocks the binding of Cas9 to the DNA (Figure 3(b)). The fluorescent protein was inserted into two loops in Cas9 at the REC2 domain and into the CTD domain of the NUC lobe. These loops are situated across the DNA-binding cleft and are occluded by pdDronpa dimerization. Although psCas9 produced an indel level lower than the parental Cas9 that was similar to other photoinduced two-component Cas9 systems [73], the ps-SpCas9 functions as a single chain, and the pdDronpa domain can also be used as a localization and expression marker. The ps-SpCas9 was also engineered to regulate gene transcription, for which the mutations D10A and H841A were introduced in Cas9 to create ps-dSpCas9, which was then fused to a VP64-p65-Rta (VPR) transactivation module [57] at the N-terminus. The pdDronpa domain was replaced to a new version of the protein called pdDronpa1.2 with reduced basal activity in the absence of light [72]. This new protein, denominated as VPR-ps-dSpCas9, showed a 58-fold induction of the reporter gene (mCherry) after light illumination. This is higher than an optimized multiple-chain light-activated Cas9 effector (LACE) system [40, 56] and comparable with chemically inducible systems [57].

A recent study also used a photomodulated dimerizing protein known as RsLOV, a photoreceptor from *Rhodobacter sphaeroides*, which in the dark is a homodimer and upon illumination by blue light dissociates to its monomeric form [74]. The RsLOV was inserted using flexible linkers at 231 positions throughout the dCas9 protein via multiplex-inverse PCR (Figure 3(c)). Chimeras were selected by fluorescence-activated cell sorting (FACS) based on the ability to modulate RFP expression in the presence of blue light (470 nm), and two photoactivatable RsLOV-Cas9 variants demonstrated repression activity enhancement under blue light. A temperature-sensitive variant, denoted as tsRC9, also was isolated and presented activity at 29°C but negligible activity at 37°C [74].

2.8. Exploring the Limits of Insertional Fusion with Cas9. In an elegant approach, Oakes et al. [75] have characterized SpCas9 tolerance to domain insertion by random insertion of an engineered Mu transposon flanked by BsaI endonuclease restriction sites into dCas9. Deep sequencing analysis showed that the transposon was inserted in >70% of all possible amino acid sites of dCas9, and this library was used as a starting point for insertion of the 86 aa PDZ domain. The library containing the plasmid carrying dCas9 and Mu transposon was digested with BsaI, and the transposon was replaced by the BsaI-digested gene encoding the PDZ domain with flanking amino acid linkers. After PDZ insertion, FACS was used to identify dCas9 variants that could repress RFP expression, and 127 positions were identified as tolerant to PDZ insertion. These sites tended to cluster around flexible loops, at solvent-exposed residues and the ends of helices, and preferential sites for domain insertion were found mainly within the helical II recognition lobe (REC), in the RuvC-III region and throughout the CTD domain. Insertions into vital motifs, such as sgRNA-binding grooves, the bridge helix, the PAM-binding pocket, and the DNA/RNA heteroduplex annealing channel, resulted in impaired Cas9 functionality. Subsequently, eight insertion sites were chosen to introduce a SH3 domain, of which six showed repression levels similar to parental dCas9 [75]. Furthermore, 2–4 insertions of multiple PDZ and SH3 domains were introduced into dCas9 at validated insertion sites, and many of these constructs were capable of repressing expression to a degree comparable with the parental dCas9. Moreover, with the aim of developing a switch Cas9, the well-characterized human estrogen receptor-α ligand-binding domain (ER-LBD) was inserted into the naive dCas9 transposition library (Figure 3(d)). The ER-LBD binds 4-HT, and a dCas9 variant carrying an ER-LBD insertion (denominated as darC9:231) capable of modulating gene repression in the presence of 4-HT was identified. The catalytic residues (D10 and H840) were reintroduced in darC9 to produce arcC9:231. In the presence of 4-HT, arC9 increased chromosomal cleavage 100- and 24-fold in *E. coli* and human cells, respectively [75]. These results demonstrate that the domain insertion profiling of Cas9 can generate important functional information, which can be used to facilitate protein engineering.

3. Structure-Based Cas9 Engineering

The crystal structure of Cas9 from *S. pyogenes* [76] provides the basis for rational engineering and modification of the protein. Using structure-based design principles, it is possible to enhance specific properties of Cas9, reduce nonspecific activity, switch specific regions of the protein, and obtain enzymes that are suitable for specific genome-engineering applications.

Nonspecific cleavage of DNA is the consequence of imperfect complementarity between the RNA guide and a genomic site, leading to off-target gene editing. The stabilization of the nontarget DNA by Cas9 is through the positively charged groove located between the HNH-, RuVC-, and PAM-interacting residues in the CTD domain. Engineering of this region holds the promise of reducing off-target edition, and the substitution of the 32 positively charged residues by alanine in this groove leads to the identification of five residues that decreased off-target cleavage [77]. The combination of these mutations generated three Cas9 variants with normal on-target activity and decreased off-target indel formation. Using a similar approach, alanine substitution of four polar or charged residues located in the same groove resulted in undetectable levels of off-target indels [78]. Structural analysis revealed that these residues participate in nonspecific interactions with the phosphate backbone of the target DNA strand.

The cleavage of DNA by Cas9 is dependent of the recognition of the protospacer-adjacent motif (PAM). The recognition of a specific PAM sequence by Cas9 is one of the limitations of using the wild-type SpCas9. Sequence databases presently contain over 1000 Cas9 orthologs whose different PAM specificities could provide insights to engineer the SpCas9 and alter its specificity [79]. The CTD domain (containing the PAM-interacting motif) from the SpCas9 has been exchanged for the Cas9 orthologous CRISPR-3 from *Streptococcus thermophilus* [76], which altered the recognition of the 5′-NGG-3′ to the 5′-NGGNG-3′ PAM from *S. thermophilus*.

A limitation of SpCas9 in gene therapy is its size (~4.3 kb), which limits its efficient gene-based delivery via recombinant adenoassociated virus (rAAV) [80, 81]. rAAV is a viral vector that has been successfully used in therapeutic gene editing. However, the size limit of an insert in this vector is ~4.7 kb, which is problematic for packaging SpCas9, sgRNA, and control elements [80, 81]. To reduce the size of Cas9, the effects of deletion of specific Cas9 domains on protein activity were investigated [76]. It was observed that the enzyme with the REC2 domain deleted (Δ175–307) maintained 50% activity as compared to its wild-type counterpart, indicating that this domain is not critical for DNA cleavage and could potentially be targeted to reduce the size of Cas9 [76].

Using another approach, the nuclease (M1 to E57-GSS linker-G729 to D1368) and the α-helical lobes (G56 to S714) of Cas9 were expressed separately [82], and although neither of the two fragments presented activity, both parts readily reassemble to an active form on contact with the sgRNA, albeit with reduced editing efficiency. In another study, the Cas9 coding sequence was divided into two segments, and each was fused to an intein moiety (N-moiety: M1 to E573 and C-moiety: C574 to D1368 or N-moiety: M1–K637 and C-moiety: T638–D1368) [83]. After the sequences were introduced into the cell by a recombinant adenoassociated virus, protein splicing efficiently reconstituted the Cas9 as a single polypeptide with full activity (Figure 3(e)), suggesting a means to use CRISPR for gene therapy via adenovirus release of genetic material. Splitting Cas9 into two segments (N-moiety: D2–V713 and C-moiety: S714–D1.368) and fusing each segment to a photodimerization domain (pMag and nMag) [84] resulted in reassociation of the two fragments on irradiation with blue light and restoration of Cas9 structure and activity, suggesting the possibility for optical regulation of Cas9 activity. Based on crystal structure analysis, the Cas9 was split into eleven sites [85], and in each case, the fragments were fused to FKBP or FRB that dimerize in the presence of rapamycin. In addition, the N-terminal fragment was fused to a nuclear export sequence (NES) and C-terminal fragment to a nuclear localization sequence (NLS). Thus, split Cas9 can be reassembled upon rapamycin induction and acted in the nucleus without detectable off-target activity. Moreover, a split dCas9 version was fused to the VP64 transactivation domain, and this construction was able to activate transcription of target genes in the presence of rapamycin.

The crystal structure of the Cas9 shows that the K866 undergoes substantial conformational change upon sgRNA binding, allowing the proper positioning of the target DNA. With the aim of modulation of Cas9, a site-specific photocaged lysine was introduced at these positions to create optochemical control of Cas9 [71]. The photocaged lysine at this position deactivated Cas9, and the caging group could be removed through light exposure. The photocaged Cas9 showed minimal background activity in the absence of UV light and reached parental Cas9 levels after light irradiation.

4. Engineering PAM Specificity by Directed Evolution

The sequences recognized by Cas9 are limited by the requirement of a specific protospacer-adjacent motif (PAM) [24]. One strategy to improve CRISPR-Cas9 is by changing PAM specificity using directed evolution techniques [86]. Based on the crystal structure of Cas9 [76], the PAM-interacting domain of SpCas9 was subjected to random mutagenesis, and a library of mutated sequences was screened against the 5′-NGA-3′ PAM-target site. Three Cas9 variants called VQR (D1135V/R1335Q/T1337R), EQR (D1135E/R1335Q/T1337R), and VRER (D1135V/G1218R/R1335E/T1337R) were obtained, whose PAM specificity changed from 5′-NGG-3′ observed in the wild-type Cas9 to 5′-NGA-3′, 5′-NGAG-3′, and 5′-NGCG-3′, respectively, thereby broadening the target range of the Cas9 [86].

This approach is limited because it needs to evolve each variant separately with a potential PAM sequence, and this limitation is exacerbated for Cas9 orthologs that specify longer PAMS. As an alternative, variants with relaxed specificities within the PAM could be evolved [87], and the predicted PI-domain from SaCas9 was randomly mutagenized and tested for PAM specificity, resulting in a Cas9 variant called KKH (E782K/N968K/R1015H) which showed the same DNA-cleavage specificity when compared to its wild-type counterpart (5′-NNGRRT-3′). Nevertheless, the KKH enzyme can also identify nonspecific sites such as 5′-NNARRT-3′, 5′-NNCRRT-3′, and 5′-NNTRRT-3′, which increases its targeting range. The crystal structures of the three variants [76] revealed that PAM-reprogramming of these enzymes is based on a synergistic effect of the mutations in the displacement of the PAM duplex, enabling different nucleotide sequences to be recognized by these new PAMs.

In order to enhance acquisition of spacer sequences flanked by non-NGG PAM motifs, SpCas9 variants were created by error-prone PCR, and after selection, a Cas9 variant I473F was identified with higher specificity for the PAM sequence 5′-NAG-3′ [88]. This variant caused an enhanced immune response against viruses due to higher rates of spacer acquisition, showing the importance of the enzyme not only in DNA-cleavage events but also in gaining new spacer sequences to facilitate the CRISPR immune response in bacteria.

A recent strategy that has been employed to broaden PAM specificity was to use phage-assisted continuous evolution (PACE) to evolve a SpCas9 variant (xCas9) [89]. Using PACE, hundreds of generations of directed evolution could be implemented, and screening used a bacterial one-hybrid selection in which dCas9 was fused to the ω subunit of a bacterial RNA polymerase which on binding to a PAM causes phage propagation. It was proposed that the replication of variants with broader PAM specificity would be favored. After continuous *in vivo* protein evolution, several variants were enriched in the pool, including R324L, S409I, and M694I, which are located near the DNA-sgRNA interface in the crystal structure. Restoring the catalytic residues showed that the variants were able to cleave the DNA with five PAM sequences (5′-NG-3′, 5′-NNG-3′, 5′-GAA-3′, 5′-GAT-3′, and 5′-CAA-3′), and off-target tests showed a significant off-target activity reduction as compared to the parental SpCas9.

5. Future Perspectives

Cas9 has a tremendous utility for the regulation and modification of complex biological systems. However, overcoming the limitations of the system is paramount to realize its full potential. An ideal Cas9-based tool should bind and/or cleave a single specific target in a complex genome without generating off-targets as side products. In addition, it would be highly desirable to develop Cas9 technologies endowed with precise spatiotemporal control, rapid responses to inducers, lack of toxicity, ease of customization, and high efficiency of *in vitro* and *in vivo* delivery. The studies described here showed that over recent years, new ways to improve Cas9 functions have been developed. These protein-engineering efforts have made significant contributions to the improvement of gene therapy, genomic imaging, and the emerging field of synthetic biology.

However, new Cas9-based tools are required in order to overcome a number of features that are still poorly explored. For example, most of the systems described in this review still present low delivery efficiency and off-target activity, are relatively expensive, and are not readily adapted for multiplexing. This suggests that there is still plenty of scope for engineering Cas9. For example, some inducible Cas9 systems use expensive steroids that have short half-lives in solution and relatively high toxicity for both prokaryote and eukaryote cells [90–92]. Engineering Cas9 to respond to a range of inexpensive and nontoxic signals would result in CRISPR applications that are more flexible and economically feasible on a larger scale. In addition, many of these studies are specific to mammalian cells. The adaptability of these tools to other platforms such as bacteria, fungus, and plant cells would greatly increase its biotechnological impact. Moreover, little is known about the allergenic potential of Cas9 in gene therapies in humans, and protein engineering plays a key role for studying and solving this potential issue. Protein engineering of catalytic and biophysical properties is as yet little explored, and Cas9 with improved catalytic efficiency or able to work in extreme conditions could be useful for engineering extremophilic organisms. In addition to Cas9 engineering, sgRNA engineering also has been used to enhance CRISPR functionality, and the combination of both approaches could be used to expand Cas9 applications [4, 93]. Given the rapid development of diverse engineered SpCas9s, it is likely that in the future, a wide range of tailored Cas9, working as simple or multiplex tools, will be available for a wide variety of genome-editing applications, including gene therapy and treatment of disease.

Abbreviations

SpCas9: Cas9 from *Streptococcus pyogenes*
SaCas9: Cas9 from *Staphylococcus aureus*
PAM: Protospacer-adjacent motif
NLS: Nuclear localization signal
CRY2: Light-inducible heterodimerizing cryptochrome 2
CIB1: Calcium- and integrin-binding protein 1
4-HT: 4-Hydroxytamoxifen
VPR: Chimeric activation domain composed of the activation domainsVP64, P65, and Rta
pdDronpa: Engineered GFP that dimerizes in the dark
LOV: Photomodulated dimerizing protein
ER-LBD: Ligand-binding domain of the estrogen receptor.

Conflicts of Interest

The authors declare that there is no conflict of interest regarding the publication of this paper.

Acknowledgments

The authors are grateful for support from Fundação de Amparo à Pesquisa do Estado de São Paulo (FAPESP) (LFR: 2016/18827-7; LFCR: 2016/20358-5). The authors thank Dr. M. Ostermeier for the helpful insights. The authors sincerely apologize to the authors whose work could not be included in this manuscript due to space limitation.

References

[1] P. Mali, K. M. Esvelt, and G. M. Church, "Cas9 as a versatile tool for engineering biology," *Nature Methods*, vol. 10, no. 10, pp. 957–963, 2013.

[2] P. D. Hsu, E. S. Lander, and F. Zhang, "Development and applications of CRISPR-Cas9 for genome engineering," *Cell*, vol. 157, no. 6, pp. 1262–1278, 2014.

[3] M. F. Copeland, M. C. Politz, and B. F. Pfleger, "Application of TALEs, CRISPR/Cas and sRNAs as trans-acting regulators in prokaryotes," *Current Opinion in Biotechnology*, vol. 29, pp. 46–54, 2014.

[4] R. Barrangou and J. A. Doudna, "Applications of CRISPR technologies in research and beyond," *Nature Biotechnology*, vol. 34, no. 9, pp. 933–941, 2016.

[5] R. Barrangou and P. Horvath, "A decade of discovery: CRISPR functions and applications," *Nature Microbiology*, vol. 2, article 17092, 2017.

[6] S. L. Morgan, N. C. Mariano, A. Bermudez et al., "Manipulation of nuclear architecture through CRISPR-mediated chromosomal looping," *Nature Communications*, vol. 8, 2017.

[7] N. Hao, K. E. Shearwin, and I. B. Dodd, "Programmable DNA looping using engineered bivalent dCas9 complexes," *Nature Communications*, vol. 8, no. 1, p. 1628, 2017.

[8] M. Jinek, K. Chylinski, I. Fonfara, M. Hauer, J. A. Doudna, and E. Charpentier, "A programmable dual-RNA-guided DNA endonuclease in adaptive bacterial immunity," *Science*, vol. 337, no. 6096, pp. 816–821, 2012.

[9] F. J. M. Mojica, C. Díez-Villaseñor, J. García-Martínez, and C. Almendros, "Short motif sequences determine the targets of the prokaryotic CRISPR defence system," *Microbiology*, vol. 155, no. 3, pp. 733–740, 2009.

[10] D. Rath, L. Amlinger, A. Rath, and M. Lundgren, "The CRISPR-Cas immune system: biology, mechanisms and applications," *Biochimie*, vol. 117, pp. 119–128, 2015.

[11] I. Fonfara, A. le Rhun, K. Chylinski et al., "Phylogeny of Cas9 determines functional exchangeability of dual-RNA and Cas9 among orthologous type II CRISPR-Cas systems," *Nucleic Acids Research*, vol. 42, no. 4, pp. 2577–2590, 2014.

[12] K. M. Esvelt, P. Mali, J. L. Braff, M. Moosburner, S. J. Yaung, and G. M. Church, "Orthogonal Cas9 proteins for RNA-guided gene regulation and editing," *Nature Methods*, vol. 10, no. 11, pp. 1116–1121, 2013.

[13] F. J. M. Mojica and F. Rodriguez-Valera, "The discovery of CRISPR in archaea and bacteria," *The FEBS Journal*, vol. 283, no. 17, pp. 3162–3169, 2016.

[14] F. J. M. Mojica and L. Montoliu, "On the origin of CRISPR-Cas technology: from prokaryotes to mammals," *Trends in Microbiology*, vol. 24, no. 10, pp. 811–820, 2016.

[15] D. Burstein, L. B. Harrington, S. C. Strutt et al., "New CRISPR-Cas systems from uncultivated microbes," *Nature*, vol. 542, no. 7640, pp. 237–241, 2017.

[16] F. Jiang and J. A. Doudna, "CRISPR–Cas9 structures and mechanisms," *Annual Review of Biophysics*, vol. 46, no. 1, pp. 505–529, 2017.

[17] H. Chen, J. Choi, and S. Bailey, "Cut site selection by the two nuclease domains of the Cas9 RNA-guided endonuclease," *The Journal of Biological Chemistry*, vol. 289, no. 19, pp. 13284–13294, 2014.

[18] G. Gasiunas, R. Barrangou, P. Horvath, and V. Siksnys, "Cas9-crRNA ribonucleoprotein complex mediates specific DNA cleavage for adaptive immunity in bacteria," *Proceedings of the National Academy of Sciences*, vol. 109, no. 39, pp. E2579–E2586, 2012.

[19] C. Anders, O. Niewoehner, A. Duerst, and M. Jinek, "Structural basis of PAM-dependent target DNA recognition by the Cas9 endonuclease," *Nature*, vol. 513, no. 7519, pp. 569–573, 2014.

[20] M. Jinek, F. Jiang, D. W. Taylor et al., "Structures of Cas9 endonucleases reveal RNA-mediated conformational activation," *Science*, vol. 343, no. 6176, article 1247997, 2014.

[21] S. H. Sternberg, S. Redding, M. Jinek, E. C. Greene, and J. A. Doudna, "DNA interrogation by the CRISPR RNA-guided endonuclease Cas9," *Nature*, vol. 507, no. 7490, pp. 62–67, 2014.

[22] M. D. Szczelkun, M. S. Tikhomirova, T. Sinkunas et al., "Direct observation of R-loop formation by single RNA-guided Cas9 and cascade effector complexes," *Proceedings of the National Academy of Sciences*, vol. 111, no. 27, pp. 9798–9803, 2014.

[23] L. Cong, F. A. Ran, D. Cox et al., "Multiplex genome engineering using CRISPR/Cas systems," *Science*, vol. 339, no. 6121, pp. 819–823, 2013.

[24] W. Jiang, D. Bikard, D. Cox, F. Zhang, and L. A. Marraffini, "RNA-guided editing of bacterial genomes using CRISPR-Cas systems," *Nature Biotechnology*, vol. 31, no. 3, pp. 233–239, 2013.

[25] V. Pattanayak, S. Lin, J. P. Guilinger, E. Ma, J. A. Doudna, and D. R. Liu, "High-throughput profiling of off-target DNA cleavage reveals RNA-programmed Cas9 nuclease specificity," *Nature Biotechnology*, vol. 31, no. 9, pp. 839–843, 2013.

[26] G. J. Gibson and M. Yang, "What rheumatologists need to know about CRISPR/Cas9," *Nature Reviews Rheumatology*, vol. 13, no. 4, pp. 205–216, 2017.

[27] J. Luo, "CRISPR/Cas9: from genome engineering to cancer drug discovery," *Trends in Cancer*, vol. 2, no. 6, pp. 313–324, 2016.

[28] S. Q. Tsai and J. K. Joung, "Defining and improving the genome-wide specificities of CRISPR-Cas9 nucleases," *Nature Reviews Genetics*, vol. 17, no. 5, pp. 300–312, 2016.

[29] J. A. Doudna and C. A. Gersbach, "Genome editing: the end of the beginning," *Genome Biology*, vol. 16, no. 1, p. 292, 2015.

[30] D. B. T. Cox, R. J. Platt, and F. Zhang, "Therapeutic genome editing: prospects and challenges," *Nature Medicine*, vol. 21, no. 2, pp. 121–131, 2015.

[31] R. Aroul-Selvam, T. Hubbard, and R. Sasidharan, "Domain insertions in protein structures," *Journal of Molecular Biology*, vol. 338, no. 4, pp. 633–641, 2004.

[32] M. Wang and G. Caetano-Anollés, "The evolutionary mechanics of domain organization in proteomes and the rise of modularity in the protein world," *Structure*, vol. 17, no. 1, pp. 66–78, 2009.

[33] L. F. Ribeiro, T. D. Warren, and M. Ostermeier, "Construction of protein switches by domain insertion and directed evolution," *Methods in Molecular Biology*, vol. 1596, pp. 43–55, 2017.

[34] J. Tullman, N. Nicholes, M. R. Dumont, L. F. Ribeiro, and M. Ostermeier, "Enzymatic protein switches built from paralogous input domains," *Biotechnology and Bioengineering*, vol. 113, no. 4, pp. 852–858, 2016.

[35] L. F. Ribeiro, N. Nicholes, J. Tullman et al., "Insertion of a xylanase in xylose binding protein results in a xylose-stimulated xylanase," *Biotechnology for Biofuels*, vol. 8, no. 1, p. 118, 2015.

[36] L. F. Ribeiro, J. Tullman, N. Nicholes et al., "A xylose-stimulated xylanase-xylose binding protein chimera created by random nonhomologous recombination," *Biotechnology for Biofuels*, vol. 9, no. 1, p. 119, 2016.

[37] M. Ostermeier, "Engineering allosteric protein switches by domain insertion," *Protein Engineering, Design and Selection*, vol. 18, no. 8, pp. 359–364, 2005.

[38] V. Stein and K. Alexandrov, "Synthetic protein switches: design principles and applications," *Trends in Biotechnology*, vol. 33, no. 2, pp. 101–110, 2015.

[39] B. Chen, L. A. Gilbert, B. A. Cimini et al., "Dynamic imaging of genomic loci in living human cells by an optimized CRISPR/Cas system," *Cell*, vol. 155, no. 7, pp. 1479–1491, 2013.

[40] L. R. Polstein and C. A. Gersbach, "A light-inducible CRISPR-Cas9 system for control of endogenous gene activation," *Nature Chemical Biology*, vol. 11, no. 3, pp. 198–200, 2015.

[41] M. F. Bolukbasi, A. Gupta, S. Oikemus et al., "DNA-binding-domain fusions enhance the targeting range and precision of Cas9," *Nature Methods*, vol. 12, no. 12, pp. 1150–1156, 2015.

[42] B. Maji, C. L. Moore, B. Zetsche et al., "Multidimensional chemical control of CRISPR-Cas9," *Nature Chemical Biology*, vol. 13, no. 1, pp. 9–11, 2017.

[43] M. Berrondo, M. Ostermeier, and J. J. Gray, "Structure prediction of domain insertion proteins from structures of individual domains," *Structure*, vol. 16, no. 4, pp. 513–527, 2008.

[44] L. F. Ribeiro, G. P. Furtado, M. R. Lourenzoni et al., "Engineering bifunctional laccase-xylanase chimeras for improved catalytic performance," *The Journal of Biological Chemistry*, vol. 286, no. 50, pp. 43026–43038, 2011.

[45] B. Pierre, J. W. Labonte, T. Xiong et al., "Molecular determinants for protein stabilization by insertional fusion to a thermophilic host protein," *Chembiochem*, vol. 16, no. 16, pp. 2392–2402, 2015.

[46] C. S. Kim, B. Pierre, M. Ostermeier, L. L. Looger, and J. R. Kim, "Enzyme stabilization by domain insertion into a thermophilic protein," *Protein Engineering Design & Selection*, vol. 22, no. 10, pp. 615–623, 2009.

[47] X. H. Zhang, L. Y. Tee, X. G. Wang, Q. S. Huang, and S. H. Yang, "Off-target effects in CRISPR/Cas9-mediated genome engineering," *Molecular Therapy - Nucleic Acids*, vol. 4, article e264, 2015.

[48] S. Q. Tsai, N. Wyvekens, C. Khayter et al., "Dimeric CRISPR RNA-guided FokI nucleases for highly specific genome editing," *Nature Biotechnology*, vol. 32, no. 6, pp. 569–576, 2014.

[49] J. P. Guilinger, D. B. Thompson, and D. R. Liu, "Fusion of catalytically inactive Cas9 to FokI nuclease improves the specificity of genome modification," *Nature Biotechnology*, vol. 32, no. 6, pp. 577–582, 2014.

[50] A. K. Singh Gautam, S. Balakrishnan, and P. Venkatraman, "Direct ubiquitin independent recognition and degradation of a folded protein by the eukaryotic proteasomes-origin of intrinsic degradation signals," *PLoS One*, vol. 7, no. 4, p. e34864, 2012.

[51] A. C. Komor, Y. B. Kim, M. S. Packer, J. A. Zuris, and D. R. Liu, "Programmable editing of a target base in genomic DNA without double-stranded DNA cleavage," *Nature*, vol. 533, no. 7603, pp. 420–424, 2016.

[52] N. M. Gaudelli, A. C. Komor, H. A. Rees et al., "Programmable base editing of A•T to G•C in genomic DNA without DNA cleavage," *Nature*, vol. 551, no. 7681, pp. 464–471, 2017.

[53] H. A. Rees, A. C. Komor, W. H. Yeh et al., "Improving the DNA specificity and applicability of base editing through protein engineering and protein delivery," *Nature Communications*, vol. 8, 2017.

[54] Y. B. Kim, A. C. Komor, J. M. Levy, M. S. Packer, K. T. Zhao, and D. R. Liu, "Increasing the genome-targeting scope and precision of base editing with engineered Cas9-cytidine deaminase fusions," *Nature Biotechnology*, vol. 35, no. 4, pp. 371–376, 2017.

[55] J. M. Gerhke, O. R. Cervantes, M. Kendell Clement, L. Pinello, and J. Keith Joung, "High-precision CRISPR-Cas9 base editors with minimized bystander and off-target mutations," *bioRxiv*, article 273938, 2018.

[56] Y. Nihongaki, S. Yamamoto, F. Kawano, H. Suzuki, and M. Sato, "CRISPR-Cas9-based photoactivatable transcription system," *Chemistry & Biology*, vol. 22, no. 2, pp. 169–174, 2015.

[57] Y. Gao, X. Xiong, S. Wong, E. J. Charles, W. A. Lim, and L. S. Qi, "Complex transcriptional modulation with orthogonal and inducible dCas9 regulators," *Nature Methods*, vol. 13, no. 12, pp. 1043–1049, 2016.

[58] K. I. Liu, M. N. B. Ramli, C. W. A. Woo et al., "A chemical-inducible CRISPR-Cas9 system for rapid control of genome editing," *Nature Chemical Biology*, vol. 12, no. 11, pp. 980–987, 2016.

[59] A. Amabile, A. Migliara, P. Capasso et al., "Inheritable silencing of endogenous genes by hit-and-run targeted epigenetic editing," *Cell*, vol. 167, no. 1, pp. 219–232.e14, 2016.

[60] P. Stepper, G. Kungulovski, R. Z. Jurkowska et al., "Efficient targeted DNA methylation with chimeric dCas9-Dnmt3a-Dnmt3L methyltransferase," *Nucleic Acids Research*, vol. 45, no. 4, pp. 1703–1713, 2017.

[61] A. Vojta, P. Dobrinić, V. Tadić et al., "Repurposing the CRISPR-Cas9 system for targeted DNA methylation," *Nucleic Acids Research*, vol. 44, no. 12, pp. 5615–5628, 2016.

[62] J. I. McDonald, H. Celik, L. E. Rois et al., "Reprogrammable CRISPR/Cas9-based system for inducing site-specific DNA methylation," *Biology Open*, vol. 5, no. 6, pp. 866–874, 2016.

[63] T. Xiong, G. E. Meister, R. E. Workman et al., "Targeted DNA methylation in human cells using engineered dCas9-methyltransferases," *Scientific Reports*, vol. 7, no. 1, p. 6732, 2017.

[64] N. A. Kearns, H. Pham, B. Tabak et al., "Functional annotation of native enhancers with a Cas9-histone demethylase fusion," *Nature Methods*, vol. 12, no. 5, pp. 401–403, 2015.

[65] I. B. Hilton, A. M. D'Ippolito, C. M. Vockley et al., "Epigenome editing by a CRISPR-Cas9-based acetyltransferase activates genes from promoters and enhancers," *Nature Biotechnology*, vol. 33, no. 5, pp. 510–517, 2015.

[66] T. Anton, S. Bultmann, H. Leonhardt, and Y. Markaki, "Visualization of specific DNA sequences in living mouse embryonic stem cells with a programmable fluorescent CRISPR/Cas system," *Nucleus*, vol. 5, no. 2, pp. 163–172, 2014.

[67] H. Ma, A. Naseri, P. Reyes-Gutierrez, S. A. Wolfe, S. Zhang, and T. Pederson, "Multicolor CRISPR labeling of chromosomal loci in human cells," *Proceedings of the National Academy of Sciences*, vol. 112, no. 10, pp. 3002–3007, 2015.

[68] N. H. Shah and T. W. Muir, "Inteins: nature's gift to protein chemists," *Chemical Science*, vol. 5, no. 2, pp. 446–461, 2014.

[69] S. H. Peck, I. Chen, and D. R. Liu, "Directed evolution of a small-molecule-triggered intein with improved splicing properties in mammalian cells," *Chemistry & Biology*, vol. 18, no. 5, pp. 619–630, 2011.

[70] K. M. Davis, V. Pattanayak, D. B. Thompson, J. A. Zuris, and D. R. Liu, "Small molecule-triggered Cas9 protein with improved genome-editing specificity," *Nature Chemical Biology*, vol. 11, no. 5, pp. 316–318, 2015.

[71] X. X. Zhou, X. Zou, H. K. Chung et al., "A single-chain photoswitchable CRISPR-Cas9 architecture for light-inducible gene editing and transcription," *ACS Chemical Biology*, vol. 13, no. 2, pp. 443–448, 2017.

[72] X. X. Zhou, L. Z. Fan, P. Li, K. Shen, and M. Z. Lin, "Optical control of cell signaling by single-chain photoswitchable kinases," *Science*, vol. 355, no. 6327, pp. 836–842, 2017.

[73] G. T. Hess, L. Frésard, K. Han et al., "Directed evolution using dCas9-targeted somatic hypermutation in mammalian cells," *Nature Methods*, vol. 13, no. 12, pp. 1036–1042, 2016.

[74] F. Richter, I. Fonfara, B. Bouazza et al., "Engineering of temperature- and light-switchable Cas9 variants," *Nucleic Acids Research*, vol. 44, no. 20, pp. 10003–10014, 2016.

[75] B. L. Oakes, D. C. Nadler, A. Flamholz et al., "Profiling of engineering hotspots identifies an allosteric CRISPR-Cas9 switch," *Nature Biotechnology*, vol. 34, no. 6, pp. 646–651, 2016.

[76] H. Nishimasu, F. A. Ran, P. D. Hsu et al., "Crystal structure of Cas9 in complex with guide RNA and target DNA," *Cell*, vol. 156, no. 5, pp. 935–949, 2014.

[77] I. M. Slaymaker, L. Gao, B. Zetsche, D. A. Scott, W. X. Yan, and F. Zhang, "Rationally engineered Cas9 nucleases with improved specificity," *Science*, vol. 351, no. 6268, pp. 84–88, 2016.

[78] B. P. Kleinstiver, V. Pattanayak, M. S. Prew et al., "High-fidelity CRISPR-Cas9 nucleases with no detectable genome-wide off-target effects," *Nature*, vol. 529, no. 7587, pp. 490–495, 2016.

[79] V. Siksnys and G. Gasiunas, "Rewiring Cas9 to target new PAM sequences," *Molecular Cell*, vol. 61, no. 6, pp. 793-794, 2016.

[80] E. Zinn and L. H. Vandenberghe, "Adeno-associated virus: fit to serve," *Current Opinion in Virology*, vol. 8, pp. 90–97, 2014.

[81] K. Chamberlain, J. M. Riyad, and T. Weber, "Expressing transgenes that exceed the packaging capacity of adeno-associated virus capsids," *Human Gene Therapy Methods*, vol. 27, no. 1, pp. 1–12, 2016.

[82] A. V. Wright, S. H. Sternberg, D. W. Taylor et al., "Rational design of a split-Cas9 enzyme complex," *Proceedings of the National Academy of Sciences*, vol. 112, no. 10, pp. 2984–2989, 2015.

[83] D. J. J. Truong, K. Kühner, R. Kühn et al., "Development of an intein-mediated split-Cas9 system for gene therapy," *Nucleic Acids Research*, vol. 43, no. 13, pp. 6450–6458, 2015.

[84] Y. Nihongaki, F. Kawano, T. Nakajima, and M. Sato, "Photoactivatable CRISPR-Cas9 for optogenetic genome editing," *Nature Biotechnology*, vol. 33, no. 7, pp. 755–760, 2015.

[85] B. Zetsche, S. E. Volz, and F. Zhang, "A split-Cas9 architecture for inducible genome editing and transcription modulation," *Nature Biotechnology*, vol. 33, no. 2, pp. 139–142, 2015.

[86] B. P. Kleinstiver, M. S. Prew, S. Q. Tsai et al., "Engineered CRISPR-Cas9 nucleases with altered PAM specificities," *Nature*, vol. 523, no. 7561, pp. 481–485, 2015.

[87] B. P. Kleinstiver, M. S. Prew, S. Q. Tsai et al., "Broadening the targeting range of Staphylococcus aureus CRISPR-Cas9 by modifying PAM recognition," *Nature Biotechnology*, vol. 33, no. 12, pp. 1293–1298, 2015.

[88] R. Heler, A. V. Wright, M. Vucelja, D. Bikard, J. A. Doudna, and L. A. Marraffini, "Mutations in Cas9 enhance the rate of acquisition of viral spacer sequences during the CRISPR-Cas immune response," *Molecular Cell*, vol. 65, no. 1, pp. 168–175, 2017.

[89] J. H. Hu, S. M. Miller, M. H. Geurts et al., "Evolved Cas9 variants with broad PAM compatibility and high DNA specificity," *Nature*, vol. 556, no. 7699, pp. 57–63, 2018.

[90] F. Atroshi, A. Rizzo, T. Westermarck, and T. Ali-Vehmas, "Effects of tamoxifen, melatonin, coenzyme Q10, and L-carnitine supplementation on bacterial growth in the presence of mycotoxins," *Pharmacological Research*, vol. 38, no. 4, pp. 289–295, 1998.

[91] X. Liu, E. Pisha, D. A. Tonetti et al., "Antiestrogenic and DNA damaging effects induced by tamoxifen and toremifene metabolites," *Chemical Research in Toxicology*, vol. 16, no. 7, pp. 832–837, 2003.

[92] P. W. Fan, F. Zhang, and J. L. Bolton, "4-Hydroxylated metabolites of the antiestrogens tamoxifen and toremifene are metabolized to unusually stable quinone methides," *Chemical Research in Toxicology*, vol. 13, no. 1, pp. 45–52, 2000.

[93] C. M. Nowak, S. Lawson, M. Zerez, and L. Bleris, "Guide RNA engineering for versatile Cas9 functionality," *Nucleic Acids Research*, vol. 44, pp. 9555–9564, 2016.

7

Systems Pharmacology-Based Approach of Connecting Disease Genes in Genome-Wide Association Studies with Traditional Chinese Medicine

Jihye Kim,[1] Minjae Yoo,[1] Jimin Shin,[1] Hyunmin Kim,[1] Jaewoo Kang,[2] and Aik Choon Tan[1]

[1]*Translational Bioinformatics and Cancer Systems Biology Laboratory, Division of Medical Oncology, Department of Medicine, School of Medicine, University of Colorado Anschutz Medical Campus, Aurora, CO 80045, USA*
[2]*Department of Computer Science and Engineering, Interdisciplinary Graduate Program in Bioinformatics, Korea University, Seoul, Republic of Korea*

Correspondence should be addressed to Aik Choon Tan; aikchoon.tan@ucdenver.edu

Academic Editor: Qin Feng

Traditional Chinese medicine (TCM) originated in ancient China has been practiced over thousands of years for treating various symptoms and diseases. However, the molecular mechanisms of TCM in treating these diseases remain unknown. In this study, we employ a systems pharmacology-based approach for connecting GWAS diseases with TCM for potential drug repurposing and repositioning. We studied 102 TCM components and their target genes by analyzing microarray gene expression experiments. We constructed disease-gene networks from 2558 GWAS studies. We applied a systems pharmacology approach to prioritize disease-target genes. Using this bioinformatics approach, we analyzed 14,713 GWAS disease-TCM-target gene pairs and identified 115 disease-gene pairs with q value < 0.2. We validated several of these GWAS disease-TCM-target gene pairs with literature evidence, demonstrating that this computational approach could reveal novel indications for TCM. We also develop TCM-Disease web application to facilitate the traditional Chinese medicine drug repurposing efforts. Systems pharmacology is a promising approach for connecting GWAS diseases with TCM for potential drug repurposing and repositioning. The computational approaches described in this study could be easily expandable to other disease-gene network analysis.

1. Introduction

Traditional Chinese medicine (TCM) has been practiced over thousands of years in China and other Asian countries for treating various symptoms and diseases [1]. The drug development process for the modern (or Western) medicine is usually initiated by developing small molecules that target a specific biological target or pathway [2]. In this drug development process, it is much easier to study the mechanism of action (MoA) of these small molecules in a context-specific disease state. For example, vemurafenib was developed to specifically inhibit BRAF p.V600E-mutant cancer cells [3]. Clinically, vemurafenib has demonstrated efficacy in treating BRAF-mutant melanoma patients, eventually received Food and Drug Administration (FDA) approval in 2011 for the treatment of late-stage melanoma [4]. Due to the known MoA (e.g., binding to BRAF mutants and inhibiting the downstream signaling pathway) and the target population (e.g., melanoma with BRAF p.V600E), it is much easier to develop predictive biomarker (e.g., BRAF p.V600E) for personalize treatment using modern drug [4]. However, it remains a challenge to elucidate the molecular mechanisms of a particular TCM in inhibiting biological process in a context-specific disease state because of the "multicomponent, multitarget" nature of TCM.

The polypharmacology property of TCM could be exploited for drug repurposing and repositioning. For example, artemisinin, an active component extracted from *Artemisia annua* L. in the early 1970s in China, has been the "magic bullet" for malaria therapy [5]. In late last year, through a high-throughput chemical biology screen, a group of researchers identified artemisinin can induce pancreatic alpha cells to transdifferentiate into insulin producing beta-like cells [6]. This finding suggests that artemisinin could be repurposed as type 1 diabetes therapy. It also opens up other opportunities to use modern technologies to reveal the unexpected interactions between TCM and diseases for drug repurposing.

With the advancement of high-throughput genotyping technologies (e.g., microarray and sequencing), it is now possible to perform genome-wide association studies (GWAS) to identify candidate genes associated with various complex diseases [7]. One of the challenges in GWAS is to translate the candidate genes into clinic for disease management and treatment. Similarly, microarray and sequencing technologies have been applied in chemical biology research to deconvolute target genes and to elucidate molecular mechanisms [8, 9]. These modern biotechnologies open up new research avenues for the investigation of TCM in a wide range of diseases. To gain insights, novel computational algorithms and bioinformatics methods are needed to analyze and interpret these massive "omics" data. Systems pharmacology is a promising research approach in translational medicine that integrates computational and experimental methods to elucidate, validate, and apply new pharmacological concepts to the development and use of small molecule and biologic drugs [10, 11]. Approaches in systems pharmacology can provide a scientifically rational for defining optimal multidrug regimens, identifying responsive patient populations, identifying translational biomarkers, drug repurposing, and designing clinical trials [10–13].

In this study, we employed systems pharmacology-based approach of connecting GWAS disease genes with traditional Chinese medicine as drug repurposing and repositioning strategy. We obtained TCM-induced gene expression study from the Gene Expression Omnibus and performed differentially expressed gene analysis to define TCM-target genes. We constructed a drug-gene network based on the TCM-target genes. We downloaded candidate genes associated with GWAS diseases from GWAS catalog [7] and constructed the GWAS disease-gene network for this study. By using a systems pharmacology-based analysis, we prioritized GWAS disease-TCM gene pairs from the two networks. We performed literature and molecular mechanisms evidence by querying PubMed and the Connectivity Map to support the GWAS disease-TCM gene pairs. We also implemented TCM-Disease web app for users to query new connections between disease and TCM-target genes. Our overall research strategy of this study is illustrated in Figure 1.

2. Materials and Methods

2.1. TCM Gene Expression Analysis. We downloaded the raw microarray data of 102 TCM component-induced gene expression studies from the Gene Expression Omnibus (GSE85871) [14]. These TCM molecules are active ingredients in Chinese herbs and TCM. Briefly, TCM molecules (either 1 μM or 10 μM) and DMSO were treated on MCF7 breast cancer cell line, and total RNA was extracted and subjected for Affymetrix HG U133A 2.0 microarray gene expression profiling in duplicate. Raw data were downloaded and normalized by Robust Multiarray Average using Affymetrix Power Tools. For data analysis, we collapsed the probe sets representing the same gene using the maximum expression value of these probe sets to improve power to detect any changes in different isoforms. Gene expression values from the duplicates were averaged, and fold change of 1.5 was used as the thresholds to select differentially expressed genes for a particular TCM compound against DMSO. We used the list of differentially expressed genes as the gene signatures as the TCM-target genes. The list of 102 TCM components is listed in Supplementary Table 1.

2.2. Genome-Wide Association Studies Disease Genes. We obtained the latest collection of genome-wide association studies (GWAS) data from the NHGRI-EBI GWAS catalog of published GWAS (https://www.ebi.ac.uk/gwas/home, data version: v1.0, gwas-catalog_v1.0-associations_e89_r2017-06-19.tsv) [7]. The GWAS catalog of GWAS contains 1987 diseases/traits collected from 2558 studies. We used the genes associated with the diseases in this study.

2.3. Mapping Disease Terms to MeSH. GWAS diseases obtained from the NHGRI-EBI GWAS catalog were mapped to the Medical Subject Headings (MeSH) terms. MeSH is the NIH/NLM/NCBI controlled vocabulary thesaurus used for indexing articles for PubMed, and it follows a hierarchical tree structure for indexing these terms. We mapped the GWAS disease names using the MeSH Browser (https://meshb.nlm.nih.gov/search), such that we could group similar diseases using the MeSH terminology tree. For this study, we focused on diseases that mapped to the MeSH terms in the "Disease [C]" category, using the third term level in the MeSH tree [CX.X.X]. We grouped all the genes associated with the same MeSH term and used as the genes associated with that disease term.

2.4. Protein-Protein Interactions Network. To construct the GWAS disease-gene network, we queried STRING to retrieve additional genes (proteins) associated with a particular set of disease genes [15]. We restrict the protein-protein interactions from STRING and only include not more than 100 addition genes that have high-confidence interactions score (score > 0.9). We build the GWAS disease-gene network using the resultant protein-protein interactions.

2.5. Statistical Framework for Prioritizing TCM-Disease Connection. To prioritize TCM-Disease connection, we performed a permutation test to compute the statistical significance of a TCM to be prioritized for treating a specific disease [16]. Our hypothesis is that the polypharmacology nature of TCM has a potential to treat a particular disease if its targets are associated with that disease-gene network. We performed a permutation based testing as below to obtain the nominal p value (P) similar to the approach of Connectivity Map [8, 9]:

$$P = \frac{\#\{G_{\mathrm{TCM}}(p) > G_{\mathrm{TCM}}\}}{\#\mathrm{permutations}}, \tag{1}$$

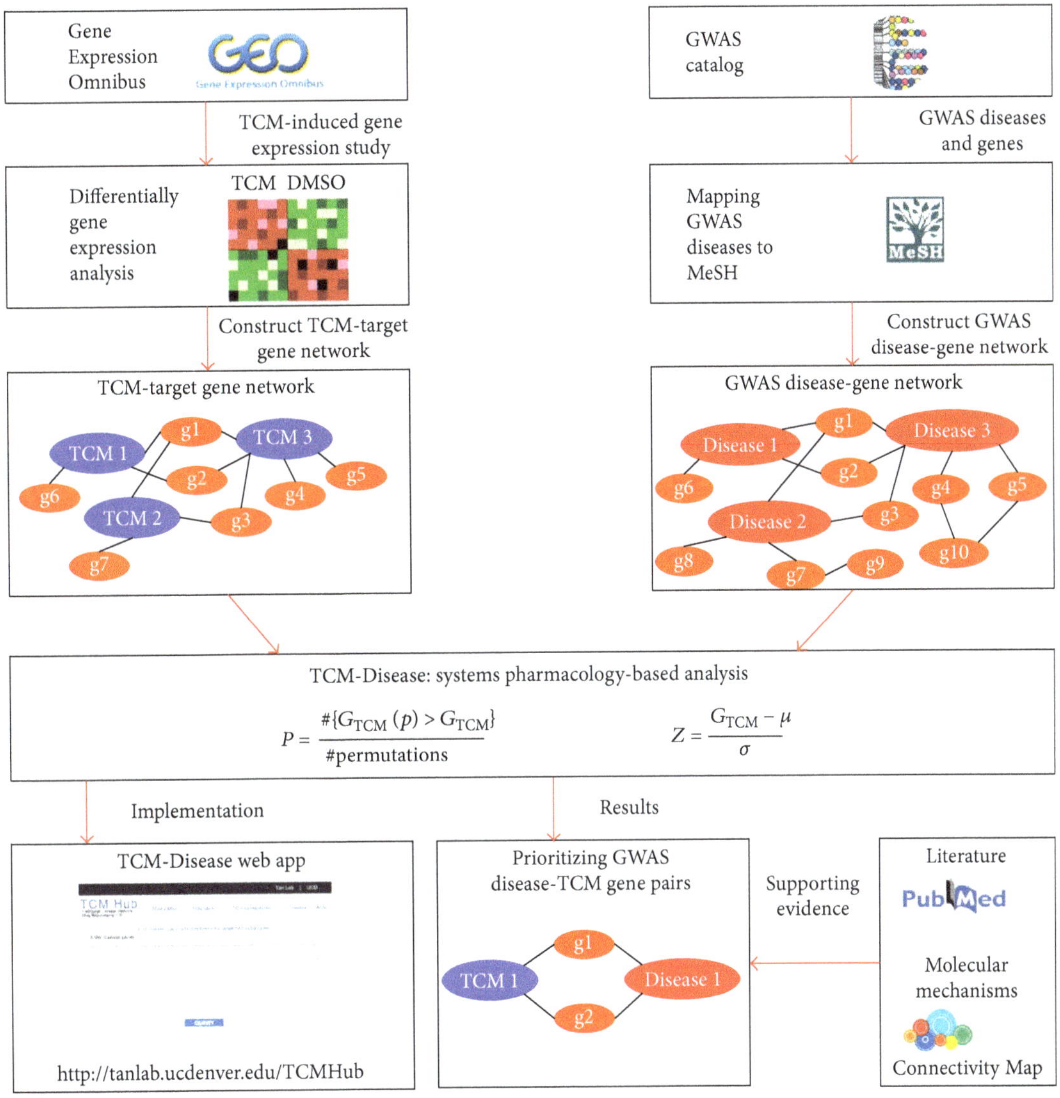

FIGURE 1: Overall research strategy of systems pharmacology-based analysis of connecting GWAS diseases to TCM-target genes.

where G_{TCM} represents the number of GWAS disease genes targeted by a TCM. $G_{TCM}(p)$ represents a permutation that GWAS disease genes were targeted by a TCM. We performed 100,000 permutations by randomly selecting the same number of genes in a GWAS disease from a genome-wide scale (we used 20,462 genes as the number of genes) in this study. A nominal p value (P) is calculated for each TCM by counting the number of permutations that have $G_{TCM}(p)$ greater than G_{TCM}, divided by the total number of permutations. Next, we performed Benjamini-Hochberg multiple testing correction method to obtain the adjusted p values (q values) for the TCM-GWAS disease genes.

To compare the TCM-target genes against each GWAS disease, we calculated a Z-score for each disease as:

$$Z = \frac{G_{TCM} - \mu}{\sigma}, \quad (2)$$

where G_{TCM} denotes the number of GWAS disease genes targeted by a TCM, μ denotes the mean number of GWAS disease genes targeted by a given TCM, and σ represents the standard deviation during the 100,000 permutations. The permutation test and Z-score calculation are inspired by the approach used in [16].

2.6. TCM-Disease Web Application Implementation. We developed TCM-Disease, the statistical framework described in this study, as a user-friendly web application. The website is implemented using Python 2.7.9, Python-CGI script deployed in a Mac OSX Server. The web application is part of the traditional Chinese medicine drug repurposing hub (TCM hub), which is freely accessible at http://tanlab.ucdenver.edu/TCMHub.

2.7. Network Visualization. To visualize the connections between GWAS disease-TCM-target genes, we utilized the Cytoscape network visualization tool (version 3.5.1) [17, 18].

2.8. Connectivity Map Analysis. We utilized the latest Connectivity Map (CMap) CLUE web application (https://clue.io)

TABLE 1: Summary statistics for the TCM gene network and GWAS disease-gene network.

TCM gene network			
Number of compounds	Number of TCM-target genes	Average number of genes per TCM compound	Number of TCM-gene interactions
102	7380	430	43,839
GWAS disease-gene network			
Number of diseases	Number of disease genes	Average number of genes per disease	Number of disease-gene interactions
293	7417	134	39,262

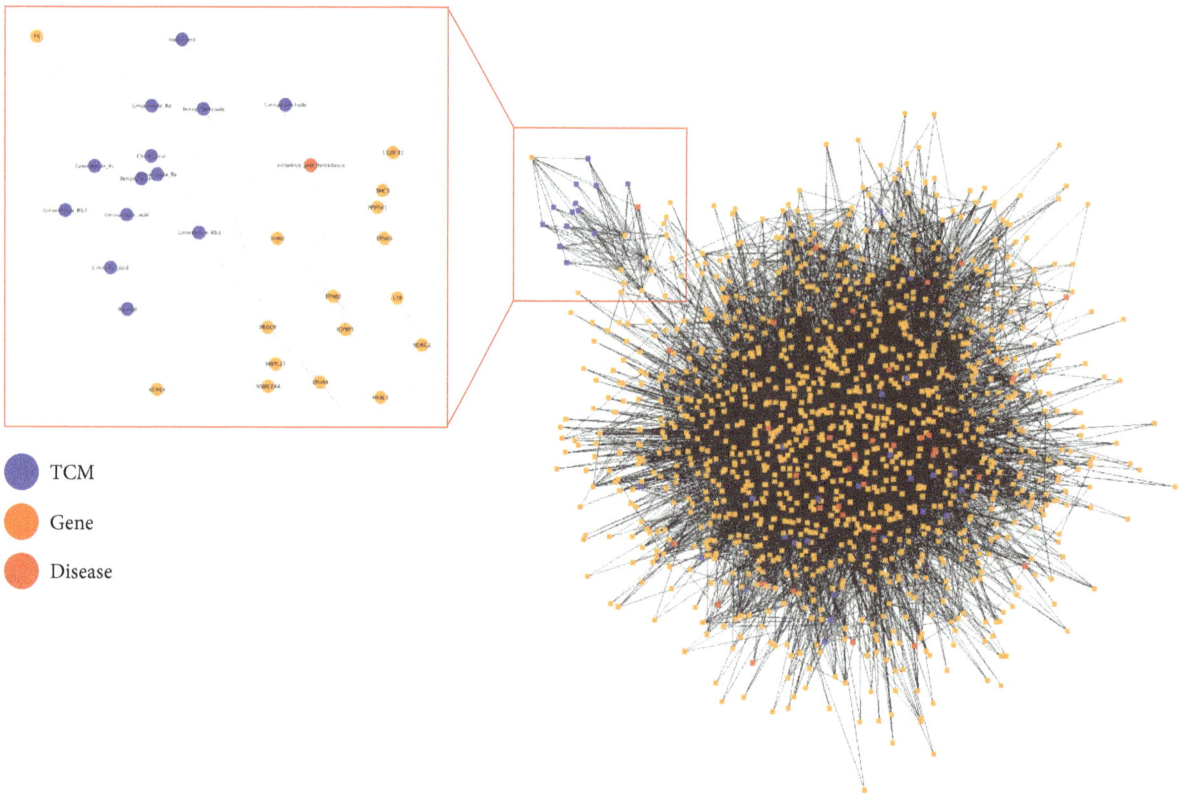

FIGURE 2: GWAS disease-TCM gene network. This is the GWAS disease-TCM gene network.

resource for identifying MoA for these TCM components [8, 9]. The latest CMap resource contains more than 1.3 million gene expression signatures that were collected from genetic or small molecule perturbations on a panel of cell lines. The CMap version (Data version 1.0.1.1. and Software Version 1.0.1.8) contains 2836, 3799, and 2160 signatures for compounds, gene knockdown, and gene overexpression, respectively. We queried the top 150 upregulated and downregulated genes for each TCM against CLUE App to retrieve the connected signatures as potential molecular mechanisms of TCM components. The results of the Connectivity Map for these 102 TCM components are available in TCM hub (http://tanlab.ucdenver.edu/TCMHub).

3. Results

3.1. Construction of the TCM-Target Gene Network. We constructed the TCM-target gene network from the analyzing microarray gene expression profiles of 102 TCM components. The final TCM-target gene network consists of 43,839 interaction pairs, which connect 102 TCM components and 7380 target genes. Table 1 summarizes the statistics of the TCM-target gene network. From the TCM-target gene network, it suggests that TCM could be repurposed for various diseases as indicated by the polypharmacology properties of these compounds.

3.2. Construction of the GWAS Disease-Gene Network. To study the disease-gene interactions, we constructed the disease-gene network from 2558 GWAS studies. We grouped the GWAS diseases using MeSH terms and only focused on the diseases in the MeSH "Disease [C]" vocabulary term in this study. The final GWAS disease-gene network consists of 39,262 interaction pairs, which connect 293 diseases and 7417 disease genes. Table 1 summarizes the statistics of the GWAS disease-gene network.

TABLE 2: GWAS disease-TCM gene pairs with $q < 0.2$ identified from this study.

TCM name	Diseases (*Z*-score)
1β-Hydroxyalantolactone	Ovarian cysts (3.952), ovarian diseases (3.365)
Ainsliadimer A	Neoplasms, glandular and epithelial (4.051)
Artemisinin	Diabetes mellitus, type 1 (4.526)
Benzoylaconitine	Embolism and thrombosis (5.058)
Benzyl benzoate	Embolism and thrombosis (6.157)
Berberine hydrochloride	Central nervous system infections (4.626) hemorrhage (5.090), hypertension, pregnancy-induced (4.347), prion diseases (5.433)
Borneol	Embolism and thrombosis (4.754)
Britanin	Central nervous system infections (5.323), prion diseases (4.554)
Bruceine D	Intestinal diseases (3.126), lip diseases (4.480), lymphoproliferative disorders (3.190), mouth abnormalities (4.471), skin diseases, papulosquamous (4.167)
Bufalin	Lip diseases (12.748), mouth abnormalities (12.730), skin diseases, papulosquamous (6.010), spinal diseases (5.987)
Chelerythrine	Autoimmune diseases of the nervous system (3.526)
Cholic acid	Embolism and thrombosis (4.536)
Cinnamaldehyde	Embolism and thrombosis (4.559)
Cinnamic acid	Embolism and thrombosis (5.398)
Curculigoside	Hypersensitivity, immediate (3.676)
Daidzin	Autoimmune diseases of the nervous system (4.358), bone marrow diseases (5.015), demyelinating autoimmune diseases, cns (3.744)
Deoxycholic acid	Embolism and thrombosis (5.072)
Four mixture	Autoimmune diseases of the nervous system (3.198), breast diseases (4.159), breast neoplasms (4.137), demyelinating autoimmune diseases, cns (3.263), gastrointestinal neoplasms (3.648), lymphoproliferative disorders (4.073), neoplasms, glandular and epithelial (3.244)
Gamabufatalin	Gastroenteritis (4.344), skin diseases, papulosquamous (5.028)
Gentiopicroside	Breast diseases (4.249), breast neoplasms (4.251)
Ginsenoside Rb2	Embolism and thrombosis (4.604)
Ginsenoside Rb3	Embolism and thrombosis (5.758)
Ginsenoside Rc	Embolism and thrombosis (6.291)
Ginsenoside Rd	Embolism and thrombosis (5.214)
Ginsenoside Re	Embolism and thrombosis (5.722)
Ginsenoside Rg1	Skin diseases, papulosquamous (4.279)
Isoborneol	Embolism and thrombosis (4.441)
Japonicone A	Acneiform eruptions (3.354), arthritis, rheumatoid (3.591), breast diseases (4.482), breast neoplasms (4.473), central nervous system infections (4.303), cocaine-related disorders (3.625), digestive system neoplasms (3.370), gastrointestinal neoplasms (5.216), intestinal diseases (3.592), leukemia (3.687), lymphoma (4.369), lymphoproliferative disorders (4.066), neoplasms, glandular and epithelial (3.772), ovarian diseases (3.535), prion diseases (5.863), rheumatic diseases (3.599), sebaceous gland diseases (3.357), urinary bladder diseases (3.722), urologic neoplasms (3.506)
Liquiritin	Autoimmune diseases of the nervous system (3.704), glucose metabolism disorders (4.245), hypersensitivity, immediate (3.993)
Narciclasine	Bone marrow diseases (3.974), gastroenteritis (3.703), intestinal diseases (3.734), lymphoma (3.848), lymphoproliferative disorders (4.190), osteitis deformans (3.623), peripheral vascular diseases (3.707), prion diseases (4.772), skin diseases, papulosquamous (4.086)
Nitidine chloride	Arthritis (3.319), bone marrow diseases (3.701), breast diseases (5.287), breast neoplasms (5.259), diabetes mellitus, type 2 (5.288), digestive system neoplasms (3.605), gastrointestinal neoplasms (3.712), glucose metabolism disorders (5.115), intestinal diseases (3.572), lymphoproliferative disorders (3.936), neoplasms, glandular and epithelial (4.427)

Table 2: Continued.

TCM name	Diseases (*Z*-score)
Notoginsenoside R1	Ovarian cysts (4.713), ovarian diseases (4.526)
Oleanic acid	Breast diseases (3.698), breast neoplasms (3.681), digestive system neoplasms (3.838), gastrointestinal neoplasms (4.394), muscular diseases (4.916), myositis (4.915), neoplasms (5.028)
Osthole	Neoplasms (5.137)
Salidroside	Autoimmune diseases of the nervous system (3.615), hypersensitivity, immediate (3.817), neoplasms (4.994)
Sanguinarine	Breast diseases (4.521), breast neoplasms (4.535)
Santonin	Breast neoplasms (3.506), gastrointestinal neoplasms (3.576)
Schisantherin A	Hypersensitivity, immediate (3.660)
Tanshinone IIA	Body weight (3.665), central nervous system infections (4.831), prion diseases (5.608)
Tetrahydropalmatine	Autoimmune diseases of the nervous system (4.315), demyelinating autoimmune diseases, cns (3.999), hypersensitivity, immediate (3.792)

3.3. Systems Pharmacology-Based Analysis of Connecting GWAS Disease-TCM-Target Genes. We performed the statistical systems pharmacology analysis to determine the connections between GWAS disease genes and TCM-target genes. We performed 14,713 connections between GWAS disease-TCM-target genes comparison using the described statistical framework as described in the Method section. Here, we used the q value < 0.2 as the threshold and identified 115 GWAS disease-TCM-target gene pairs from the analysis. Figure 2 shows the GWAS disease-TCM gene network. This list of GWAS disease-TCM-target gene pairs consists of 40 TCM and 42 GWAS diseases as summarized in Table 2. Figure 3 illustrates the heatmap of the 40 TCM and 42 GWAS diseases.

3.4. Case Study of GWAS Disease-TCM Gene Pairs. We illustrate the following GWAS disease-TCM gene pairs identified from the bioinformatics analysis and provide experimental validations supported by published literature. We also queried the TCM gene expression signatures to the latest version of Connectivity Map for potential mechanism of action (MoA) evidence of the TCM against GWAS diseases.

3.4.1. Japonicone A as Lymphoma Treatment. We identified japonicone A as a potential treatment for lymphoma (q value = 0.06) (Figure 4(a)). Japonicone A is a natural product extracted from the aerial part of *Inula japonica* Thunb (Asteraceae family) [19] and has been demonstrated to possess activities in anti-inflammatory, antidiabetes, and anticancer [19, 20]. By querying the TCM hub, japonicone A is connected with NF-κB pathway inhibitors as one of the mechanisms of action (MoA). Interestingly, japonicone A has been recently demonstrated to inhibit the growth of Burkitt lymphoma via inhibiting the NF-κB pathway [21]. This provides experimental support of japonicone A that could be repurposed as the lymphoma treatment.

3.4.2. Artemisinin as Type 1 Diabetes Mellitus Treatment. From the systems pharmacology-based analysis, we identified artemisinin, an antimalaria drug, that is connected with type 1 diabetes mellitus (q value = 0.11) (Figure 4(b)). Artemisinin is a compound extracted from *Artemisia annua* L. (Chinese name: qinghao). The discovery of artemisinin by Tu Youyou and her colleague has been the breakthrough therapy for malaria [5], and this discovery has led to the 2015 Nobel Prize of Physiology or Medicine awarded Tu Youyou. Notably, in a recent study, it has been demosntrated that treating antimalarial drug artemisinin in pancreatic alpha cells could transform into functional beta-like cells through enhanced GABA signaling [6]. From the gene expression analysis, GABA is one of the top overexpressed genes after artemisinin treatment. This demonstrates that this systems pharmacology approach could identify artemisinin as drug repurposed for diabetes mellitus treatment.

3.4.3. Sanguinarine as Breast Cancer Treatment. Another GWAS disease-TCM gene pair identified from this study is sanguinarine as breast cancer treatment (q value = 0.07). Sanguinarine is a compound extract from bloodroot (*Sanguinaria canadensis*) and has been used to treat different diseases [22]. Recent study has demonstrated that sanguinarine has anticancer therapy in inhibiting cell migration and the induction of apoptosis in basal-like breast cancer. The authors also identified that the sanguinarine is a potent dihydrofolate reductase (DHFR) inhibitor [23]. This provides evidence that sanguinarine could be used as anticancer drug for the treatment of basal-like breast cancer through systems pharmacology analysis.

3.4.4. Berberine Hydrochloride as Preeclampsia Treatment. From the GWAS disease-TCM genes analysis, we identified berberine hydrochloride as a potential pregnancy-induced hypertension (preeclampsia) (q value = 0.13). Berberine hydrochloride is an extract from the Chinese goldthread *Coptis chinensis*. This TCM compound has been shown to antioxidant, anti-inflammatory, and antimicrobial activities [24, 25]. It has been recently demonstrated that berberine hydrochloride could alleviate preeclampsia by regulating IL2/IL10 and BCL2/BAX expressions in preeclampsia rat

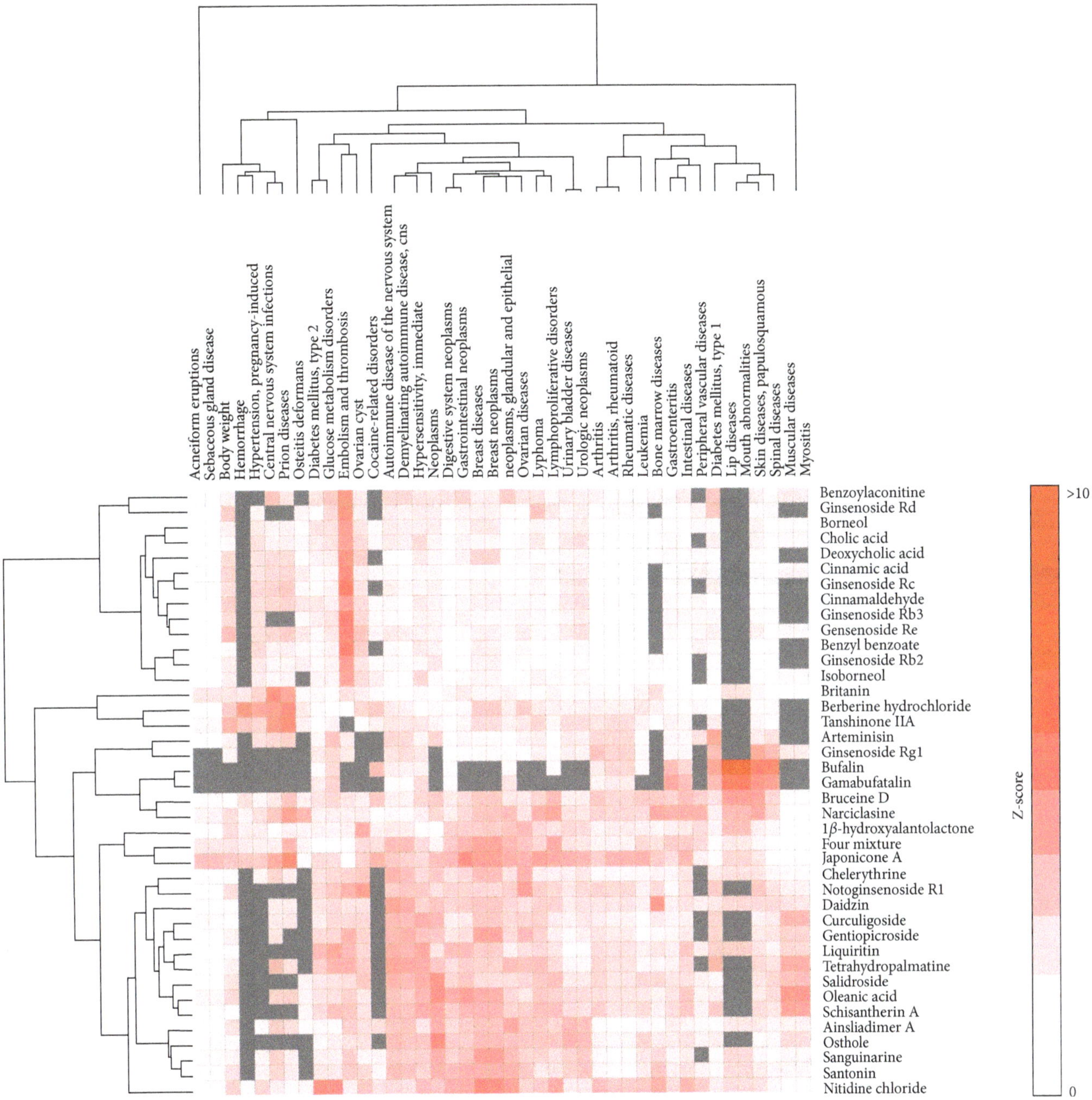

FIGURE 3: Heatmap showing the statistical significance GWAS disease-TCM gene pairs by Z-score.

models [26]. This warrants further investigation of using berberine hydrochloride for the treatment of preeclampsia.

3.4.5. Tanshinone IIA as Obesity Treatment. Finally, the systems pharmacology analysis identifies tanshinone IIA as treatment for obesity (q value = 0.19). Tanshinone IIA, a commonly used TCM for antioxidant and anti-inflammatory, is a compound extracted from *Salvia miltiorrhiza* (Chinese: Danshen) [27]. It has been recently demonstrated that tanshinone IIA could be used for obesity treatment through peroxisome proliferator-activated receptor (PPAR) gamma antagonism [28]. This provides a drug repurposing of tanshinone IIA for obesity treatment.

Taken together, we showed that the GWAS disease-TCM gene pairs identified in this study could be validated by literature evidence. Furthermore, the MoA of these TCM against the GWAS disease could be supported by Connectivity Map analysis.

3.5. TCM-Disease Web Application. To facilitate the translation of connecting TCM-target genes with GWAS disease genes, we implemented the statistical framework as TCM-Disease web application. For the TCM-Disease web application, users will submit the list of GWAS disease genes and perform the statistical test for identifying disease-TCM gene pairs. The web application returns the list of disease-TCM

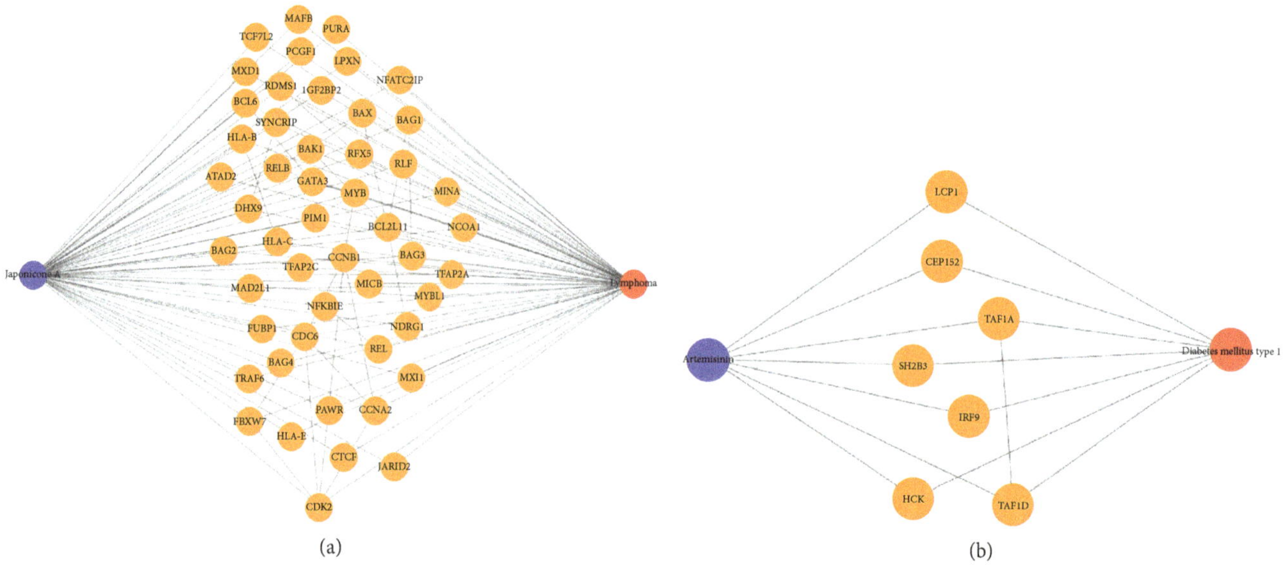

Figure 4: Example of GWAS disease-TCM gene networks. (a) Japonicone A as lymphoma treatment. (b) Artemisinin as type 1 diabetes mellitus treatment.

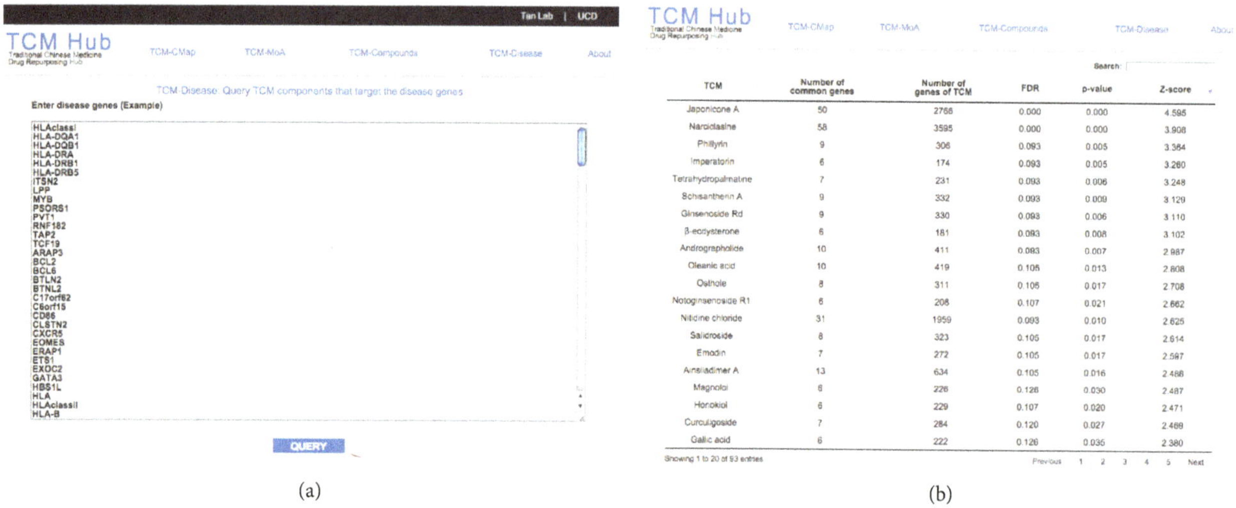

Figure 5: Screenshot of the TCM-Disease web application. (a) A list of disease-related genes are entering into the query box. (b) The web application performs the systems pharmacology permutation test and Z-score and returns the TCM component sorted by p value. This example is querying the lymphoma disease genes, and japonicone A is returned as the top TCM connected to this disease.

gene pairs (currently the database consists of 102 TCM) ranked by the permutation p values and Z-scores. Figure 5 shows the screenshot of the TCM-Disease web application.

4. Discussion

TCM has been successfully applied to treat various diseases for the past several thousand years, due to its polypharmacology nature. In this study, we performed a systems pharmacology-based approach to connect GWAS disease genes with TCM-target genes. We derived 102 TCM components and their target genes by analyzing microarray gene expression experiments. We constructed disease-gene networks from 2558 GWAS studies. We applied a systems pharmacology framework to prioritize disease-target genes as potential repurposing and repositioning TCM for these diseases. Using this bioinformatics approach, we analyzed 14,713 GWAS disease-TCM-target gene pairs and identified 115 disease-gene pairs with q value < 0.2. We validated several of these GWAS disease-TCM-target gene pairs with literature evidence, demonstrating that this computational approach could reveal novel indications for TCM.

Systems pharmacology and network biology approaches have been applied to study disease-gene interactions [16, 29–34]. For example, Fang et al. employed systems pharmacology-based approach to prioritize natural products that target somatic-mutated genes in cancer [16]. They introduced the statistical framework applied in this study, and they demonstrated that by using this approach they could identify clinically actionable alterations in cancer using natural products [16]. Kim et al. performed a systematic analysis of structural similarities between traditional oriental medicine (TOM) compounds and human metabolites for studying the mechanisms of action and suggest approaches to reduce toxicity [35]. They identified that TOM compounds could provide the complementary action, neutralizing action, facilitating action, and pharmacokinetic potentiation in 38 different synergistic combinations [35]. This provides an opportunity to use TCM/TOM in combination with other modern therapeutics. All of these studies rely on using systems pharmacology or network biology approaches to elucidate the disease-TCM gene interactions. This supports the use of systems pharmacology-based approach in the current study of identifying GWAS disease-TCM-target gene pairs.

TCM-Disease web application is part of the TCM hub—a bioinformatics resource for traditional Chinese medicine drug repurposing study (http://tanlab.ucdenver.edu/TCMHub). TCM hub currently consists of four web applications: (1) TCM-CMap, (2) TCM-MoA, (3) TCM-Compounds, and (4) TCM-Disease. TCM-CMap is the implementation of the "connectivity map" concept based on 102 TCM compounds. The TCM-CMap web app allows users to connect a gene signature (up and down genes) to one of the 102 TCM compounds in the reference database. TCM-MoA web app allows users to search the molecular mechanisms of TCM compounds. TCM-Compounds web app provides the CLUE connection results for the 102 TCM compounds. The CLUE connection results include connecting the TCM compound gene signatures to 2836, 3799, and 2160 signatures for compounds (CP), gene knockdown (KD), and gene overexpression (OE), respectively. TCM-Disease web app allows users to identify disease-TCM gene pairs for potential drug repurposing and repositioning. We believe that TCM hub represents a unique bioinformatics resource for in silico study of traditional Chinese medicine.

There are several limitations of the current GWAS disease-TCM gene networks study. First, the number of TCM is limited; currently, this study is only assessing 102 TCM components based on gene expression changes, which may limit the coverage of TCM-target gene networks. Second, the number of diseases investigated in the current study is limited to 293 GWAS diseases, which may limit the coverage of disease-gene networks. Lastly, our current validation of the associations of GWAS disease-TCM genes is based on literature evidence with experimental results. To address these limitations, as part of our future work, we plan to increase the number of TCM components by mining Gene Expression Omnibus (GEO) for additional TCM microarray studies. We will complement the gene expression signatures with TCM-target genes by mining established drug-target databases such as TCMID [36], TCM-MeSH [37], PharmDB-K [38], or DSigDB [39]. By integrating the gene expression targets and manually curated targets, we can expand the coverage of TCM-target gene networks. In parallel, we plan to expand the disease-gene networks by using data curated from DisGeNet [40], currently contains 561,119 gene-disease associations and more than 9000 diseases. By constructing the disease-gene network from this large database, it will increase the coverage of the diseases and potentially increased the value in revealing novel disease-TCM genes.

5. Conclusions

In summary, we present a systems pharmacology-based approach for connecting GWAS diseases with TCM for potential drug repurposing and repositioning. We employed several unique bioinformatics and computational approaches to identify and prioritize GWAS disease-TCM gene pairs. The computational approaches described in this study could be easily expandable to other disease-gene network analysis. We also develop TCM-Disease web application to facilitate the traditional Chinese medicine drug repurposing efforts.

Conflicts of Interest

The authors declare that they have no competing interests.

Authors' Contributions

Jihye Kim and Minjae Yoo contributed equally to this work.

Acknowledgments

The authors thank the Tan Lab members for their useful comments on this manuscript. This work is partly supported by the David F. and Margaret T. Grohne Family Foundation.

References

[1] F. Cheung, "TCM: made in China," *Nature*, vol. 480, no. 7378, pp. S82–S83, 2011.

[2] J. Knowles and G. Gromo, "A guide to drug discovery: target selection in drug discovery," *Nature Reviews Drug Discovery*, vol. 2, no. 1, pp. 63–69, 2003.

[3] G. Bollag, P. Hirth, J. Tsai et al., "Clinical efficacy of a RAF inhibitor needs broad target blockade in BRAF-mutant melanoma," *Nature*, vol. 467, no. 7315, pp. 596–599, 2010.

[4] G. Kim, A. E. McKee, Y. M. Ning et al., "FDA approval summary: vemurafenib for treatment of unresectable or metastatic melanoma with the BRAFV600E mutation," *Clinical Cancer Research*, vol. 20, no. 19, pp. 4994–5000, 2014.

[5] Y. Tu, "The discovery of artemisinin (qinghaosu) and gifts from Chinese medicine," *Nature Medicine*, vol. 17, no. 10, pp. 1217–1220, 2011.

[6] J. Li, T. Casteels, T. Frogne et al., "Artemisinins target GABA A receptor signaling and impair α cell identity," *Cell*, vol. 168, no. 1-2, pp. 86–100.e15, 2017.

[7] J. MacArthur, E. Bowler, M. Cerezo et al., "The new NHGRI-EBI catalog of published genome-wide association studies

(GWAS catalog)," *Nucleic Acids Research*, vol. 45, no. D1, pp. D896–D901, 2017.

[8] A. Subramanian, R. Narayan, S. M. Corsello et al., "A next generation connectivity map: L1000 platform and the first 1,000,000 profiles," *Cell*, vol. 171, no. 6, pp. 1437–1452.e17, 2017.

[9] J. Lamb, E. D. Crawford, D. Peck et al., "The Connectivity Map: using gene-expression signatures to connect small molecules, genes, and disease," *Science*, vol. 313, no. 5795, pp. 1929–1935, 2006.

[10] S. I. Berger and R. Iyengar, "Network analyses in systems pharmacology," *Bioinformatics*, vol. 25, no. 19, pp. 2466–2472, 2009.

[11] S. Zhao and R. Iyengar, "Systems pharmacology: network analysis to identify multiscale mechanisms of drug action," *Annual Review of Pharmacology and Toxicology*, vol. 52, no. 1, pp. 505–521, 2012.

[12] Z. Wang and Y. Y. Wang, "Modular pharmacology: deciphering the interacting structural organization of the targeted networks," *Drug Discovery Today*, vol. 18, no. 11-12, pp. 560–566, 2013.

[13] M. Kibble, N. Saarinen, J. Tang, K. Wennerberg, S. Mäkelä, and T. Aittokallio, "Network pharmacology applications to map the unexplored target space and therapeutic potential of natural products," *Natural Product Reports*, vol. 32, no. 8, pp. 1249–1266, 2015.

[14] C. Lv, X. Wu, X. Wang et al., "The gene expression profiles in response to 102 traditional Chinese medicine (TCM) components: a general template for research on TCMs," *Scientific Reports*, vol. 7, no. 1, p. 352, 2017.

[15] D. Szklarczyk, J. H. Morris, H. Cook et al., "The STRING database in 2017: quality-controlled protein-protein association networks, made broadly accessible," *Nucleic Acids Research*, vol. 45, no. D1, pp. D362–D368, 2017.

[16] J. Fang, C. Cai, Q. Wang, P. Lin, Z. Zhao, and F. Cheng, "Systems pharmacology-based discovery of natural products for precision oncology through targeting cancer mutated genes," *CPT: Pharmacometrics & Systems Pharmacology*, vol. 6, no. 3, pp. 177–187, 2017.

[17] P. Shannon, A. Markiel, O. Ozier et al., "Cytoscape: a software environment for integrated models of biomolecular interaction networks," *Genome Research*, vol. 13, no. 11, pp. 2498–2504, 2003.

[18] M. S. Cline, M. Smoot, E. Cerami et al., "Integration of biological networks and gene expression data using Cytoscape," *Nature Protocols*, vol. 2, no. 10, pp. 2366–2382, 2007.

[19] J. J. Shan, M. Yang, and J. W. Ren, "Anti-diabetic and hypolipidemic effects of aqueous-extract from the flower of Inula japonica in alloxan-induced diabetic mice," *Biological & Pharmaceutical Bulletin*, vol. 29, no. 3, pp. 455–459, 2006.

[20] J. J. Qin, H. Z. Jin, J. J. Fu et al., "Japonicones A–D, bioactive dimeric sesquiterpenes from Inula japonica Thunb," *Bioorganic & Medicinal Chemistry Letters*, vol. 19, no. 3, pp. 710–713, 2009.

[21] X. Li, X. Yang, Y. Liu et al., "Japonicone A suppresses growth of Burkitt lymphoma cells through its effect on NF-κB," *Clinical Cancer Research*, vol. 19, no. 11, pp. 2917–2928, 2013.

[22] A. Croaker, G. J. King, J. H. Pyne, S. Anoopkumar-Dukie, and L. Liu, "Sanguinaria canadensis: traditional medicine, phytochemical composition, biological activities and current uses," *International Journal of Molecular Sciences*, vol. 17, no. 9, p. 1414, 2016.

[23] C. Kalogris, C. Garulli, L. Pietrella et al., "Sanguinarine suppresses basal-like breast cancer growth through dihydrofolate reductase inhibition," *Biochemical Pharmacology*, vol. 90, no. 3, pp. 226–234, 2014.

[24] Z. Li, Y. N. Geng, J. D. Jiang, and W. J. Kong, "Antioxidant and anti-inflammatory activities of berberine in the treatment of diabetes mellitus," *Evidence-based Complementary and Alternative Medicine*, vol. 2014, Article ID 289264, 12 pages, 2014.

[25] H. H. Yu, K. J. Kim, J. D. Cha et al., "Antimicrobial activity of berberine alone and in combination with ampicillin or oxacillin against methicillin-resistant Staphylococcus aureus," *Journal of Medicinal Food*, vol. 8, no. 4, pp. 454–461, 2005.

[26] A. Wang, Q. Liu, J. Zhang, and R. Zheng, "Berberine alleviates preeclampsia possibly by regulating the expression of interleukin-2/interleukin-10 and Bcl-2/Bax," *International Journal of Clinical and Experimental Medicine*, vol. 8, no. 9, pp. 16301–16307, 2015.

[27] H.-Q. Yin, Y.-S. Kim, Y.-J. Choi et al., "Effects of tanshinone IIA on the hepatotoxicity and gene expression involved in alcoholic liver disease," *Archives of Pharmacal Research*, vol. 31, no. 5, pp. 659–665, 2008.

[28] Z. Gong, C. Huang, X. Sheng et al., "The role of tanshinone IIA in the treatment of obesity through peroxisome proliferator-activated receptor gamma antagonism," *Endocrinology*, vol. 150, no. 1, pp. 104–113, 2009.

[29] G.-B. Zhang, Q.-Y. Li, Q.-L. Chen, and S.-B. Su, "Network pharmacology: a new approach for Chinese herbal medicine research," *Evidence-based Complementary and Alternative Medicine*, vol. 2013, Article ID 621423, 9 pages, 2013.

[30] P. Li, J. Chen, W. Zhang, B. Fu, and W. Wang, "Transcriptome inference and systems approaches to polypharmacology and drug discovery in herbal medicine," *Journal of Ethnopharmacology*, vol. 195, pp. 127–136, 2017.

[31] X. Zhang, J. Gu, L. Cao et al., "Network pharmacology study on the mechanism of traditional Chinese medicine for upper respiratory tract infection," *Molecular BioSystems*, vol. 10, no. 10, pp. 2517–2525, 2014.

[32] J. Gu, H. Zhang, L. Chen, S. Xu, G. Yuan, and X. Xu, "Drug-target network and polypharmacology studies of a traditional Chinese medicine for type II diabetes mellitus," *Computational Biology and Chemistry*, vol. 35, no. 5, pp. 293–297, 2011.

[33] H. Liang, H. Ruan, Q. Ouyang, and L. Lai, "Herb-target interaction network analysis helps to disclose molecular mechanism of traditional Chinese medicine," *Scientific Reports*, vol. 6, no. 1, article 36767, 2016.

[34] W. Zhou, J. Wang, Z. Wu, C. Huang, A. Lu, and Y. Wang, "Systems pharmacology exploration of botanic drug pairs reveals the mechanism for treating different diseases," *Scientific Reports*, vol. 6, no. 1, article 36985, 2016.

[35] H. U. Kim, J. Y. Ryu, J. O. Lee, and S. Y. Lee, "A systems approach to traditional oriental medicine," *Nature Biotechnology*, vol. 33, no. 3, pp. 264–268, 2015.

[36] R. Xue, Z. Fang, M. Zhang, Z. Yi, C. Wen, and T. Shi, "TCMID: traditional Chinese medicine integrative database for herb molecular mechanism analysis," *Nucleic Acids Research*, vol. 41, no. D1, pp. D1089–D1095, 2013.

[37] R.-Z. Zhang, S.-J. Yu, H. Bai, and K. Ning, "TCM-Mesh: the database and analytical system for network pharmacology

analysis for TCM preparations," *Scientific Reports*, vol. 7, no. 1, p. 2821, 2017.

[38] J.-H. Lee, K. M. Park, D.-J. Han et al., "PharmDB-K: integrated bio-pharmacological network database for traditional Korean medicine," *PLoS One*, vol. 10, no. 11, article e0142624, 2015.

[39] M. Yoo, J. Shin, J. Kim et al., "DSigDB: drug signatures database for gene set analysis," *Bioinformatics*, vol. 31, no. 18, pp. 3069–3071, 2015.

[40] J. Piñero, À. Bravo, N. Queralt-Rosinach et al., "DisGeNET: a comprehensive platform integrating information on human disease-associated genes and variants," *Nucleic Acids Research*, vol. 45, no. D1, pp. D833–D839, 2017.

8

The Challenges and Opportunities in the Clinical Application of Noncoding RNAs: The Road Map for miRNAs and piRNAs in Cancer Diagnostics and Prognostics

Preethi Krishnan[1,2] **and Sambasivarao Damaraju**[1,2]

[1]*Department of Laboratory Medicine and Pathology, University of Alberta, Edmonton, Canada T6G 1Z2*
[2]*Cross Cancer Institute, Edmonton, Canada T6G 1Z2*

Correspondence should be addressed to Sambasivarao Damaraju; sdamaraj@ualberta.ca

Academic Editor: Wilfred van IJcken

Discoveries on nonprotein-coding RNAs have induced a paradigm shift in our overall understanding of gene expression and regulation. We now understand that coding and noncoding RNA machinery work in concert to maintain overall homeostasis. Based on their length, noncoding RNAs are broadly classified into two groups—long (>200 nt) and small noncoding RNAs (<200 nt). These RNAs perform diverse functions—gene regulation, splicing, translation, and posttranscriptional modifications. MicroRNAs (miRNAs) and PIWI-interacting RNAs (piRNAs) are two classes of small noncoding RNAs that are now classified as master regulators of gene expression. They have also demonstrated clinical significance as potential biomarkers and therapeutic targets for several diseases, including cancer. Despite these similarities, both these RNAs are generated through contrasting mechanisms, and one of the aims of this review is to cover the distance travelled since their discovery and compare and contrast the various facets of these RNAs. Although these RNAs show tremendous promise as biomarkers, translating the findings from bench to bedside is often met with roadblocks. The second aim of this review therefore is to highlight some of the challenges that hinder application of miRNA and piRNA as in guiding treatment decisions.

1. MicroRNAs

1.1. Introduction. For a long time, the field of molecular biology has been governed by the central dogma, which can simply be explained as DNA transcribes to RNA and RNA translates to protein. While this still holds true, recent discoveries have added additional layers to this principle. A group of RNAs termed noncoding RNAs have been found to play a role in regulating transcription and translation. These RNAs were previously not recognized to have significant roles in human health and disease, but we now understand that they are involved in diverse roles ranging from gene regulation to alternative splicing to protein translation. Noncoding RNAs are broadly classified into two groups based on their size: long noncoding RNAs (lncRNAs, https://lncipedia.org/, http://www.noncode.org/), which are generally >200 nucleotides, and small noncoding RNAs (sncRNAs), which are generally <200 nucleotides, and both these groups of RNAs function as regulators of gene expression. sncRNAs include subclasses such as microRNAs (miRNAs, http://www.mirbase.org/), PIWI-interacting RNAs (piRNAs, http://genome.ucsc.edu/), transfer RNAs (tRNAs, http://gtrnadb2009.ucsc.edu/), small nucleolar RNAs (snoRNAs, https://www-snorna.biotoul.fr/), small nuclear RNAs (snRNAs, https://www.ensembl.org/info/genome/genebuild/ncrna.html), and small interfering RNAs (siRNAs, https://www.ncbi.nlm.nih.gov/projects/genome/rnai/). Of these, miRNAs, siRNAs, and piRNAs are classified as regulatory RNAs. Several differences and similarities exist between these three RNAs [1, 2]. For instance, siRNAs regulate gene expression largely by degrading the target mRNA, whereas miRNAs and piRNAs regulate gene expression either by degrading the target mRNA or by inhibiting translation. While siRNAs perform autosilencing (genes that siRNAs target and the origins of the siRNA are from the same gene), miRNAs and piRNAs predominantly perform heterosilencing (targets are different from the genes from which they originate). Likewise, similarities and differences exist in the biogenesis pathway of siRNAs and miRNAs, which are detailed in the review by Bartel [2]. The characteristic feature of these three types of RNAs is their interaction with Argonaute (AGO) proteins to guide target-specific gene regulation. Two classes of AGO proteins exist: AGO and PIWI (P-element-induced Wimpy testis) [3]. While miRNAs and siRNAs interact with the AGO class of proteins, piRNAs are found to interact with the PIWI class of proteins.

1.2. Discovery of miRNAs. MicroRNAs are small (~22 nt), noncoding, regulatory RNAs that control gene expression posttranscriptionally by binding to the 3′ UTR of mRNA and promote mRNA degradation or inhibit protein translation [2, 4–6]. miRNAs are the most studied class of sncRNAs and are well explored for their roles in various physiological and pathological conditions.

The first miRNA was discovered by Lee et al. and Wightman et al. in 1993, in the context of studies on the growth pattern of *C. elegans* [4, 5]. Growth of *C. elegans* depends on heterochronic genes such as *lin-4* that turn on and off during the transition to different developmental stages. Loss-of-function mutation resulted in the absence of adult structures in the worm but reiteration of characteristics specific to early stages; that is, in place of adult structures, the nematode develops early stage structures. By contrast, mutations in another gene termed *lin-14* caused the opposite effect—the worms matured prematurely, suggesting that the two genes imparted opposite effects in the development of *C. elegans*. It was later discovered that *lin-4* repressed the activity of *lin-14*; however, the mechanism behind this regulation remained elusive. While the Ambros lab had identified that *lin-4* generated two transcripts, one short (~22 nt) and one long transcript, Ruvkun's lab had found the complementary-binding site for the shorter transcript in the 3′ untranslated region of *lin-14*. They also discovered that *lin-4* binds directly to *lin-14* and suppresses the latter's expression, forming the explanation for the molecular mechanism of small RNAs. These RNAs were initially known as small temporal RNAs as their roles were largely believed to be restricted to the temporal development of nematodes. A few years later, Hamilton and Baulcombe observed a similar silencing mechanism in plants, thus expanding the role of these small RNAs to other organisms [7]. Around the same time, Ambros and his team had discovered that the regulation of *lin-4* was not limited to *lin-14* alone but could regulate another gene called *lin-28*, suggesting possible implications on overall gene regulation [8]. In 2000, Reinhart et al. and Pasquinelli et al. reported the discovery of a second miRNA—let-7—and had made a crucial observation that miRNAs were evolutionarily conserved and not specific to nematodes [9, 10]. This marked the beginning of a surge of reports on miRNAs—discovery, biogenesis, roles, and their significance in different conditions—normal and diseased.

1.3. Genomic Location of miRNAs. miRNA genes are found across all chromosomes. They may originate from the intergenic regions or genic regions (both protein coding and non-protein coding) [2, 11–14]. Further, some miRNAs may be in a distant location from other miRNAs, while some others may be in proximity and may exist as clusters. A cluster, as defined by miRBase (http://www.mirbase.org/), is a group of miRNAs that are located within 10 kb of each other [15]. miRNAs belonging to the same cluster may either be cotranscribed or transcribed independently [16–18].

1.4. Biogenesis of miRNAs. The biogenesis of miRNA begins in the nucleus where miRNA genes are transcribed by RNA polymerase II or III into long primary transcripts (pri-miRNAs) that are polyadenylated at the 3′ end and capped at the 5′ end [19–21]. miRNAs arising from intergenic regions are transcribed by polymerase II or III, whereas miRNAs originating from intronic regions are transcribed by polymerase II [22]. Pri-miRNAs contain stem loop structures and are cleaved at the stem of the hairpin structure by a cellular RNase class II endonuclease III enzyme called Drosha along with DGCR8/Pasha into hairpin structures called precursor miRNAs (pre-miRNAs) which are approximately 70–120 nt long [20]. The pre-miRNA harbors a 5′ phosphate and a 2-nucleotide overhang at the 3′ end, characteristic of endonuclease III cleavage. Pre-miRNA is then transported to the cytoplasm with the help of Exportin 5 along with Ran-GTP, which is then processed by the cytoplasmic dsRNase III Dicer into approximately 22 nt miRNA: miRNA duplex with 2 nt overhanging at its 3′ end [2, 11]. The duplex is unwound by helicase, and only one mature strand (~20 nt long) enters the multicomponent complex called RNA-induced silencing complex (RISC), which harbors AGO, and the complementary strand is degraded [2]. Most often, the strand with low internal stability at the 5′ end is retained within RISC [23, 24]. The mature miRNA mediates gene expression by binding to the complementary sequence in the 3′ untranslated region of the target messenger RNA (mRNA). Depending on the complementarity shared, the target mRNAs may be degraded (if the two RNAs are perfectly complementary to each other) or the protein translation may be inhibited (if they share imperfect complementarity) [25]. Excellent schematics of canonical pathways of miRNA biogenesis have been published elsewhere [2, 16].

Alternatively, mirtrons (miRNAs present in the introns) may be processed through splicing mechanisms [26]. In this process, pre-miRNAs are generated by splicing of the introns, which are released as lariat structures. Debranching enzymes act upon these lariat structures for their linearization and to form the short hairpin-like forms. The pre-miRNAs thus formed are transported to the cytoplasm and are cleaved by Dicer to form mature miRNAs. An alternate pathway of miRNA biogenesis is provided in the review by Naqvi et al. [27]. This mechanism was discovered in flies and worms and was later observed in different organisms.

1.5. Mechanisms of Action. The interaction between miRNA and target mRNAs predominantly occur at the miRNA seed region (2–8 nt in the 5′ end of the miRNA) and the 3′ UTR of mRNA through sequence complementarity [28]. Several in silico databases such as TargetScan (http://www.targetscan.org/) and miRanda (http://www.microRNA.org/) predict the targets of miRNA based on sequence complementarity. Since these are in silico predictions, two approaches are commonly adopted to gain confidence in the predictions—(i) targets which are predicted in multiple databases are used for functional validation or (ii) targets which are in silico predicted are overlapped with differentially expressed genes identified in tissues. The direct interaction of miRNA with its cognate targets is first confirmed using luciferase assays. Functional validations of the targets are performed using gene overexpression or knockout models in cell lines to observe the effect on the phenotype. miRNA-mRNA interactions include both direct and indirect effects on translation [25]. In the direct effects, initiation of translation or postinitiation of translation is inhibited. While in the former the association of ribosomes with target mRNAs is prevented, the latter includes premature ribosome fall-off, reduced/stalled elongation, or cotranslational protein degradation. Indirect effects of miRNA-mRNA interaction include deadenylation, resulting in degradation or increased turnover. These effects occur in the cytoplasm, predominantly in the processing bodies (P-bodies), which are enriched for factors involved in mRNA degradation. The mRNAs whose protein formation is prevented (by direct or indirect effects) may be sequestered in the P-bodies, which can be used later for translation or can be degraded.

Expression of miRNAs is also regulated at different levels—transcription, Drosha processing, transfer of pre-miRNAs from the nucleus to cytoplasm, and Dicer processing and loading into the RISC complex. Regulation of miRNAs are clearly explained in the reviews by Davis and Hata [29], Winter et al. [16], and Slezak-Prochazka et al. [30]. In addition, miRNAs are affected by the presence of single nucleotide variants and mutations in the miRNA regions or in the binding site of miRNAs leading to improper biogenesis of miRNAs or improper regulation of mRNA targets. Thus, a complex but coordinated association between the regulation of and by miRNAs determines the balance between normal and diseased conditions.

1.6. Detection and Quantification of miRNAs. Platforms that have been used for gene expression profiling have also been adopted for small RNAs. However, small RNAs pose several challenges for development of a profiling platform [31–34]—(i) the small size of these RNAs makes it difficult to design a complementary probe or a traditional primer, where, often, the size of a probe/primer is equal to or more than the size of small RNAs; (ii) miRNA isoforms can differ by a single nucleotide and this distinction is difficult to obtain unless the platform is highly sensitive to detect even one nucleotide difference; (iii) the GC content of miRNAs vary greatly, thus making it difficult to standardize the melting temperatures for annealing reactions in a genome-wide study; and (iv) rapid rate of discovery, making it difficult to reuse the data generated on platforms using preprinted probes, based on the existing annotation.

Nevertheless, three main platforms used for profiling small noncoding RNAs include quantitative reverse transcription polymerase chain reaction (qRT-PCR), microarray, and next-generation sequencing (NGS) [35]. Every platform has its own merits and limitations, and a summary of these platforms is provided below.

1.6.1. Quantitative Reverse Transcription Polymerase Chain Reaction (qRT-PCR). One of the commonly adopted techniques is qRT-PCR that relies on cDNA generated from RNA, followed by real-time PCR. However, due to the small size of miRNAs, conversion of miRNA to cDNA is challenging. Therefore, either poly(A) tail is added or a stem loop primer is added, depending on the method adopted for qRT-PCR. Once cDNAs are generated, PCR reactions are performed. The amplified products are usually detected either by SYBR green or using TaqMan probes. While qRT-PCR offers the advantage of being highly sensitive and specific with a high dynamic range (6 orders of magnitude), this technique has low throughput, and only a limited number of RNAs can be interrogated on this platform [31] due to assay design limitations. It is a labor-intensive method, especially when the experiment includes higher number of samples and the optimal reaction conditions may vary according to sequence-specific differences. The difficulty in designing optimal probes for detecting small RNAs is higher when the miRNAs differ only by one or two nucleotides. Although profiling of multiple miRNAs on a single plate is frequently seen in the literature [36, 37], there is a limitation on the number of miRNAs that can be detected, and this approach may still not cover the entire genome. A major problem with qRT-PCR method is the choice of internal reference used [38]. There is no standard reference gene available yet, and the choice of reference gene thus varies from study to study. One of the common normalizers used is RNU6 [39]. For these reasons, this platform may be best suited to validate or probe for candidate molecules and may be especially useful when the sample amount is limiting.

1.6.2. Microarray. This platform is based on hybridization of probe and target sequences, which are complementary. The technique of printing hundreds to thousands of probes on a chip is perfected and has evolved over the past 15 years. The small size of RNAs poses challenges in this technique as well [34]. Microarrays are less sensitive when compared to qRT-PCR [31, 33], and as with qRT-PCR, the analysis might be restricted to a single class of molecules (e.g., miRNAs alone), in a specific genome build. A platform with low sensitivity will generate many false negative calls, whereas higher sensitivity and reduced specificity would result in a higher number of false positives. The chance of identifying novel RNAs is minimal, and capturing of RNAs with single nucleotide differences is challenging. The dynamic range of this platform is moderate (4 orders of magnitude), but it allows profiling of higher number of miRNAs (compared to qRT-PCR) [31, 40, 41]. The lack of ability to perform absolute quantification of molecules renders it more suitable for comparing relative abundances of molecules that fall within the dynamic range of the platform between two conditions such as normal and diseased. Microarray-based profiling techniques require specialized equipment for hybridization, fluidics stations, incubators, and scanners. The data from high-density arrays also require knowledge of appropriate bioinformatics platforms. Commercially available informatics platforms, for example, the Partek Genomics Suite (http://www.partek.com/pgs), are well designed for use by biologists.

Although both qRT-PCR and microarray have their own advantages, these methods rely on a predetermined set of RNAs based on a specific genome build, thereby leaving us blinded to the functions of other RNAs that cannot be captured because of platform limitations.

1.6.3. Next-Generation Sequencing (NGS). NGS refers to the sequencing of millions of reads in parallel, yielding higher throughput [40] and more coverage, necessitating the use of powerful computing skills and algorithms for analysis [32]. Several platforms are available for NGS including Illumina (https://www.illumina.com/), Roche (http://sequencing.roche.com/), and Ion Torrent (https://www.thermofisher.com/). These platforms incorporate the sequencing-by-synthesis method that helps in detecting known and unknown RNAs.

The advantage of NGS is that it offers absolute quantification of molecules, higher coverage, and high sensitivity and specificity [35]. It does not require the knowledge of genomic annotation (prerequisite for qRT-PCR and microarray), and the reads can be assembled de novo [42]. This platform overcomes the problems of hybridization encountered in the sequencing technique, is capable of capturing reads with even a single nucleotide difference, and is useful for identifying novel RNAs. It exhibits high dynamic range (>10 orders of magnitude), enabling quantification of low amounts of molecules, and allows parallel quantification of multiple RNA types not restricted to one particular class. Since NGS does not depend on any particular genome build, reanalysis of the existing data based on the current or future annotations of the genome (genome build) is possible. However, sequencing biases may be introduced due to the number of steps involved in sample preparation; data analysis and interpretation is complex due to the large volumes of data being generated. As with all high-throughput techniques, NGS is also not free from batch effects and is discussed later in the review. Of the three profiling platforms, NGS is also the most expensive, but it offsets the costs by allowing mining of all small RNA classes which is not possible on microarray or qRT-PCR platforms.

With several user-friendly bioinformatics platforms now available for data analysis, complexity of data and its mining, once considered a limitation for NGS, has now been overcome [32]. However, different researchers employ different analytical steps starting from aligning the reads to normalization to statistical tests used for identifying differentially expressed RNAs. For instance, several commonly used normalization methods and packages include RPKM [43], DESeq2 [44], edgeR [45, 46], and TMM [47]. There are advantages and disadvantages for each of these, and there is no clear consensus on the best method for NGS data. Similarly, the statistical test employed for identifying differentially expressed RNAs also remains debatable, mainly because of the distribution of reads. While most of the microarray data are log transformed, mainly to bring the data to normal distribution, the use of log transformation is questionable for NGS data. NGS data is count data, and when normalized, it does not fall into normal distribution. Therefore, several methods have to be tried to gain confidence in the results, and regardless of any method that is used, one needs to validate the signatures using a different platform and in different cohorts, preferably using external (geographically distinct regions and diverse population ancestries) validation datasets.

In addition to the three profiling methods mentioned above, other methods are continuously being developed for profiling miRNAs. Some of these methods include NanoString (https://www.nanostring.com/), digital PCR, and multiplex arrays.

1.7. miRNAs as Contributors to Tumorigenesis. Since their discovery in 1993, miRNAs have been studied in great depth [48] for their role as key players in normal developmental processes [49], including cell growth and apoptosis [50, 51], hematopoietic cell lineage differentiation [52], muscle cell proliferation and differentiation [53], and also in cancer [54, 55] as well as cardiovascular [56, 57], autoimmune [58, 59], and neurodegenerative diseases [60, 61]. miRNA deregulation in cancer was first demonstrated by Calin et al. [62], wherein the authors observed a deletion at the 13q14 locus in chronic lymphocytic leukemia, which also harbored the miR-15/16 cluster. Reduced expression of miR-15/16 in tumors and their potential to target antiapoptotic target Bcl2 suggested the role of miRNAs as potential tumor suppressors. Conversely, some miRNAs are upregulated in cancers. For example, miR-155 is overexpressed in human B cell lymphoma. Further, miR-155 transgenic mice were observed to favor B cell proliferation, followed by B cell malignancy [63]. Inhibition of miR-21 (a miRNA that is commonly overexpressed in most tumors) led to regression of tumors and overexpression of miR-21 led to the B cell malignant phenotype [64]. These early studies suggested the possible role of miRNAs as tumor suppressors and oncogenes and thus paved the way for a surge of reports confirming these roles. Some miRNAs such as miR-221 have also been found to have dual roles of both oncogenes and tumor suppressors depending on the cancer type in which they are expressed [65–67]. It is now estimated that miRNAs regulate approximately 60% of the protein-coding genes [68, 69]. With an increasing number of studies devoted to miRNAs, the role of miRNAs began to expand, and we now recognize their critical contribution to tumorigenesis—they are important in every hallmark of cancer [70, 71], including evading growth suppressors, avoiding immune destruction, enabling replicative immortality, tumor-promoting inflammation, activating invasion and metastasis, resisting cell death, and sustaining proliferative signaling.

1.8. miRNAs as Cancer Biomarkers. A biomarker is defined as "a characteristic that is objectively measured and evaluated as an indicator of normal biological processes, pathogenic processes, or pharmacologic responses to a therapeutic intervention" by the NIH Biomarkers Definitions Working Group [72]. Biomarkers that can aid in disease identification are termed diagnostic biomarkers, whereas biomarkers that guide treatment decisions can be broadly classified as either prognostic or predictive biomarkers. While the former class predicts progression or outcome in an untreated individual, the latter provides information on the likely benefit or detriment from treatment. Therefore, prognostic factors are helpful in identifying patients who are at risk for recurrence and/or death, which may eventually help in modifying treatment modalities, whereas predictive factors may aid in identifying patients who are likely to respond to a treatment. Some of the characteristics of an ideal biomarker include [73] the following: (i) unique expression profile reflecting the diseased condition and altered expression levels in the disease tissue, when compared to the normal tissue, (ii) reliable in identifying the condition, (iii) high stability, and (iv) highly specific and sensitive. miRNAs are highly stable, even in formalin-fixed and paraffin-embedded (FFPE) specimens preserved for over 20 years [74]. Further, miRNAs are stable in body fluids [75] such as plasma [76], serum [75], saliva [77], and urine [78]. They are known for their tissue specificity and can be easily detected by multiple profiling platforms. All these factors make a strong case of why miRNAs are suitable candidates for biomarkers.

1.8.1. miRNAs as Diagnostic Biomarkers. miRNAs are capable of distinguishing between different tumor subtypes. For instance, miRNAs were found to be differentially expressed between the basal and luminal subtypes of breast cancer, and these findings were also validated in an independent data set [79]. miRNAs have been identified as classifiers of histological subtypes of breast cancer. Nine miRNAs (miR-210, let-7d, miR-181a, miR-221, miR-10b, miR-126, miR-218, miR-335-3p, and miR-143) together as a composite signature were found to distinguish ductal carcinoma in situ and invasive ductal carcinoma [80]. Similarly, miRNAs were found to distinguish between different subtypes of renal cell carcinoma (RCC) [81]. Using a decision tree classifier, the identified miRNA signature showed a sensitivity of >95% for distinguishing normal from RCC and papillary RCC and showed 100% sensitivity for clear cell RCC and in distinguishing oncocytoma from chromophobe RCC subtype. Crossvalidation showed 90% accuracy. One of the important roles for miRNA signature is to identify tissues of origin for metastatic cancers of unknown primary site. Ferracin et al. identified a 47-miRNA signature to estimate the tissue of origin probability [82]. This signature identified primary cancers with 100% accuracy and metastatic cancers with 86% accuracy. Independent validation of this signature resulted in 86% accuracy in identifying metastatic cases, and prediction of cancer of unknown primary tumors reached probabilities higher than 90%.

1.8.2. miRNAs as Prognostic Biomarkers. miRNAs have been found to be associated with various clinicopathological factors such as tumor size and stage and as well as metastasis. Research focusing on this aspect of miRNAs has intensified over the years, and now we have mounting evidence associating the expression of miRNAs to clinical outcomes, suggesting their role in cancer prognostication. Let-7 was the first miRNA discovered to have prognostic significance [83]. Expression of let-7 was significantly lower in lung cancer patients, and those patients with low expression of let-7 had shorter survival. Further, let-7 was also shown to be an independent prognostic factor after adjusting for clinicopathological factors. In one of the early studies [84] conducted using microarray profiling, a 13-miRNA signature distinguished (i) chronic lymphocytic patients exhibiting high expression of ZAP-70 (prognostic factor) from patients with low levels of ZAP-70 and (ii) between cases with mutated and unmutated IgV_H. This signature was associated with disease progression. The observation that the 13-miRNA signature was associated with conventional prognostic factors and was independently associated with disease progression suggested their potential role as independent prognostic factors. In lung cancer, high expression of miR-155 was associated with reduced survival in multivariate analysis [85]. Based on qRT-PCR profiling, Yu et al. identified a five-miRNA signature as an independent prognostic factor for non-small cell lung cancer and also validated this in a validation dataset [86]. A risk score was constructed using the five miRNAs which were significant in univariate Cox analysis. The data was split into low- and high-risk groups based on the median value. Possible prognostic relevance of miRNAs in breast cancer began with the study by Iorio et al. who had identified miRNAs that correlated with breast cancer clinicopathological features, for example, receptor status, tumor stage, vascular invasion, and proliferation index [87]. Similarly, miRNAs discriminated between long-term and short-term survivors of pancreatic cancer [88], showed association with clinicopathological factors of this disease [89]. Likewise, using RT-PCR assays, a seven-miRNA signature associated with overall survival and relapse-free survival of gastric cancer patients [90]. Following some of these early studies and with the intervention of newer and advanced technological platforms to profile miRNAs, several studies reported prognostic markers for various cancer types. Recently, studies have shown the utility of sequencing platforms to identify miRNAs showing clinical relevance. For instance, we have sequenced miRNAs using the Illumina Genome Analyzer platform to identify prognostic markers for breast cancer [39]. Similarly, Roth et al. have identified prognostic miRNAs in the blood of primary lymphoma patients using Illumina HiSeq [91]. Although this is encouraging, we still need to travel a long path in identifying clinically applicable miRNA signatures. Some of the challenges and pitfalls in identifying robust biomarkers are discussed in a later section in this review.

1.9. miRNAs in Cancer Therapeutics. Although treatment options for cancer are fairly well developed, there is always a constant need for alternative treatment regimens to further improve survival and quality of life. miRNAs mainly function either as tumor suppressors or as oncogenes. Therefore, the motto for designing therapies using miRNAs would either be to restore tumor suppressive miRNAs using miRNA mimics or to suppress oncogenic miRNAs using inhibitors.

1.9.1. miRNA Mimics. miRNA mimics are chemically synthesized double-stranded miRNA-like molecules designed for gene-silencing purposes [92]. These are generally ~22 nucleotides long and are modified slightly from endogenous miRNAs to enhance stability and loading of miRNA in a RISC complex [93, 94]. One of the ways through which miRNA mimics can be delivered is through the use of nanoparticles. To achieve target specificity, these nanoparticles are coated with antibodies that can recognize tumor-specific antigens, thus binding specifically to tumor cells and delivering the harbored miRNA mimics to tumor cells. One of the well-studied examples is miRNA-34a in neuroblastoma. miR-34a is a proapoptotic tumor-suppressive miRNA that targets multiple genes including MYCN and BCL2 that was encapsulated in a nanoparticle coated with antibodies to disialoganglioside GD2. Delivery of miR-34a through this mode showed multiple pathway regulation such as increased apoptosis and decreased angiogenesis [95]. Intranasal administration of let-7 has also shown to decrease tumor formation in lung cancer-positive mouse models exhibiting mutation for the K-ras oncogene [96]. Expression vector-based gene regulation has also emerged as a means of inducing higher copies of synthetic miRNAs for effective target regulation. Chung et al. demonstrated this using the RNA polymerase II expression vector based on the noncoding RNA BIC [97]. miR-155 precursor sequences are found within BIC, and it was found that the expression of the third exon of BIC is adequate for generating miR-155 copies. Using this premise, a vector capable of expressing two different miRNAs from a single transcript that can effectively inhibit at least two different mRNA targets was constructed. Recently, Esposito et al. identified aptamer-miRNA conjugates as novel tools for *in vivo* targeted delivery of miRNAs with therapeutic benefit [98]. They demonstrated this potential using a tumor suppressor miRNA: let-7g which was conjugated to an aptamer that can selectively bind to tumor cells expressing oncogenic receptor tyrosine kinase Axl. Delivery of this conjugate resulted in marked reduction of tumor cell survival, migration, and growth in a xenograft mouse model of lung cancer. Another approach for delivering miRNA mimics is through the use of liposomes [99]. This was demonstrated in lung cancer models for delivering miR-29b through cationic lipoplexes. miR-29b is a tumor suppressor that is downregulated in cancer. By increasing the expression of miR-29b, lung cancer growth was reduced by about 60%.

1.9.2. miRNA Inhibitors. Inhibitory strategy works mainly based on antisense oligonucleotides (also called as anti-miRs) that include locked nucleic acids (LNA) anti-miRNAs, antagomiRs, miRNA sponges, miRNA masks, and small molecule inhibitors of miRNAs. Anti-miRs work by binding with the mature miRNAs and thereby blocking their interaction with their respective targets. Importantly, anti-miRs possessing 2′-O-methoxyethyl and phosphorothioate modification have been demonstrated to block let-7 activity in *C. elegans* [100]. A classic example of LNA is the inhibitory strategy developed against miR-122 for hepatitis C infection. This miRNA is in clinical trials currently and has so far showed positive results, in terms of reducing the viral load in a dose-dependent manner, thus reducing the likelihood of developing hepatocellular carcinoma [101, 102].

Multi-miRNA therapy, where two or more miRNAs act as therapeutic agents, can be used to modulate a mixture of mRNAs and thereby modulate multiple pathways. Similarly, miRNAs can be combined with siRNAs to target a particular gene and boost the inhibition since both these molecules work on similar principles [48]. miRNA sponges are artificially designed multiple complementary target sites for specific miRNA binding, thus preventing the binding of miRNAs to naturally occurring complementary sites in mRNAs. The advantage of this method is that since sponges have multiple miRNA-binding sites, the designed RNAs can sponge the effect of a family of miRNAs sharing the same seed sequence [103]. Naturally occurring RNAs that act as sponge RNAs are the long noncoding RNAs and some transcribed and processed pseudogenes [104, 105].

1.9.3. miRNA-Based Alternative Therapeutic Options. A unique strategy to target miRNAs is through the use of small molecule inhibitors of miRNAs. These agents regulate the transcription or the biogenesis of miRNAs rather than inhibit the interaction of miRNA with its target. A specific example is enoxacin, which by binding to TRBP (the human immunodeficiency virus-transactivating response RNA-binding protein) enhances the production of tumor suppressor miRNAs and inhibits tumor growth in a broad spectrum of cancer types [106]. Likewise, polylysine and trypaflavine inhibit miRNA-mediated gene regulation and thereby attenuate tumorigenic conditions. While the former molecule inhibits Dicer-mediated processing, the latter acts by moderating the association of miRNA with Argonaute proteins [107]. Another strategy to reduce the effect of miRNA on tumor growth or tumor-associated systemic effects is through blocking the production of exosomes or microvesicles and/or blocking the miRNAs harbored within these vesicles. miRNAs act as effective mediators of communication between different cells or tissues. miR-21 and miR-29a released from tumor cells in exosomes can act as ligands that can bind to TLR7 receptors present on the surface of macrophages residing in the tumor microenvironment. This binding triggers an inflammatory reaction that intensifies tumor growth. Therefore, blocking these secreted miRNAs can emerge as a potential therapeutic option [48].

2. PIWI-Interacting RNAs (piRNAs)

2.1. Introduction. piRNAs are a recently discovered (2006) class of small noncoding regulatory RNAs that are slightly longer (25–32 nt) than the 18–25 nt long miRNAs [108–111]. Similar to miRNAs, piRNAs also interact with the AGO family of RNA-binding proteins to guide target-specific gene regulation, specifically the PIWI subclass of AGO proteins. In contrast to the AGO proteins, which are ubiquitously present in all the tissues, PIWI proteins are predominantly present in the germline cells. In 1997, Lin and Spradling [112] observed an arrest in the self-renewing capacity of germline stem cells present in *Drosophila*, discovering the indispensable role of PIWI proteins in germline maintenance. This observation was confirmed by Cox et al. [113], who also showed that *PIWI* is present both in somatic cells and in germ cells. While *PIWI* expression in somatic cells adjacent to germ cells is required for germline stem cell maintenance, *PIWI* in the germ cells are essential for embryogenesis. Further, they also demonstrated that in humans, *C. elegans*, and *Arabidopsis*, *PIWI* codes for proteins necessary for stem cell maintenance, suggesting an evolutionarily conserved role for PIWI proteins.

2.2. Discovery of piRNAs

2.2.1. Discovery of piRNAs in Mouse. In 2006, four groups [108–111] independently identified another important class of small RNAs from mouse testes—the PIWI-interacting RNAs (piRNAs). These RNAs were found to be more abundant than any other class of small RNAs, and it was estimated that every spermatid would contain approximately one million piRNAs. Since the mouse-specific PIWI proteins were expressed in a temporal manner during mouse spermatogenesis, it is highly likely that the RNAs interacting with these proteins also show a time-dependent expression. Based on the length, two classes of RNAs were found to be interacting with PIWI proteins: the length of one class of RNAs ranged from 26 to 28 nt and the second class had a size range of 29–32 nt. While the shorter group of RNAs are expressed in the early stages of spermatogenesis (from spermatogonia until the round spermatid appears), the longer class of RNAs appeared later and were more abundant in the adult testes. Several features of this class of RNAs were reported: (i) largely present in germ cells; (ii) expressed in a developmentally regulated manner; (iii) mapping to the genome in clusters; (iv) strong strand bias, showing a preference for Uridine at the 5′ end; (v) show sense and antisense orientation; (vi) map to the intergenic regions, protein-coding regions, and repetitive sequences; (vii) did not show any stem loop structures, as seen in the primary-miRNAs; (viii) had 5′ phosphate group and 3′ OH group; (ix) showed uneven genomic distribution; and (x) showed poor conservation with distant species but were found to be well conserved within closely related genomes such as mouse and rat. Initially, they were believed to be part of repeat-associated siRNAs (rasiRNAs) as they showed regulation of repetitive elements such as transposons in the germline. Dicer-independent biogenesis of piRNAs and their interaction with PIWI proteins distinguished them from both miRNAs and siRNAs and were thus named as PIWI-interacting RNAs or piRNAs.

2.2.2. Discovery of piRNAs in Drosophila. There are two subclades of PIWI proteins in *Drosophila*—aubergine and PIWI—which play important roles in germ cell formation and in their development. These proteins were also found to be essential for silencing the transposons. The mechanism through which PIWI proteins promote gene silencing was not known until 2006 when Saito et al. [114] demonstrated that PIWI proteins associated with a set of small RNAs (25–29 nt) in the ovary of *Drosophila*. These small RNAs were distinct from miRNAs and siRNAs and were found to be derived from repetitive sequences. Further, these RNAs were present in both sense and antisense orientation and possessed a 5′ Uridine bias, suggesting their processing by RNase III enzyme. Immunoprecipitation and northern blot analyses revealed that PIWI proteins did not partner with miRNAs or siRNAs and their association with rasiRNAs was specific. An interesting observation was that PIWI proteins exhibited slicer activity in the nucleus of the cell, in contrast to the AGO protein which exhibited slicer activity in the cell's cytoplasm. In all, Saito et al. demonstrated that PIWI proteins perform gene-silencing activity in the nucleus through their specific association with rasiRNAs. These rasiRNAs, associated with PIWI proteins, were later called as piRNAs.

Brennecke et al. [115] performed a detailed study to identify and understand the small RNAs associated with PIWI proteins. They purified the PIWI protein complexes obtained from the ovary and sequenced the small RNAs found within the complex. Several observations were made: (i) the length of the small RNAs ranged between 23 and 29 nt, (ii) these RNAs showed strong preference for Uridine at the 5′ end, and (iii) most of the sequences could be aligned to annotated transposons or their remnants. An important finding was the identification of genomic origin of piRNAs. piRNAs were found to be present as clusters in discrete loci in the genome, and these were observed to be bidirectionally transcribed. These clusters were often found in the regions of high transposonal activity such as in the pericentromeric and telomeric heterochromatin and in the euchromatin regions [115].

2.2.3. Discovery of piRNAs in Humans. Immunoprecipitation and northern blot analyses revealed the presence of piRNAs in fetal and adult human ovaries and testes [116]. The identified piRNAs exhibited all the characteristics found in piRNAs identified from mice. While more than half of the piRNAs were found to arise from unidirectional-promoter-containing genes, the remaining piRNAs were found to arise from bidirectional-promoter-containing genes. The piRNAs were widely distributed across the autosomes and sex chromosomes. piRNAs were found in higher amounts in adult testes and ovaries and fetal ovaries, when compared to fetal testes. Sexually dimorphic patterns of piRNAs were observed in terms of length, distribution, and expression. In general, piRNAs appeared to be slightly longer in adult testes, when compared to the RNAs obtained from the ovaries. In contrast to what was observed in *Drosophila* or mice, piRNAs in humans mapped to the intergenic regions predominantly and only 2% of the identified piRNAs mapped to the transposable elements. This suggests that the primary function of piRNAs in humans may not include regulating transposons. Ha et al. profiled the piRNAs from human adult testis [117]. All the features of piRNAs reported in humans and other model organisms were confirmed in their study. In addition, genic regions of protein-coding genes were reported as the possible origin and genomic location of piRNAs in humans. Genic regions include 3′ and 5′ untranslated regions (UTRs) and the coding regions. piRNAs were found to be significantly enriched in the 3′ UTRs. However, the expression of these piRNAs did not correlate strongly with the expression of its corresponding host genes, from where the piRNA originated, indicating an alternative mechanism for piRNA generation from protein-coding genes. piRNA transcripts were also observed to be generated from long noncoding RNAs which are >200 nt in length. In addition, we [118, 119] and others [120–123] have suggested that transfer RNAs and small nucleolar RNAs may also be the sources of piRNAs as we observed some of the piRNA genes to exist within the boundaries of transfer RNAs and small nucleolar RNAs [118–123]. However, the process through which piRNA transcripts are generated remains elusive.

2.3. Biogenesis of piRNAs. The main difference between piRNAs and miRNAs biogenesis is that piRNAs do not require a double-stranded precursor molecule nor do they require the Dicer molecule for their biogenesis.

The biogenesis pathway of piRNA remains elusive, and the majority of our understanding concerns the *Drosophila* pathway. Often, piRNA biogenesis is also associated with silencing of target genes and requires all three PIWI proteins—PIWI, AUB, and AGO3. While PIWI proteins are found predominantly in the nuclei of germ cells and the somatic cells [124], AUB and AGO3 are found in the cytoplasmic regions of the germ cells [115]. In general, PIWI and AUB show preference for sequences with a 5′ Uridine bias mapping antisense to transposons, whereas AGO3 prefers sequences with adenine at the 10th position, mapping sense to transposons [115]. piRNAs take two routes for their processing: primary synthesis pathway and the secondary pathway/ping-pong amplification [125–128]. It is believed that the primary biogenesis pathway is necessary to initiate PIWI pathways, while the secondary pathway is necessary for maintaining the piRNA levels as well as for target silencing. While we only observe the primary pathway in somatic cells, we observe both primary and secondary pathways in germ cells [129].

Primary synthesis begins with the transcription from piRNA clusters by RNA polymerase II. The exact machinery involved in the processing is not known. However, it has been proposed that Zucchini, an endonuclease enzyme, cleaves the long single-stranded transcript, generating a 5′ phosphate residue. These piRNA-like molecules thus generated are loaded onto PIWI proteins (most often, PIWI and AUB proteins). The 3′ ends, which generally possess extra bases, are trimmed upon their association with PIWI proteins, and the length of a mature piRNA is established at this stage. Factors contributing to 3′ end trimming are not known yet. However, this is followed by 2′-O-methylation by Hen1 to confer stability to the piRNA and generate mature piRNA.

The secondary mechanism, known as the amplification cycle, involves only AUB and AGO3 and not PIWI proteins. piRNAs generated from the primary pathway may enter into the secondary pathway and subsequently bind with Aub. Using piRNA as a guide, the piRNA-AUB pair recognizes a complementary sequence from the transposons. AUB cleaves the complementary sequence, 10 nucleotides from the 5′ end of the primary piRNA. This generates a new piRNA with the 5′ end, which then binds with AGO3, and the same cycle of pairing with target sequencing, trimming, and generation of piRNA continues. In this cycle, the binding of piRNAs to AUB and AGO3 alternate with each other and the sequences that bind to these proteins are complementary to each other. This secondary cycle which serves the dual purpose of generating piRNAs and silencing transposons is thus called as amplification cycle or ping-pong cycle.

A similar biogenesis pathway has been observed in mice. Although there are striking differences in the manner in which piRNAs are generated in *C. elegans*, many of the factors identified in flies and mouse have also been observed in *C. elegans* (reviewed in [130]). Although there are no conclusive reports on the biogenesis pathway of piRNAs in humans, there are clues to suggest that it may be similar to the process observed in *Drosophila* or mouse [116, 131].

2.4. Functions of PIWI Proteins and piRNAs. Functions of piRNAs and PIWI proteins can be classified under two categories—developmental and regulatory (reviewed in [125]). The development functions can be further categorized into germline and somatic functions.

2.4.1. Contribution to Developmental Processes. Our understanding on the development functions of PIWI proteins originates from *Drosophila*, mice, *C. elegans*, and other model organisms. The major roles of PIWI proteins in the germline function include the formation of germ cell [132–134], maintenance of germline stem cells [112], meiosis [135], spermiogenesis [136], and oogenesis [112, 137]. Gene knockout and knockin experiments demonstrate the contribution of PIWI proteins in these functions. The significance of PIWI proteins has expanded beyond germ cells to somatic tissues. For instance, they are known to mediate epigenetic regulation and stem cell maintenance in *Drosophila* as well as maintenance of neoblast cells in planaria [138, 139]. The development of ciliates involves germline micronuclei and somatic macronuclei [140]. A certain amount of DNA sequences found in the somatic macronucleus has to be eliminated during sexual reproduction, and PIWI proteins are known to play a major role in DNA elimination.

2.4.2. Contribution to Gene Regulation. PIWI proteins may serve as epigenetic suppressors or activators [141–143], depending on the recruitment of specific proteins. It has also been observed that transposon-coding genes are not methylated in the absence of PIWI proteins, reflecting to a loss of epigenetic control (reviewed in [125]). piRNAs and PIWI proteins may both serve as upstream mediators of epigenetic control and may also be involved in transcriptional gene silencing. The role of piRNAs and PIWI proteins in silencing transposonal activities is well studied. It is believed that piRNAs occur as clusters. A specific example is the flamenco region in flies, which harbors one of the largest piRNA clusters [115, 144]. A disruption in the flamenco region interrupts the production of piRNAs, with a simultaneous increase in transposon activity. Also, the biogenesis pathway of piRNAs also serves a dual purpose—to generate piRNAs and to mediate gene silencing. Cofractionation of PIWI proteins and piRNAs with polysomes has hinted at the possibility of a potential role for PIWI proteins and piRNAs in translational control [110, 145]. One of the important observations is the role of piRNAs in posttranscriptional gene silencing. Although the mechanism still remains unclear, it is believed that piRNAs may act in a manner similar to that of miRNAs. A study from Esposito et al. [146] showed that piR_015520 negatively regulates its host gene, melatonin receptor 1A (*MTNR1A*), offering new functions for piRNAs, similar to miRNAs. Other studies have also confirmed the relationship between piRNAs and its corresponding target mRNAs [147–150], even though it is not known if the piRNAs have any seed sequence that determines its complementary binding with the target mRNA.

2.5. Detection and Quantification of piRNAs. Traditionally, piRNAs bound to PIWI proteins were identified by coimmunoprecipitation, followed by sequencing or RT-PCR of RNAs. The piRNAs thus identified were further validated using northern blot analyses and in situ hybridization [108–111, 151]. Similar to miRNAs, piRNAs can also be detected using microarray, sequencing [152], and qRT-PCR. One of the challenges with sequencing piRNAs lies in the binding of adapters to the 3′ end as these RNAs are characterized by the presence of 2′-O-methylation [153]. Therefore, traditional sequencing often leads to underrepresentation of piRNAs. One of the ways to overcome this challenge is through the inclusion of treatment with periodate before adding adaptors during library preparation [153]. This is often followed by confirmation with piRNA databases and/or coimmunoprecipitation with PIWI proteins.

Recently, piRNAs have also been profiled using microarray platforms. The most commonly used microarray platform is the Arraystar platform (Arraystar, Rockville, MD) that has been designed to profile more than 23,000 piRNAs in humans [154–156]. Often studies conducted using microarrays first identify piRNAs of interest and then perform crossplatform validation using qRT-PCR. To date, a qPCR array for piRNAs has not been developed, and investigators need to put in efforts to design individual assays. So far, candidate piRNA molecules have been validated by qRT-PCR using either the SYBR Green method or the TaqMan method using stem loop primers.

2.6. PIWI Proteins and piRNAs as Contributors to Tumorigenesis. The contribution of PIWI-piRNA to cancer comes in different layers. First, PIWI-piRNA complexes maintain genomic integrity by silencing transposons through DNA methylation. Transposons are mobile elements that can get transferred and interspersed between different genomic locations, generating insertions, deletions, and other chromosomal aberrations. All of these aberrations are critical for tumor development. Second, the role of PIWI-piRNA in conferring epigenetic modification has been demonstrated in model organisms. PIWI is guided by the piRNA to the genomic region where the piRNA shares complementarity. This interaction helps in recruiting factors necessary for imparting methylation and/or acetylation, which are essential for transcriptional regulation of gene expression [157]. One example of this mode of gene regulation is the silencing of killer immunoglobulin-like receptors (*KIRs*) [158]. Antisense transcripts of *KIR* are processed to piRNA-like molecules. The piRNA-PIWI complex binds to the promoter region and favors recruitment of methylation factors, enabling transcriptional silencing of the gene.

While the role of piRNAs in regulating transposons and in mediating epigenetic regulation is well documented [159], the role of somatic piRNAs in cancer biology is increasingly being recognized [160]. Our knowledge on the significance of piRNAs in cancer originated from delineating their role in germ cell cancers [161]. Although when compared to germline piRNAs, the number of piRNAs are lesser in somatic cells (both level of expression and number of distinct piRNAs), their contribution to tumorigenesis is significant. While germline piRNAs are known for their regulation of transposons, recent studies suggest their role as regulators of gene expression in the somatic cells [148, 162]. Similar to miRNAs, piRNAs, along with their cognate partner (PIWI), have been shown to bind to complementary target sequences and promote mRNA deadenylation or degradation. Similar to PIWI, piRNAs are dysregulated in cancers, when compared to normal tissues. Further, the expression patterns of piRNAs have also been reported to vary across different tissue types, suggesting a possible tissue-specific expression of piRNAs [163]. Likewise, similar to PIWI, recent studies have shown that up- or downregulation of piRNAs may induce imbalance in the cell machinery, causing increased cell proliferation [164], invasion and metastasis [165], decreased apoptosis [166], and so on. With these diverse functions, piRNAs are now classified as "master regulators of gene expression."

2.7. piRNAs as Diagnostic and Prognostic Markers of Cancer. piRNAs are slowly garnering attention as diagnostic and prognostic biomarkers for various cancer types. For example, Cui et al. observed significantly lower levels of piR-651 and piR-823 in the peripheral blood of gastric cancer patients and their association with clinicopathological features [167]. They also obtained an area under the curve value of 0.841, 0.812, and 0.860 for piR-651, piR-823, and piR-651, respectively, combined with piR-823, indicating the diagnostic potential of these piRNAs. Two studies demonstrated association of piRNAs with renal cell carcinoma. In the study conducted by Busch et al., three piRNAs (piR-30924, piR-38756, and piR-57125) were associated with tumor recurrence and overall survival. Two of these (piR-30924 and piR-57125) also emerged as independent prognostic markers [154]. In another study conducted by Li et al., high expression of three piRNAs (piR-32051, piR-39894, and piR-43607) was associated with metastasis, clinical stage, and survival in renal cell carcinoma [168]. In gastric cancer, one piRNA (FR222326) and three piRNAs (FR290353, FR064000, and FR387750/FR157678) associated with overall survival and recurrence free survival, respectively [169]. Similarly, in breast cancer, four piRNAs (hsa_piR_009051, hsa_piR_021032, hsa_piR_015249, and hsa_piR_020541) associated with overall survival and six piRNAs (hsa_piR_017061, hsa_piR_009051, hsa_piR_021032, hsa_piR_004153, hsa_piR_017716, and hsa_piR_019914) showed association with recurrence-free survival [162]. Analogous to miRNAs, piRNAs are stable in biofluids such as plasma and blood [170]. piRNAs isolated from these biofluids were found to be deregulated in patients with colorectal cancer, prostate cancer, and pancreatic cancer [171]. More studies confirming the prognostic potential of circulating piRNAs may add a new dimension to the existing roles of piRNAs.

2.8. PIWI-piRNAs and Cancer Therapeutics. It is becoming evident that PIWI proteins and piRNAs are integral components of tumor development. These molecules operate at various levels including transcriptional, posttranscriptional, and epigenetic. Altered expressions of both these molecules have proved to alter cell growth, bringing about increased cell proliferation, decreased apoptosis, and increased invasiveness and motility, all leading to transformation of a normal cell to a malignant cell. Similar to miRNAs, silencing of piRNAs and PIWI may reverse these phenotypes. Dysregulation of PIWI proteins also contribute to altered levels of piRNAs in a cell. Therefore, regulating the expression of PIWI would serve dual purposes—control the expression of piRNAs which otherwise may be dysregulated in a cancer cell and in restoring a cell's balance. Regulation of mRNAs by piRNAs is currently being addressed. While the exact mechanism remains elusive, piRNAs are most likely to interact with mRNAs through complementary sequences and bring about mRNA degradation or translation inhibition, similar to miRNA-mediated regulation. Although not well studied, piRNAs and PIWI tend to share a feedback regulatory mechanism. While the control of PIWI over the expression of piRNAs is quite well known, possible regulation of PIWI expression by piRNAs, in a manner similar to miRNA-mediated regulation, has only been proposed. We had, in our earlier study, observed that (i) hsa_piR_021032 shares a complementary sequence with PIWIL2 and (ii) the expressions of both these molecules were in opposite directions [162]. These observations suggest that piRNAs may directly regulate the expression of PIWI, a mechanism that has not yet been experimentally validated. So far, we do not have any therapeutics developed using these molecules. Silencing or inducing the expression of PIWI/piRNA can alter the recruitment of epigenetic factors and can modify the overall expression levels of PIWI/piRNAs and mRNAs via transcription or posttranscription, thereby influencing the overall development and progression of cancer. The field of piRNAs is relatively young but promising. However, all these observations and evidence highlight the critical role that these molecules play in overall tumorigenesis and emphasize the need to comprehensively understand the mechanisms of these molecules and explore their therapeutic potential.

3. Challenges Involved in Biomarker Development

miRNAs belong to a well-studied class of small noncoding RNAs. The role of miRNAs as biomarkers has also been well explored, and several investigators are focusing on comprehensive understanding and utility of these RNAs in clinics. Other classes of small RNAs such as piRNAs have also shown promise as biomarkers, and extensive studies are needed in this direction to translate this finding to clinics. However, many challenges exist and thus hinder the immediate application of these molecules in clinics. Some of these challenges are outlined below.

3.1. Choice of Normal Samples. The choice of normal samples chosen for comparison with tumor tissues remains debatable. One school of thought encourages the use of tumor-adjacent normal tissues, as these are scored also for the presence of tumor cells [172, 173]. Since these tissues do not show the presence of tumor cells, these are considered to be a safe choice for comparison. However, another school of thought dismisses the idea of using adjacent normal tissues for two reasons—(a) lack of molecular differences between tumor and adjacent normal tissues against the presence of difference between tumor and normal tissues obtained from apparently healthy individuals [174, 175] and (b) lack of consensus on the distance from the tumor that a tissue can be designated as normal [174, 176, 177]. Studies have been conducted to understand the molecular differences between tumor-adjacent normal tissues, normal tissues obtained from healthy individuals, and tumor tissues. For instance, in a study conducted by Sanz-Pamplona et al. [174], mRNA levels from 98-paired adjacent normal mucosa were compared with colorectal cancer tissues and with 50 colon mucosa from healthy donors. It was observed that a number of genes were activated in the tumor-adjacent mucosa, similar to the tumor tissues, and the activation of these genes was not observed in the normal tissues from healthy donors. In fact, principal component analysis of the three tissue sets identified different clusters for the three tissue types, suggesting that considerable differences existed between the tumor-adjacent "normal" tissues and healthy normal tissues. Further, a total of 895 genes were found to be differentially expressed between the two normal tissues. Functional enrichment analysis and network analysis of the activated genes from adjacent normal tissues identified pathways significant for tumorigenesis, implying that the adjacent normal tissues may mimic the effects of tumorigenesis. The same was validated in publicly available datasets, consistent with the view that adjacent stromal cells crosstalk with tumor cells within this microenvironment (called as "the field effect"). Therefore, use of adjacent normal tissues as a reference may result in misleading interpretations. This bystander effect of tumors on adjacent normal tissues has been observed in other cancer types as well, such as prostate cancer [175, 178, 179] and breast cancer. Similar anomalous patterns of expression were reported even for noncoding RNAs when tumor-adjacent normal versus reduction mammoplasty normal tissues were compared [180]. In support of this general premise, differential miRNA expressions were reported between reduction mammoplasty and tumor-adjacent normal tissues. Differentially expressed long noncoding RNAs were absent when tumor-adjacent normal tissues versus tumor tissues were compared [181]. However, differentially expressed transcripts were identified when tumor tissues were compared with reduction mammoplasty specimens. It is safe to infer from the above examples that where possible, use of normal tissues obtained from apparently healthy individuals may be more appropriate tissue for comparison with tumor tissues [181].

3.2. Profiling Platform. miRNAs can be profiled using different platforms including microarray, qRT-PCR, and NGS. Each platform possesses its own advantages and disadvantages, and therefore, the choice of profiling platform is critical for discovery and validation studies. Going forward, while NGS may be the preferred choice for profiling small noncoding RNAs, it may not be easily applicable in the clinics, and therefore, one needs to optimize other platforms such as NanoString to validate the expression of miRNAs obtained from NGS and their applicability to clinic.

3.3. Sample Size and Study Design. The number of samples chosen in a discovery cohort is crucial. A biomarker study should represent the population. The higher the sample size, the higher is the statistical power of that study. For instance, identifying differentially expressed RNAs from five samples in each group may give rise to false positive associations and may lead to dimensionality problem, where the numbers of markers are much higher than the number of samples used. Therefore, multiple rounds of validation of signatures in different populations are required to gain confidence in the identified signature.

3.4. Study Design and Analysis Methods. Different factors need to be considered when designing a biomarker study. Many studies have used the case-control approach [182, 183]. It is common practice to identify differentially expressed RNAs between the two groups and test the prognostic relevance of differentially expressed RNAs. While it is important that a biomarker shows distinct expression patterns between normal and tumor tissues, it is not necessary that only differentially expressed RNAs will show prognostic relevance. With this objective, biomarker studies have also been conducted using only the tumor tissues and without focusing on the differentially expressed RNAs—a case-only study [86, 90]. To understand the differences and similarities between these two methods, recent studies from our group compared the results obtained from both the approaches and observed that the case-only study is an unbiased approach that identifies more numbers of clinically significant RNAs [39, 118, 119, 162]. The case-only approach not only identifies all the RNAs identified in the case-control approach but also identifies newer molecules that were not differentially expressed.

3.5. Batch Effects. Batch effects should be corrected for. Analyzing the data in the presence of batch effects may lead to spurious associations. While it may not be possible to contain for all the source of variations, it is more appropriate to acknowledge, detect, and contain as many sources of technical variations as possible. One of the ways to detect for batch effects would be using principal component analysis, and one of the ways to quantify and remove batch effects would be using ANOVA. Leek et al. clearly explain the causes, effects of, and methods to detect and correct for nonbiological sources of variation [184].

3.6. Expression of RNAs. Handling of high-throughput data, especially NGS data, is challenging. A large number of RNAs may be expressed only in one or two samples and that too with very low expression, with one read count. This single count could be due to technical artifacts. Therefore, it may be a good idea to remove the RNAs which have low expression. Some investigators consider the total read count in the dataset, whereas some others consider per-sample read count [185–189]. Based on the objective of the study, one has to determine the cut-off for read counts. For instance, one of the characteristics of biomarkers is that a biomarker should be expressed in most of the samples. In this case, choosing a molecule that is expressed in >90% of the samples may be appropriate [39, 118, 119, 162].

3.7. Data Normalization. The choice of normalization method is vital. Several methods exist for normalizing the data generated from microarray as well as from NGS. While most of the microarray studies now employ quantile normalization, the most appropriate method for normalizing NGS data remains elusive. Several choices include RPKM, and TMM. Each of these methods and/or packages has been used by researchers to successfully identify RNAs. However, we do not yet have consensus on which method is more suitable to handle NGS data.

3.8. Single versus Panel Markers. The choice of single markers versus combination markers is important. Biology is driven by multiple interacting modules, and the increasing use of panels of mRNA markers for prognosis is recognized; use of a single marker (for instance, miRNA, which has the intrinsic property of pleiotropy) may or may not capture the higher level of complexity inherent to biology. A single miRNA may regulate multiple mRNAs, and multiple miRNAs may target a single mRNA. While some argue the use of single markers as more effective markers, a combined signature of various RNAs such as miRNAs, piRNAs, and mRNAs may more closely reflect the overall biology of the underlying disease.

The confidence in a biomarker study largely depends on its reproducibility in external datasets with large sample sizes. This adds strength to the initial study findings and may eventually lead to generalizing the initial study findings. However, it is not an easy task to obtain an external dataset because several factors need to be considered. These include (a) number of samples available, (b) histological and molecular subtypes of the cancer studied, (c) profiling platform used for the study, (d) number of events (dead/alive, recurrence), (e) the amount of clinical information available and attempts to adjust for confounding factors including sex, (f) number of years of follow-up, and (g) the population ancestry; these may all influence the replicability of findings across laboratories.

4. Conclusion

Despite progress in the miRNA field and utility of miRNAs as diagnostic, prognostic, or predictive markers, their adaptation to the clinic is slow, and that is to be expected. More than 100 clinical trials (https://clinicaltrials.gov/) are currently underway for use of miRNAs in cancer and other diseases. Compared to miRNA, the piRNA field is in its infancy, and hence, applications to the clinic will not be seen immediately. The identified and characterized miRNA markers need several independent validations with large sample sizes. Biomarker acceptance will be high if applicable to all ethnic groups. These biomarkers should also undergo extensive validations in prospective cohorts and with clear benefit in terms of positive predictive value and cost benefit over the existing markers currently in use. Last but not the least, change in clinical practice guidelines meets with considerable resistance as this requires extensive education and expertise to accept new diagnostic or prognostic tools across physicians, administrators, and laboratory personnel. The field of genomics is fast evolving, and integrating genomics courses and bioinformatics at the undergraduate level will build the awareness and competence as the younger generation scientists and physicians are well versed in genomics and do not shy away from embracing newer disciplines. There are large efforts underway internationally in universities and affiliated teaching hospitals to create precision medicine or precision oncology disciplines, and it should not be long before the new generation of biomarkers finds its place in health care.

Disclosure

The authors acknowledge that some sections of the manuscript are based on the document that served as a partial fulfillment of the doctoral thesis requirements at the University of Alberta submitted by Preethi Krishnan (https://era.library.ualberta.ca/files/cxw42n810z/Krishnan_Preethi_201607_PhD.pdf).

Conflicts of Interest

The authors declare no conflicts of interests.

Acknowledgments

Funding was provided by the Canadian Breast Cancer Foundation (CBCF)—Prairies/NWT chapter (to Sambasivarao Damaraju). The authors thank Dr. Vickie Baracos for critical reading of the manuscript and edits.

References

[1] R. W. Carthew and E. J. Sontheimer, "Origins and mechanisms of miRNAs and siRNAs," *Cell*, vol. 136, no. 4, pp. 642–655, 2009.

[2] D. P. Bartel, "MicroRNAs: genomics, biogenesis, mechanism, and function," *Cell*, vol. 116, no. 2, pp. 281–297, 2004.

[3] M. A. Carmell, Z. Xuan, M. Q. Zhang, and G. J. Hannon, "The Argonaute family: tentacles that reach into RNAi, developmental control, stem cell maintenance, and tumorigenesis," *Genes & Development*, vol. 16, no. 21, pp. 2733–2742, 2002.

[4] R. C. Lee, R. L. Feinbaum, and V. Ambros, "The *C. elegans* heterochronic gene *lin-4* encodes small RNAs with antisense complementarity to *lin-14*," *Cell*, vol. 75, no. 5, pp. 843–854, 1993.

[5] B. Wightman, I. Ha, and G. Ruvkun, "Posttranscriptional regulation of the heterochronic gene *lin-14* by *lin-4* mediates temporal pattern formation in *C. elegans*," *Cell*, vol. 75, no. 5, pp. 855–862, 1993.

[6] R. C. Lee and V. Ambros, "An extensive class of small RNAs in *Caenorhabditis elegans*," *Science*, vol. 294, no. 5543, pp. 862–864, 2001.

[7] A. J. Hamilton and D. C. Baulcombe, "A species of small antisense RNA in posttranscriptional gene silencing in plants," *Science*, vol. 286, no. 5441, pp. 950–952, 1999.

[8] E. G. Moss, R. C. Lee, and V. Ambros, "The cold shock domain protein LIN-28 controls developmental timing in *C. elegans* and is regulated by the *lin-4* RNA," *Cell*, vol. 88, no. 5, pp. 637–646, 1997.

[9] B. J. Reinhart, F. J. Slack, M. Basson et al., "The 21-nucleotide *let-7* RNA regulates developmental timing in *Caenorhabditis elegans*," *Nature*, vol. 403, no. 6772, pp. 901–906, 2000.

[10] A. E. Pasquinelli, B. J. Reinhart, F. Slack et al., "Conservation of the sequence and temporal expression of *let-7* heterochronic regulatory RNA," *Nature*, vol. 408, no. 6808, pp. 86–89, 2000.

[11] F. Fazi and C. Nervi, "MicroRNA: basic mechanisms and transcriptional regulatory networks for cell fate determination," *Cardiovascular Research*, vol. 79, no. 4, pp. 553–561, 2008.

[12] M. Lagos-Quintana, R. Rauhut, W. Lendeckel, and T. Tuschl, "Identification of novel genes coding for small expressed RNAs," *Science*, vol. 294, no. 5543, pp. 853–858, 2001.

[13] N. C. Lau, L. P. Lim, E. G. Weinstein, and D. P. Bartel, "An abundant class of tiny RNAs with probable regulatory roles in *Caenorhabditis elegans*," *Science*, vol. 294, no. 5543, pp. 858–862, 2001.

[14] Y. Lee, K. Jeon, J. T. Lee, S. Kim, and V. N. Kim, "MicroRNA maturation: stepwise processing and subcellular localization," *The EMBO Journal*, vol. 21, no. 17, pp. 4663–4670, 2002.

[15] X. Zhang, G. Azhar, and J. Y. Wei, "The expression of microRNA and microRNA clusters in the aging heart," *PLoS One*, vol. 7, no. 4, article e34688, 2012.

[16] J. Winter, S. Jung, S. Keller, R. I. Gregory, and S. Diederichs, "Many roads to maturity: microRNA biogenesis pathways and their regulation," *Nature Cell Biology*, vol. 11, no. 3, pp. 228–234, 2009.

[17] J. Hausser and M. Zavolan, "Identification and consequences of miRNA-target interactions - beyond repression of gene expression," *Nature Reviews. Genetics*, vol. 15, no. 9, pp. 599–612, 2014.

[18] P. Ramalingam, J. K. Palanichamy, A. Singh et al., "Biogenesis of intronic miRNAs located in clusters by independent transcription and alternative splicing," *RNA*, vol. 20, no. 1, pp. 76–87, 2013.

[19] G. M. Borchert, W. Lanier, and B. L. Davidson, "RNA polymerase III transcribes human microRNAs," *Nature Structural & Molecular Biology*, vol. 13, no. 12, pp. 1097–1101, 2006.

[20] Y. Lee, C. Ahn, J. Han et al., "The nuclear RNase III Drosha initiates microRNA processing," *Nature*, vol. 425, no. 6956, pp. 415–419, 2003.

[21] J. Starega-Roslan, E. Koscianska, P. Kozlowski, and W. J. Krzyzosiak, "The role of the precursor structure in the biogenesis of microRNA," *Cellular and Molecular Life Sciences*, vol. 68, no. 17, pp. 2859–2871, 2011.

[22] Y. Lee, M. Kim, J. Han et al., "MicroRNA genes are transcribed by RNA polymerase II," *The EMBO Journal*, vol. 23, no. 20, pp. 4051–4060, 2004.

[23] A. Khvorova, A. Reynolds, and S. D. Jayasena, "Functional siRNAs and miRNAs exhibit strand bias," *Cell*, vol. 115, no. 2, pp. 209–216, 2003.

[24] D. S. Schwarz, G. Hutvagner, T. Du, Z. Xu, N. Aronin, and P. D. Zamore, "Asymmetry in the assembly of the RNAi enzyme complex," *Cell*, vol. 115, no. 2, pp. 199–208, 2003.

[25] T. W. Nilsen, "Mechanisms of microRNA-mediated gene regulation in animal cells," *Trends in Genetics*, vol. 23, no. 5, pp. 243–249, 2007.

[26] J. G. Ruby, C. H. Jan, and D. P. Bartel, "Intronic microRNA precursors that bypass Drosha processing," *Nature*, vol. 448, no. 7149, pp. 83–86, 2007.

[27] A. R. Naqvi, M. N. Islam, N. R. Choudhury, and Q. M. Haq, "The fascinating world of RNA interference," *International Journal of Biological Sciences*, vol. 5, no. 2, pp. 97–117, 2009.

[28] J. G. Doench and P. A. Sharp, "Specificity of microRNA target selection in translational repression," *Genes & Development*, vol. 18, no. 5, pp. 504–511, 2004.

[29] B. N. Davis and A. Hata, "Regulation of microRNA biogenesis: a miRiad of mechanisms," *Cell Communication and Signaling: CCS*, vol. 7, no. 1, p. 18, 2009.

[30] I. Slezak-Prochazka, S. Durmus, B. J. Kroesen, and A. van den Berg, "MicroRNAs, macrocontrol: regulation of miRNA processing," *RNA*, vol. 16, no. 6, pp. 1087–1095, 2010.

[31] M. Baker, "MicroRNA profiling: separating signal from noise," *Nature Methods*, vol. 7, no. 9, pp. 687–692, 2010.

[32] F. Pimentel, P. Bonilla, Y. G. Ravishankar et al., "Technology in microRNA profiling: circulating microRNAs as noninvasive cancer biomarkers in breast cancer," *Journal of Laboratory Automation*, vol. 20, no. 5, pp. 574–588, 2015.

[33] C. C. Pritchard, H. H. Cheng, and M. Tewari, "MicroRNA profiling: approaches and considerations," *Nature Reviews Genetics*, vol. 13, no. 5, pp. 358–369, 2012.

[34] A. W. Wark, H. J. Lee, and R. M. Corn, "Multiplexed detection methods for profiling microRNA expression in biological samples," *Angewandte Chemie International Edition*, vol. 47, no. 4, pp. 644–652, 2008.

[35] S. Tam, R. de Borja, M. S. Tsao, and J. D. McPherson, "Robust global microRNA expression profiling using next-generation sequencing technologies," *Laboratory Investigation*, vol. 94, no. 3, pp. 350–358, 2014.

[36] A. Drusco, G. J. Nuovo, N. Zanesi et al., "MicroRNA profiles discriminate among colon cancer metastasis," *PLoS One*, vol. 9, no. 6, article e96670, 2014.

[37] E. Nadal, A. Truini, A. Nakata et al., "A novel serum 4-microRNA signature for lung cancer detection," *Scientific Reports*, vol. 5, no. 1, article 12464, 2015.

[38] F. Marabita, P. de Candia, A. Torri, J. Tegner, S. Abrignani, and R. L. Rossi, "Normalization of circulating microRNA expression data obtained by quantitative real-time RT-PCR," *Briefings in Bioinformatics*, vol. 17, no. 2, pp. 204–212, 2016.

[39] P. Krishnan, S. Ghosh, B. Wang et al., "Next generation sequencing profiling identifies miR-574-3p and miR-660-5p as potential novel prognostic markers for breast cancer," *BMC Genomics*, vol. 16, no. 1, p. 735, 2015.

[40] L. De Cecco, M. Dugo, S. Canevari, M. G. Daidone, and M. Callari, "Measuring microRNA expression levels in oncology: from samples to data analysis," *Critical Reviews in Oncogenesis*, vol. 18, no. 4, pp. 273–287, 2013.

[41] P. J. Hurd and C. J. Nelson, "Advantages of next-generation sequencing versus the microarray in epigenetic research," *Briefings in Functional Genomics & Proteomics*, vol. 8, no. 3, pp. 174–183, 2009.

[42] A. Grada and K. Weinbrecht, "Next-generation sequencing: methodology and application," *The Journal of Investigative Dermatology*, vol. 133, no. 8, article e11, pp. 1–4, 2013.

[43] A. Mortazavi, B. A. Williams, K. McCue, L. Schaeffer, and B. Wold, "Mapping and quantifying mammalian transcriptomes by RNA-Seq," *Nature Methods*, vol. 5, no. 7, pp. 621–628, 2008.

[44] M. I. Love, W. Huber, and S. Anders, "Moderated estimation of fold change and dispersion for RNA-seq data with DESeq2," *Genome Biology*, vol. 15, no. 12, p. 550, 2014.

[45] M. D. Robinson, D. J. McCarthy, and G. K. Smyth, "edgeR: a bioconductor package for differential expression analysis of digital gene expression data," *Bioinformatics*, vol. 26, no. 1, pp. 139-140, 2010.

[46] D. J. McCarthy, Y. Chen, and G. K. Smyth, "Differential expression analysis of multifactor RNA-Seq experiments with respect to biological variation," *Nucleic Acids Research*, vol. 40, no. 10, pp. 4288–4297, 2012.

[47] M. D. Robinson and A. Oshlack, "A scaling normalization method for differential expression analysis of RNA-seq data," *Genome Biology*, vol. 11, no. 3, p. R25, 2010.

[48] I. Berindan-Neagoe, C. Monroig Pdel, B. Pasculli, and G. A. Calin, "MicroRNAome genome: a treasure for cancer diagnosis and therapy," *CA: a Cancer Journal for Clinicians*, vol. 64, no. 5, pp. 311–336, 2014.

[49] D. Sayed and M. Abdellatif, "MicroRNAs in development and disease," *Physiological Reviews*, vol. 91, no. 3, pp. 827–887, 2011.

[50] A. Cimmino, G. A. Calin, M. Fabbri et al., "miR-15 and miR-16 induce apoptosis by targeting BCL2," *Proceedings of the National Academy of Sciences of the United States of America*, vol. 102, no. 39, pp. 13944–13949, 2005.

[51] J. A. Chan, A. M. Krichevsky, and K. S. Kosik, "MicroRNA-21 is an antiapoptotic factor in human glioblastoma cells," *Cancer Research*, vol. 65, no. 14, pp. 6029–6033, 2005.

[52] C. Z. Chen, L. Li, H. F. Lodish, and D. P. Bartel, "MicroRNAs modulate hematopoietic lineage differentiation," *Science*, vol. 303, no. 5654, pp. 83–86, 2004.

[53] J. F. Chen, E. M. Mandel, J. M. Thomson et al., "The role of microRNA-1 and microRNA-133 in skeletal muscle proliferation and differentiation," *Nature Genetics*, vol. 38, no. 2, pp. 228–233, 2006.

[54] S. Volinia, G. A. Calin, C. G. Liu et al., "A microRNA expression signature of human solid tumors defines cancer gene targets," *Proceedings of the National Academy of Sciences of the United States of America*, vol. 103, no. 7, pp. 2257–2261, 2006.

[55] J. Lu, G. Getz, E. A. Miska et al., "MicroRNA expression profiles classify human cancers," *Nature*, vol. 435, no. 7043, pp. 834–838, 2005.

[56] Y. Zhao, E. Samal, and D. Srivastava, "Serum response factor regulates a muscle-specific microRNA that targets Hand2 during cardiogenesis," *Nature*, vol. 436, no. 7048, pp. 214–220, 2005.

[57] E. van Rooij, L. B. Sutherland, N. Liu et al., "A signature pattern of stress-responsive microRNAs that can evoke cardiac hypertrophy and heart failure," *Proceedings of the National Academy of Sciences of the United States of America*, vol. 103, no. 48, pp. 18255–18260, 2006.

[58] J. Stanczyk, D. M. Pedrioli, F. Brentano et al., "Altered expression of microRNA in synovial fibroblasts and synovial tissue in rheumatoid arthritis," *Arthritis & Rheumatism*, vol. 58, no. 4, pp. 1001–1009, 2008.

[59] D. Liu, H. Zhao, S. Zhao, and X. Wang, "MicroRNA expression profiles of peripheral blood mononuclear cells in patients with systemic lupus erythematosus," *Acta Histochemica*, vol. 116, no. 5, pp. 891–897, 2014.

[60] A. Zovoilis, H. Y. Agbemenyah, R. C. Agis-Balboa et al., "microRNA-34c is a novel target to treat dementias," *The EMBO Journal*, vol. 30, no. 20, pp. 4299–4308, 2011.

[61] J. Kim, K. Inoue, J. Ishii et al., "A microRNA feedback circuit in midbrain dopamine neurons," *Science*, vol. 317, no. 5842, pp. 1220–1224, 2007.

[62] G. A. Calin, C. D. Dumitru, M. Shimizu et al., "Frequent deletions and down-regulation of micro- RNA genes miR15 and miR16 at 13q14 in chronic lymphocytic leukemia," *Proceedings of the National Academy of Sciences of the United States of America*, vol. 99, no. 24, pp. 15524–15529, 2002.

[63] S. Costinean, N. Zanesi, Y. Pekarsky et al., "Pre-B cell proliferation and lymphoblastic leukemia/high-grade lymphoma in Eμ-miR155 transgenic mice," *Proceedings of the National Academy of Sciences of the United States of America*, vol. 103, no. 18, pp. 7024–7029, 2006.

[64] P. P. Medina, M. Nolde, and F. J. Slack, "OncomiR addiction in an in vivo model of microRNA-21-induced pre-B-cell lymphoma," *Nature*, vol. 467, no. 7311, pp. 86–90, 2010.

[65] E. Callegari, B. K. Elamin, F. Giannone et al., "Liver tumorigenicity promoted by microRNA-221 in a mouse transgenic model," *Hepatology*, vol. 56, no. 3, pp. 1025–1033, 2012.

[66] C. M. M. Gits, P. F. van Kuijk, M. B. E. Jonkers et al., "MiR-17-92 and miR-221/222 cluster members target KIT and ETV1 in human gastrointestinal stromal tumours," *British Journal of Cancer*, vol. 109, no. 6, pp. 1625–1635, 2013.

[67] J. Qin and M. Luo, "MicroRNA-221 promotes colorectal cancer cell invasion and metastasis by targeting RECK," *FEBS Letters*, vol. 588, no. 1, pp. 99–104, 2014.

[68] R. C. Friedman, K. K. H. Farh, C. B. Burge, and D. P. Bartel, "Most mammalian mRNAs are conserved targets of microRNAs," *Genome Research*, vol. 19, no. 1, pp. 92–105, 2008.

[69] M. Selbach, B. Schwanhausser, N. Thierfelder, Z. Fang, R. Khanin, and N. Rajewsky, "Widespread changes in protein synthesis induced by microRNAs," *Nature*, vol. 455, no. 7209, pp. 58–63, 2008.

[70] D. Hanahan and R. A. Weinberg, "The hallmarks of cancer," *Cell*, vol. 100, no. 1, pp. 57–70, 2000.

[71] D. Hanahan and R. A. Weinberg, "Hallmarks of cancer: the next generation," *Cell*, vol. 144, no. 5, pp. 646–674, 2011.

[72] Biomarkers Definitions Working Group, "Biomarkers and surrogate endpoints: preferred definitions and conceptual framework," *Clinical Pharmacology and Therapeutics*, vol. 69, no. 3, pp. 89–95, 2001.

[73] A. Etheridge, I. Lee, L. Hood, D. Galas, and K. Wang, "Extracellular microRNA: a new source of biomarkers," *Mutation Research*, vol. 717, no. 1-2, pp. 85–90, 2011.

[74] L. Bovell, C. Shanmugam, V. R. Katkoori et al., "miRNAs are stable in colorectal cancer archival tissue blocks," *Frontiers in Bioscience*, vol. E4, pp. 1937–1940, 2012.

[75] T. Blondal, S. Jensby Nielsen, A. Baker et al., "Assessing sample and miRNA profile quality in serum and plasma or other biofluids," *Methods*, vol. 59, no. 1, pp. S1–S6, 2013.

[76] P. S. Mitchell, R. K. Parkin, E. M. Kroh et al., "Circulating microRNAs as stable blood-based markers for cancer detection," *Proceedings of the National Academy of Sciences of the United States of America*, vol. 105, no. 30, pp. 10513–10518, 2008.

[77] N. J. Park, H. Zhou, D. Elashoff et al., "Salivary microRNA: discovery, characterization, and clinical utility for oral cancer detection," *Clinical Cancer Research*, vol. 15, no. 17, pp. 5473–5477, 2009.

[78] C. Mall, D. M. Rocke, B. Durbin-Johnson, and R. H. Weiss, "Stability of miRNA in human urine supports its biomarker potential," *Biomarkers in Medicine*, vol. 7, no. 4, pp. 623–631, 2013.

[79] C. Blenkiron, L. D. Goldstein, N. P. Thorne et al., "MicroRNA expression profiling of human breast cancer identifies new markers of tumor subtype," *Genome Biology*, vol. 8, no. 10, p. R214, 2007.

[80] S. Volinia, M. Galasso, M. E. Sana et al., "Breast cancer signatures for invasiveness and prognosis defined by deep sequencing of microRNA," *Proceedings of the National Academy of Sciences of the United States of America*, vol. 109, no. 8, pp. 3024–3029, 2012.

[81] Y. M. Youssef, N. M. White, J. Grigull et al., "Accurate molecular classification of kidney cancer subtypes using microRNA signature," *European Urology*, vol. 59, no. 5, pp. 721–730, 2011.

[82] M. Ferracin, M. Pedriali, A. Veronese et al., "MicroRNA profiling for the identification of cancers with unknown primary tissue-of-origin," *The Journal of Pathology*, vol. 225, no. 1, pp. 43–53, 2011.

[83] J. Takamizawa, H. Konishi, K. Yanagisawa et al., "Reduced expression of the *let-7* microRNAs in human lung cancers in association with shortened postoperative survival," *Cancer Research*, vol. 64, no. 11, pp. 3753–3756, 2004.

[84] G. A. Calin, M. Ferracin, A. Cimmino et al., "A microRNA signature associated with prognosis and progression in chronic lymphocytic leukemia," *The New England Journal of Medicine*, vol. 353, no. 17, pp. 1793–1801, 2005.

[85] N. Yanaihara, N. Caplen, E. Bowman et al., "Unique microRNA molecular profiles in lung cancer diagnosis and prognosis," *Cancer Cell*, vol. 9, no. 3, pp. 189–198, 2006.

[86] S. L. Yu, H. Y. Chen, G. C. Chang et al., "MicroRNA signature predicts survival and relapse in lung cancer," *Cancer Cell*, vol. 13, no. 1, pp. 48–57, 2008.

[87] M. V. Iorio, M. Ferracin, C. G. Liu et al., "MicroRNA gene expression deregulation in human breast cancer," *Cancer Research*, vol. 65, no. 16, pp. 7065–7070, 2005.

[88] M. Bloomston, W. L. Frankel, F. Petrocca et al., "MicroRNA expression patterns to differentiate pancreatic adenocarcinoma from normal pancreas and chronic pancreatitis," *Journal of the American Medical Association*, vol. 297, no. 17, pp. 1901–1908, 2007.

[89] C. Roldo, E. Missiaglia, J. P. Hagan et al., "MicroRNA expression abnormalities in pancreatic endocrine and acinar tumors are associated with distinctive pathologic features and clinical behavior," *Journal of Clinical Oncology*, vol. 24, no. 29, pp. 4677–4684, 2006.

[90] X. Li, Y. Zhang, Y. Zhang, J. Ding, K. Wu, and D. Fan, "Survival prediction of gastric cancer by a seven-microRNA signature," *Gut*, vol. 59, no. 5, pp. 579–585, 2010.

[91] P. Roth, A. Keller, J. D. Hoheisel et al., "Differentially regulated miRNAs as prognostic biomarkers in the blood of

primary CNS lymphoma patients," *European Journal of Cancer*, vol. 51, no. 3, pp. 382–390, 2015.

[92] H. Y. Jin, A. Gonzalez-Martin, A. V. Miletic et al., "Transfection of microRNA mimics should be used with caution," *Frontiers in Genetics*, vol. 6, p. 340, 2015.

[93] Z. Wang, "The guideline of the design and validation of MiRNA mimics," *Methods in Molecular Biology*, vol. 676, pp. 211–223, 2011.

[94] D. W. Thomson, C. P. Bracken, J. M. Szubert, and G. J. Goodall, "On measuring miRNAs after transient transfection of mimics or antisense inhibitors," *PLoS One*, vol. 8, no. 1, article e55214, 2013.

[95] A. Tivnan, W. S. Orr, V. Gubala et al., "Inhibition of neuroblastoma tumor growth by targeted delivery of microRNA-34a using anti-disialoganglioside GD2 coated nanoparticles," *PLoS One*, vol. 7, no. 5, article e38129, 2012.

[96] A. Esquela-Kerscher, P. Trang, J. F. Wiggins et al., "The *let-7* microRNA reduces tumor growth in mouse models of lung cancer," *Cell Cycle*, vol. 7, no. 6, pp. 759–764, 2008.

[97] K. H. Chung, C. C. Hart, S. Al-Bassam et al., "Polycistronic RNA polymerase II expression vectors for RNA interference based on BIC/miR-155," *Nucleic Acids Research*, vol. 34, no. 7, article e53, 2006.

[98] C. L. Esposito, L. Cerchia, S. Catuogno et al., "Multifunctional aptamer-miRNA conjugates for targeted cancer therapy," *Molecular Therapy*, vol. 22, no. 6, pp. 1151–1163, 2014.

[99] Y. Wu, M. Crawford, Y. Mao et al., "Therapeutic delivery of microRNA-29b by cationic lipoplexes for lung cancer," *Molecular Therapy - Nucleic Acids*, vol. 2, article e84, 2013.

[100] G. Hutvagner, M. J. Simard, C. C. Mello, and P. D. Zamore, "Sequence-specific inhibition of small RNA function," *PLoS Biology*, vol. 2, no. 4, article E98, 2004.

[101] M. Lindow and S. Kauppinen, "Discovering the first microRNA-targeted drug," *The Journal of Cell Biology*, vol. 199, no. 3, pp. 407–412, 2012.

[102] H. L. Janssen, H. W. Reesink, E. J. Lawitz et al., "Treatment of HCV infection by targeting microRNA," *The New England Journal of Medicine*, vol. 368, no. 18, pp. 1685–1694, 2013.

[103] M. S. Ebert, J. R. Neilson, and P. A. Sharp, "MicroRNA sponges: competitive inhibitors of small RNAs in mammalian cells," *Nature Methods*, vol. 4, no. 9, pp. 721–726, 2007.

[104] Z. Du, T. Sun, E. Hacisuleyman et al., "Integrative analyses reveal a long noncoding RNA-mediated sponge regulatory network in prostate cancer," *Nature Communications*, vol. 7, article 10982, 2016.

[105] P. M. Harrison, D. Zheng, Z. Zhang, N. Carriero, and M. Gerstein, "Transcribed processed pseudogenes in the human genome: an intermediate form of expressed retrosequence lacking protein-coding ability," *Nucleic Acids Research*, vol. 33, no. 8, pp. 2374–2383, 2005.

[106] S. Melo, A. Villanueva, C. Moutinho et al., "Small molecule enoxacin is a cancer-specific growth inhibitor that acts by enhancing TAR RNA-binding protein 2-mediated microRNA processing," *Proceedings of the National Academy of Sciences of the United States of America*, vol. 108, no. 11, pp. 4394–4399, 2011.

[107] K. Watashi, M. L. Yeung, M. F. Starost, R. S. Hosmane, and K. T. Jeang, "Identification of small molecules that suppress microRNA function and reverse tumorigenesis," *The Journal of Biological Chemistry*, vol. 285, no. 32, pp. 24707–24716, 2010.

[108] A. Aravin, D. Gaidatzis, S. Pfeffer et al., "A novel class of small RNAs bind to MILI protein in mouse testes," *Nature*, vol. 442, no. 7099, pp. 203–207, 2006.

[109] A. Girard, R. Sachidanandam, G. J. Hannon, and M. A. Carmell, "A germline-specific class of small RNAs binds mammalian piwi proteins," *Nature*, vol. 442, no. 7099, pp. 199–202, 2006.

[110] S. T. Grivna, E. Beyret, Z. Wang, and H. Lin, "A novel class of small RNAs in mouse spermatogenic cells," *Genes & Development*, vol. 20, no. 13, pp. 1709–1714, 2006.

[111] T. Watanabe, A. Takeda, T. Tsukiyama et al., "Identification and characterization of two novel classes of small RNAs in the mouse germline: retrotransposon-derived siRNAs in oocytes and germline small RNAs in testes," *Genes & Development*, vol. 20, no. 13, pp. 1732–1743, 2006.

[112] H. Lin and A. C. Spradling, "A novel group of pumilio mutations affects the asymmetric division of germline stem cells in the *Drosophila* ovary," *Development*, vol. 124, no. 12, pp. 2463–2476, 1997.

[113] D. N. Cox, A. Chao, J. Baker, L. Chang, D. Qiao, and H. Lin, "A novel class of evolutionarily conserved genes defined by piwi are essential for stem cell self-renewal," *Genes & Development*, vol. 12, no. 23, pp. 3715 3727, 1998.

[114] K. Saito, K. M. Nishida, T. Mori et al., "Specific association of piwi with rasiRNAs derived from retrotransposon and heterochromatic regions in the *Drosophila* genome," *Genes & Development*, vol. 20, no. 16, pp. 2214–2222, 2006.

[115] J. Brennecke, A. A. Aravin, A. Stark et al., "Discrete small RNA-generating loci as master regulators of transposon activity in *Drosophila*," *Cell*, vol. 128, no. 6, pp. 1089–1103, 2007.

[116] Z. Williams, P. Morozov, A. Mihailovic et al., "Discovery and characterization of piRNAs in the human fetal ovary," *Cell Reports*, vol. 13, no. 4, pp. 854–863, 2015.

[117] H. Ha, J. Song, S. Wang et al., "A comprehensive analysis of piRNAs from adult human testis and their relationship with genes and mobile elements," *BMC Genomics*, vol. 15, no. 1, p. 545, 2014.

[118] P. Krishnan, S. Ghosh, B. Wang et al., "Profiling of small nucleolar RNAs by next generation sequencing: potential new players for breast cancer prognosis," *PLoS One*, vol. 11, no. 9, article e0162622, 2016.

[119] P. Krishnan, S. Ghosh, B. Wang et al., "Genome-wide profiling of transfer RNAs and their role as novel prognostic markers for breast cancer," *Scientific Reports*, vol. 6, no. 1, article 32843, 2016.

[120] V. D. Martinez, E. A. Vucic, K. L. Thu et al., "Unique somatic and malignant expression patterns implicate PIWI-interacting RNAs in cancer-type specific biology," *Scientific Reports*, vol. 5, no. 1, article 10423, 2015.

[121] X. He, X. Chen, X. Zhang et al., "An Lnc RNA (GAS5)/SnoRNA-derived piRNA induces activation of TRAIL gene by site-specifically recruiting MLL/COMPASS-like complexes," *Nucleic Acids Research*, vol. 43, no. 7, pp. 3712–3725, 2015.

[122] S. P. Keam, P. E. Young, A. L. McCorkindale et al., "The human piwi protein Hiwi2 associates with tRNA-derived piRNAs in somatic cells," *Nucleic Acids Research*, vol. 42, no. 14, pp. 8984–8995, 2014.

[123] F. Zhong, N. Zhou, K. Wu et al., "A SnoRNA-derived piRNA interacts with human interleukin-4 pre-mRNA and induces

its decay in nuclear exosomes," *Nucleic Acids Research*, vol. 43, no. 21, pp. 10474–10491, 2015.

[124] D. N. Cox, A. Chao, and H. Lin, "Piwi encodes a nucleoplasmic factor whose activity modulates the number and division rate of germline stem cells," *Development*, vol. 127, no. 3, pp. 503–514, 2000.

[125] T. Thomson and H. Lin, "The biogenesis and function of PIWI proteins and piRNAs: progress and prospect," *Annual Review of Cell and Developmental Biology*, vol. 25, no. 1, pp. 355–376, 2009.

[126] Y. Mei, D. Clark, and L. Mao, "Novel dimensions of piRNAs in cancer," *Cancer Letters*, vol. 336, no. 1, pp. 46–52, 2013.

[127] M. J. Luteijn and R. F. Ketting, "PIWI-interacting RNAs: from generation to transgenerational epigenetics," *Nature Reviews Genetics*, vol. 14, no. 8, pp. 523–534, 2013.

[128] C. B. Assumpcao, D. Q. Calcagno, T. M. Araujo et al., "The role of piRNA and its potential clinical implications in cancer," *Epigenomics*, vol. 7, no. 6, pp. 975–984, 2015.

[129] P. D. Zamore, "Somatic piRNA biogenesis," *The EMBO Journal*, vol. 29, no. 19, pp. 3219–3221, 2010.

[130] E. M. Weick and E. A. Miska, "piRNAs: from biogenesis to function," *Development*, vol. 141, no. 18, pp. 3458–3471, 2014.

[131] E. F. Roovers, D. Rosenkranz, M. Mahdipour et al., "Piwi proteins and piRNAs in mammalian oocytes and early embryos," *Cell Reports*, vol. 10, no. 12, pp. 2069–2082, 2015.

[132] A. N. Harris and P. M. Macdonald, "*Aubergine* encodes a *Drosophila* polar granule component required for pole cell formation and related to eIF2C," *Development*, vol. 128, no. 14, pp. 2823–2832, 2001.

[133] H. B. Megosh, D. N. Cox, C. Campbell, and H. Lin, "The role of PIWI and the miRNA machinery in *Drosophila* germline determination," *Current Biology*, vol. 16, no. 19, pp. 1884–1894, 2006.

[134] T. Thomson, N. Liu, A. Arkov, R. Lehmann, and P. Lasko, "Isolation of new polar granule components in *Drosophila* reveals P body and ER associated proteins," *Mechanisms of Development*, vol. 125, no. 9-10, pp. 865–873, 2008.

[135] M. A. Carmell, A. Girard, H. J. van de Kant et al., "MIWI2 is essential for spermatogenesis and repression of transposons in the mouse male germline," *Developmental Cell*, vol. 12, no. 4, pp. 503–514, 2007.

[136] W. Deng and H. Lin, "Miwi, a murine homolog of piwi, encodes a cytoplasmic protein essential for spermatogenesis," *Developmental Cell*, vol. 2, no. 6, pp. 819–830, 2002.

[137] T. Schupbach and E. Wieschaus, "Female sterile mutations on the second chromosome of *Drosophila* melanogaster. II. Mutations blocking oogenesis or altering egg morphology," *Genetics*, vol. 129, no. 4, pp. 1119–1136, 1991.

[138] P. W. Reddien, N. J. Oviedo, J. R. Jennings, J. C. Jenkin, and A. Sanchez Alvarado, "SMEDWI-2 is a PIWI-like protein that regulates planarian stem cells," *Science*, vol. 310, no. 5752, pp. 1327–1330, 2005.

[139] D. Palakodeti, M. Smielewska, Y. C. Lu, G. W. Yeo, and B. R. Graveley, "The PIWI proteins SMEDWI-2 and SMEDWI-3 are required for stem cell function and piRNA expression in planarians," *RNA*, vol. 14, no. 6, pp. 1174–1186, 2008.

[140] K. Mochizuki, N. A. Fine, T. Fujisawa, and M. A. Gorovsky, "Analysis of a piwi-related gene implicates small RNAs in genome rearrangement in tetrahymena," *Cell*, vol. 110, no. 6, pp. 689–699, 2002.

[141] B. Brower-Toland, S. D. Findley, L. Jiang et al., "*Drosophila* PIWI associates with chromatin and interacts directly with HP1a," *Genes & Development*, vol. 21, no. 18, pp. 2300–2311, 2007.

[142] M. Pal-Bhadra, B. A. Leibovitch, S. G. Gandhi et al., "Heterochromatic silencing and HP1 localization in *Drosophila* are dependent on the RNAi machinery," *Science*, vol. 303, no. 5658, pp. 669–672, 2004.

[143] H. Yin and H. Lin, "An epigenetic activation role of piwi and a piwi-associated piRNA in *Drosophila* melanogaster," *Nature*, vol. 450, no. 7167, pp. 304–308, 2007.

[144] E. Sarot, G. Payen-Groschene, A. Bucheton, and A. Pelisson, "Evidence for a piwi-dependent RNA silencing of the gypsy endogenous retrovirus by the *Drosophila* melanogaster flamenco gene," *Genetics*, vol. 166, no. 3, pp. 1313–1321, 2004.

[145] S. T. Grivna, B. Pyhtila, and H. Lin, "MIWI associates with translational machinery and PIWI-interacting RNAs (piRNAs) in regulating spermatogenesis," *Proceedings of the National Academy of Sciences of the United States of America*, vol. 103, no. 36, pp. 13415–13420, 2006.

[146] T. Esposito, S. Magliocca, D. Formicola, and F. Gianfrancesco, "piR_015520 belongs to piwi-associated RNAs regulates expression of the human melatonin receptor 1A gene," *PLoS One*, vol. 6, no. 7, article e22727, 2011.

[147] H. Chu, G. Hui, L. Yuan et al., "Identification of novel piRNAs in bladder cancer," *Cancer Letters*, vol. 356, no. 2, Part B, pp. 561–567, 2015.

[148] A. Hashim, F. Rizzo, G. Marchese et al., "RNA sequencing identifies specific PIWI-interacting small non-coding RNA expression patterns in breast cancer," *Oncotarget*, vol. 5, no. 20, pp. 9901–9910, 2014.

[149] L. Peng, L. Song, C. Liu et al., "piR-55490 inhibits the growth of lung carcinoma by suppressing mTOR signaling," *Tumour Biology*, vol. 37, no. 2, pp. 2749–2756, 2016.

[150] T. Watanabe and H. Lin, "Posttranscriptional regulation of gene expression by piwi proteins and piRNAs," *Molecular Cell*, vol. 56, no. 1, pp. 18–27, 2014.

[151] E. J. Lee, S. Banerjee, H. Zhou et al., "Identification of piRNAs in the central nervous system," *RNA*, vol. 17, no. 6, pp. 1090–1099, 2011.

[152] Q. Yang, J. Hua, L. Wang et al., "MicroRNA and piRNA profiles in normal human testis detected by next generation sequencing," *PLoS One*, vol. 8, no. 6, article e66809, 2013.

[153] C. A. Raabe, T. H. Tang, J. Brosius, and T. S. Rozhdestvensky, "Biases in small RNA deep sequencing data," *Nucleic Acids Research*, vol. 42, no. 3, pp. 1414–1426, 2014.

[154] J. Busch, B. Ralla, M. Jung et al., "Piwi-interacting RNAs as novel prognostic markers in clear cell renal cell carcinomas," *Journal of Experimental & Clinical Cancer Research*, vol. 34, no. 1, p. 61, 2015.

[155] M. E. Reeves, M. Firek, A. Jliedi, and Y. G. Amaar, "Identification and characterization of RASSF1C piRNA target genes in lung cancer cells," *Oncotarget*, vol. 8, no. 21, pp. 34268–34282, 2017.

[156] I. S. Henaoui, C. Jacovetti, I. Guerra Mollet et al., "PIWI-interacting RNAs as novel regulators of pancreatic beta cell function," *Diabetologia*, vol. 60, no. 10, pp. 1977–1986, 2017.

[157] R. J. Ross, M. M. Weiner, and H. Lin, "PIWI proteins and PIWI-interacting RNAs in the soma," *Nature*, vol. 505, no. 7483, pp. 353–359, 2014.

[158] F. Cichocki, T. Lenvik, N. Sharma, G. Yun, S. K. Anderson, and J. S. Miller, "Cutting edge: KIR antisense transcripts are processed into a 28-base PIWI-like RNA in human NK cells," *Journal of Immunology*, vol. 185, no. 4, pp. 2009-2012, 2010.

[159] M. Halic and D. Moazed, "Transposon silencing by piRNAs," *Cell*, vol. 138, no. 6, pp. 1058–1060, 2009.

[160] K. W. Ng, C. Anderson, E. A. Marshall et al., "Piwi-interacting RNAs in cancer: emerging functions and clinical utility," *Molecular Cancer*, vol. 15, no. 1, p. 5, 2016.

[161] T. B. Rounge, K. Furu, R. I. Skotheim, T. B. Haugen, T. Grotmol, and E. Enerly, "Profiling of the small RNA populations in human testicular germ cell tumors shows global loss of piRNAs," *Molecular Cancer*, vol. 14, no. 1, p. 153, 2015.

[162] P. Krishnan, S. Ghosh, K. Graham, J. R. Mackey, O. Kovalchuk, and S. Damaraju, "Piwi-interacting RNAs and PIWI genes as novel prognostic markers for breast cancer," *Oncotarget*, vol. 7, no. 25, pp. 37944–37956, 2016.

[163] C. D. Malone, J. Brennecke, M. Dus et al., "Specialized piRNA pathways act in germline and somatic tissues of the *Drosophila* ovary," *Cell*, vol. 137, no. 3, pp. 522–535, 2009.

[164] J. Cheng, J. M. Guo, B. X. Xiao et al., "piRNA, the new non-coding RNA, is aberrantly expressed in human cancer cells," *Clinica Chimica Acta*, vol. 412, no. 17-18, pp. 1621–1625, 2011.

[165] I. Daugaard, M. T. Veno, Y. Yan et al., "Small RNA sequencing reveals metastasis-related microRNAs in lung adenocarcinoma," *Oncotarget*, vol. 8, no. 16, pp. 27047–27061, 2017.

[166] Y. Wang, T. Gable, M. Z. Ma et al., "A piRNA-like small RNA induces chemoresistance to cisplatin-based therapy by inhibiting apoptosis in lung squamous cell carcinoma," *Molecular Therapy - Nucleic Acids*, vol. 6, pp. 269–278, 2017.

[167] L. Cui, Y. Lou, X. Zhang et al., "Detection of circulating tumor cells in peripheral blood from patients with gastric cancer using piRNAs as markers," *Clinical Biochemistry*, vol. 44, no. 13, pp. 1050–1057, 2011.

[168] Y. Li, X. Wu, H. Gao et al., "Piwi-interacting RNAs (piRNAs) are dysregulated in renal cell carcinoma and associated with tumor metastasis and cancer-specific survival," *Molecular Medicine*, vol. 21, no. 1, pp. 1–8, 2015.

[169] V. D. Martinez, K. S. Enfield, D. A. Rowbotham, and W. L. Lam, "An atlas of gastric PIWI-interacting RNA transcriptomes and their utility for identifying signatures of gastric cancer recurrence," *Gastric Cancer*, vol. 19, no. 2, pp. 660–665, 2016.

[170] X. Yang, Y. Cheng, Q. Lu, J. Wei, H. Yang, and M. Gu, "Detection of stably expressed piRNAs in human blood," *International Journal of Clinical and Experimental Medicine*, vol. 8, no. 8, pp. 13353–8, 2015.

[171] T. Yuan, X. Huang, M. Woodcock et al., "Plasma extracellular RNA profiles in healthy and cancer patients," *Scientific Reports*, vol. 6, no. 1, article 19413, 2016.

[172] Y. Okugawa, Y. Toiyama, S. Toden et al., "Clinical significance of SNORA42 as an oncogene and a prognostic biomarker in colorectal cancer," *Gut*, vol. 66, no. 1, pp. 107–117, 2017.

[173] R. Wei, G. L. Huang, M. Y. Zhang et al., "Clinical significance and prognostic value of microRNA expression signatures in hepatocellular carcinoma," *Clinical Cancer Research*, vol. 19, no. 17, pp. 4780–4791, 2013.

[174] R. Sanz-Pamplona, A. Berenguer, D. Cordero et al., "Aberrant gene expression in mucosa adjacent to tumor reveals a molecular crosstalk in colon cancer," *Molecular Cancer*, vol. 13, no. 1, p. 46, 2014.

[175] U. R. Chandran, R. Dhir, C. Ma, G. Michalopoulos, M. Becich, and J. Gilbertson, "Differences in gene expression in prostate cancer, normal appearing prostate tissue adjacent to cancer and prostate tissue from cancer free organ donors," *BMC Cancer*, vol. 5, no. 1, p. 45, 2005.

[176] B. J. Braakhuis, M. P. Tabor, J. A. Kummer, C. R. Leemans, and R. H. Brakenhoff, "A genetic explanation of Slaughter's concept of field cancerization: evidence and clinical implications," *Cancer Research*, vol. 63, no. 8, pp. 1727–1730, 2003.

[177] L. Hawthorn, L. Lan, and W. Mojica, "Evidence for field effect cancerization in colorectal cancer," *Genomics*, vol. 103, no. 2-3, pp. 211–221, 2014.

[178] M. A. Troester, M. H. Lee, M. Carter et al., "Activation of host wound responses in breast cancer microenvironment," *Clinical Cancer Research*, vol. 15, no. 22, pp. 7020–7028, 2009.

[179] M. Radovich, S. E. Clare, R. Atale et al., "Characterizing the heterogeneity of triple-negative breast cancers using microdissected normal ductal epithelium and RNA-sequencing," *Breast Cancer Research and Treatment*, vol. 143, no. 1, pp. 57–68, 2014.

[180] B. N. Hannafon, P. Sebastiani, A. de las Morenas, J. Lu, and C. L. Rosenberg, "Expression of microRNA and their gene targets are dysregulated in preinvasive breast cancer," *Breast Cancer Research*, vol. 13, no. 2, p. R24, 2011.

[181] E. Wagner, Y. Liu, B. Schneider, A. M. Storniolo, J. Han, and C. He, "Abstract 3268: differences in expression of lncRNAs in breast tumor, adjacent normal-appearing breast tissue, and normal breast tissue from healthy donors," *Cancer Research.*, vol. 74, Supplement 19, p. 3268, 2014.

[182] P. Gasparini, L. Cascione, M. Fassan et al., "MicroRNA expression profiling identifies a four microRNA signature as a novel diagnostic and prognostic biomarker in triple negative breast cancers," *Oncotarget*, vol. 5, no. 5, pp. 1174–1184, 2014.

[183] K. Kleivi Sahlberg, G. Bottai, B. Naume et al., "A serum microRNA signature predicts tumor relapse and survival in triple-negative breast cancer patients," *Clinical Cancer Research*, vol. 21, no. 5, pp. 1207–1214, 2015.

[184] J. T. Leek, R. B. Scharpf, H. C. Bravo et al., "Tackling the widespread and critical impact of batch effects in high-throughput data," *Nature Reviews. Genetics*, vol. 11, no. 10, pp. 733–739, 2010.

[185] S. Eminaga, D. C. Christodoulou, F. Vigneault, G. M. Church, and J. G. Seidman, "Quantification of microRNA expression with next-generation sequencing," *Current Protocols in Molecular Biology*, vol. 4, p. 17, 2013, Chapter 4:Unit.

[186] S. Nygaard, A. Jacobsen, M. Lindow et al., "Identification and analysis of miRNAs in human breast cancer and teratoma samples using deep sequencing," *BMC Medical Genomics*, vol. 2, no. 1, p. 35, 2009.

[187] K. Schee, S. Lorenz, M. M. Worren et al., "Deep sequencing the microRNA transcriptome in colorectal cancer," *PLoS One*, vol. 8, no. 6, article e66165, 2013.

9

Ensemble Methods with Voting Protocols Exhibit Superior Performance for Predicting Cancer Clinical Endpoints and Providing More Complete Coverage of Disease-Related Genes

Runyu Jing,[1] Yu Liang,[2] Yi Ran,[3] Shengzhong Feng,[1] Yanjie Wei,[1] and Li He[3]

[1]*Shenzhen Institutes of Advanced Technology, Chinese Academy of Sciences, Shenzhen 518055, China*
[2]*College of Chemistry, Sichuan University, Chengdu 610064, China*
[3]*Biogas Appliance Quality Supervision and Inspection Center, Biogas Institute of Ministry of Agriculture, Chengdu, Sichuan, China*

Correspondence should be addressed to Yanjie Wei; yj.wei@siat.ac.cn and Li He; helibiogas@126.com

Academic Editor: Zhichao Liu

In genetic data modeling, the use of a limited number of samples for modeling and predicting, especially well below the attribute number, is difficult due to the enormous number of genes detected by a sequencing platform. In addition, many studies commonly use machine learning methods to evaluate genetic datasets to identify potential disease-related genes and drug targets, but to the best of our knowledge, the information associated with the selected gene set was not thoroughly elucidated in previous studies. To identify a relatively stable scheme for modeling limited samples in the gene datasets and reveal the information that they contain, the present study first evaluated the performance of a series of modeling approaches for predicting clinical endpoints of cancer and later integrated the results using various voting protocols. As a result, we proposed a relatively stable scheme that used a set of methods with an ensemble algorithm. Our findings indicated that the ensemble methodologies are more reliable for predicting cancer prognoses than single machine learning algorithms as well as for gene function evaluating. The ensemble methodologies provide a more complete coverage of relevant genes, which can facilitate the exploration of cancer mechanisms and the identification of potential drug targets.

1. Introduction

With the development of genetic sequencing technology, genetic information could be recorded as gene expression data. Data mining, such as using machine learning methods, is commonly used to reveal latent correlations between diseases and gene expression. Supervised or semisupervised machine learning algorithms have been proposed to predict the clinical outcomes of cancers [1–4] in the context of tumorigenesis. The support vector machine (SVM) [5–8] and artificial neural network (ANN) [9–11] algorithms were the most commonly used approaches for predicting prognoses. In addition, the Bayesian probability model [12–14] and the fuzzy neural network [15] were also used for cancer prognosis prediction. The microarray quality control (MAQC) project thoroughly investigated the performance of models for the prediction of clinical outcomes of breast cancer, multiple myeloma, and neuroblastoma and were common practices for microarray-based model construction and validation [16]. The network-based approaches have seen a recent widespread use for the identification of cancer-related genes and have revealed the molecular mechanisms of various cancers [17–22]. However, to the best of our knowledge, no studies have examined multiple algorithms with two different kinds of expression data and their ensemble performance with a limited number of samples, which might be crucial when using them in practical. In most cases, the number of available samples is restricted due to the

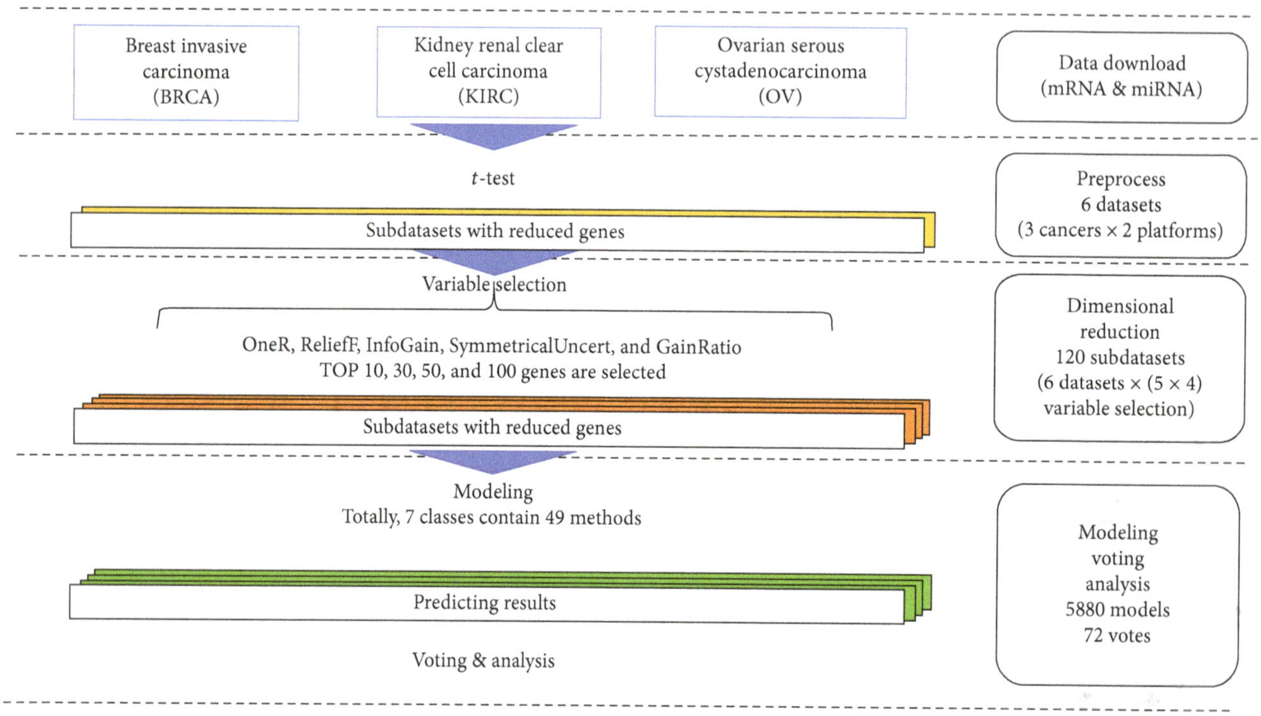

Figure 1: Work flow of the whole process. First, the datasets were downloaded from the GDC (Genomic Data Commons) database. Next, the downloaded mRNA and microRNA sequencing data are united by the usable information. The t-test was used afterwards to determine the significantly expressed genes. Five selection methods were used to select the cancer-associated genes and the subdatasets generated according to the ranks. Finally, the prediction results were integrated by a voting protocol. Note that every subdataset was divided into two pieces for cross-validation and independent test in the ratio 4 : 1 before variable selection. Only the datasets for cross-validation will be used for variable selection and modeling.

cost, privacy, and other reasons. A limited sample number causes a predictive model to be more sensitive to the dataset distribution, and the lack of prior knowledge simultaneously reduces the overall predicting performance which could be reflected by MCC (Matthews correlation coefficient). Moreover, a single machine learning algorithm provides insufficient coverage of the disease-related genes because it only uses genes that show the greatest difference of expression profiles between the phenotypic statuses compared when the genes have a similar function. Thus, identifying a stable predictive model using a limited number of samples becomes a challenge.

To address this problem, the present study thoroughly investigated the performance of single models among different datasets and proposed a strategy to combine multiple predictive models as well as the datasets into a final ensemble for clinical prediction. Compared with a single machine learning algorithm, an ensemble scheme could not only perform more reliably when predicting clinical endpoints but could also provide broader coverage of disease-related genes, which will be beneficial for further downstream analysis in such applications as the identification of potential drugs.

2. Materials and Methods

The workflow of this study is listed in Figure 1. The datasets were carefully generated such that the scale and representation of the samples in the different datasets are consistent such that the predictions were comparable. This section describes the data and methods used.

2.1. GDC Data. All data were downloaded from the NCI's Genomic Data Commons (GDC) [23] by using the official web-based API (https://gdc.cancer.gov/developers/gdc-application-programming-interface-api). The genomic data were from the official normalized microRNA and RNA sequence expression data because it was restricted by multiplatform coverage and accessibility; a portion of the cancer data was excluded from this study. For example, the number of available samples of neuroblastoma in GDC is 1127 (this number might change if the database is updated), but only approximately half of these samples have associated RNA sequence data, and no microRNA sequence data are available. Finally, the freely accessible data for breast invasive carcinoma (BRCA), ovarian serous cystadenocarcinoma (OV), and kidney renal clear cell carcinoma (KIRC) were downloaded and used for modeling. The clinical information was downloaded in the XML format. The relationships of samples from different platforms (such as mRNA, microRNA, and clinical) were identified by the official MetaData file in the JSON format. The detailed distribution can be found in Table 1.

2.2. Preprocessing. Since we wanted to make the results from two sequencing platforms (e.g., mRNA and microRNA) comparable and able for voting at last, the selection of samples was determined by the integration of different platforms.

TABLE 1: Scales of the three datasets.

Disease name	Sample number in modeling dataset	Sample number in independent test dataset	Number of kept genes	
BRCA	558	141	mRNA	24585
			miRNA	722
KIRC	112	29	mRNA	9119
			miRNA	190
OV	66	18	mRNA	4390
			miRNA	83

Only samples that contained clinical information and expression data for both mRNA and microRNA were retained for subsequent analysis.

The sample label was determined by the clinical information. For OV and KIRC, the label was determined by the survival time. A sample was deemed positive if the recorded survival time was larger than one year and if the patient was still alive (based on the clinical data). Similarly, a sample was deemed negative if the associated patient was dead and if the recorded survival time was less than one year. The sample label for BRCA was determined by the estrogen receptor (ER) status (negative or positive), which was also recorded in the clinical XML file. Therefore, if the required information for a sample could not be found in the clinical data or its status did not satisfy the criteria (e.g., the patient was alive but the survival time was less than one year), it was excluded.

The sample number was reduced by the clinical information. Student's t-test was used to subsequently reduce the mRNA and microRNA numbers. Only genes that had significant expression with a p value less than 0.05 were retained in a dataset. The ratio of positive to negative samples was kept in an appropriate range to reduce classification bias. In this study, the range of this ratio was 0.5 to 2. For example, if a dataset contained 22 positive and 50 negative samples, 6 negative samples were randomly removed to adjust the ratio so it fell within the required range. The eliminated datasets were divided into two parts for cross-validation and independent tests in a 4 : 1 ratio. The scale of the datasets is listed in Table 1. Only the datasets for cross-validation will be used for variable selection and the 5-fold cross-validation; the datasets for independent testing will not participate in modeling. And only the independent prediction will be used for further ensemble analysis and comparison.

2.3. Machine Learning Methods. 49 modeling methods in WEKA [24] (version 3.8.1) were investigated in this study. The methods were divided into seven different classes by the developers of WEKA according to specific features of the methods (Table 2). The different method classes had different features. In the *functions* class, most of the methods use a functional solution for modeling the data, such as *LibSVM* [25] and logistic regression [26], and in most cases, few mechanisms are available for ensemble learning, such as voting or resampling. However, in the *meta* class, the methods use resampling and voting for classification and regression, and the methods in the other classes are considered model units, such as *AdaBoost* [27] and *Bagging* [28]. The methods from the *bayes* class are from the probability and graph theory, and most of them, including *NaïveBayes* [29], *BayesNetwork* [30], and *BayesianLogisticRegression* [31], are sensitive to sample number. Similarly, methods in the *rules* class use rules (such as decision table) for classification [32]. The methods in the *lazy* class are instance-based and could be optimized for better efficiency using a lazy algorithm [33]. Most of the methods in the *trees* class are based on the classification and regression tree algorithm, but the way they are carried out is different. Many other mechanisms are integrated into the *trees* such as resampling used in a random forest [34]. Finally, the *misc* class contains methods for which it is difficult to assign to another class. Only two methods fell into this class in this study, namely, the *VFI*, an ensemble method based on a voting protocol [35], and the *HyperPipes*, based on an algorithm that finds similarities among attributes. Considering running time, comparability, and reducing the risk of overfitting, only default parameters are used for modeling.

Because the sample number was limited, fewer genes should be considered to avoid overfitting. In this study, five variable selection methods were used for dimensional reduction: *OneR*, *ReliefF*, *InfoGain*, *SymmetricalUncert*, and *GainRatio*. *OneR* is executed by using the *OneR* classifier, which is based on measuring the error between the attributes and the response values [36]. *ReliefF* uses a resampling mechanism for evaluating the attributes [37]. The other methods are from the information theory, and the associated formulas are

$$
\begin{aligned}
\mathrm{GainRatio} = (\mathrm{class}, \mathrm{attribute}) &= \frac{H(\mathrm{class}) - H(\mathrm{class}|\mathrm{attribute})}{H(\mathrm{attribute})}, \\
\mathrm{InfoGain}(\mathrm{class}, \mathrm{attribute}) &= H(\mathrm{class}) - H(\mathrm{class}|\mathrm{attribute}), \\
\mathrm{SymmetricalUncert}(\mathrm{class}, \mathrm{attribute}) & \\
&= 2\frac{H(\mathrm{class}) - H(\mathrm{class}|\mathrm{attribute})}{H(\mathrm{class})} \\
&\quad + H(\mathrm{attribute}),
\end{aligned}
\tag{1}
$$

where the "$H()$" in the formula is the information entropy (Shannon entropy) [38] and "class" denotes the values of a label. According to their formulas, *GainRatio* could be considered as a normalization of *InfoGain*. However, the information entropies of all of the attributes, such as the expression of mRNAs and microRNAs, are different, so both methods are used in this study. By using the ranking mechanism in WEKA, in every subdataset for cross-validation, the attributes can be ranked, and the top 10, 30, 50, and 100 ranked attributes are selected for cross-validation and modeling. Note that since the number of microRNAs in OV is limited (83, which is less than 100), the numbers of the attributes in the subdatasets are 10, 30, 50, and 83. Totally, we investigated a total of 5 variable selection methods × 4 subdatasets × 49 modeling methods = 980 predictive models.

TABLE 2: Methods used.

Class	Method names
bayes	NaiveBayes, BayesianLogisticRegression, BayesNet, ComplementNaiveBayes, DMNBtext, NaiveBayesMultinomial, NaiveBayesMultinomialUpdateable
functions	Logistic, MultilayerPerceptron, RBFNetwork, SimpleLogistic, SPegasos, VotedPerceptron, LibSVM
lazy	IB1, IBk, KStar, LWL
meta	AdaBoostM1, Bagging, Dagging, Decorate, END, FilteredClassifier, LogitBoost, MultiBoostAB, MultiClassClassifier
misc	HyperPipes, VFI
rules	ConjunctiveRule, DecisionTable, DTNB, NNge, OneR, PART, Ridor, ZeroR
trees	ADTree, BFTree, DecisionStump, FT, J48, J48graft, LADTree, LMT, NBTree, RandomForest, RandomTree, REPTree

The names, including the class names, are from WEKA. The results and discussion are based on the classes.

2.4. Integrating the Predictions by Voting. After generating hundreds of models, it is possible to combine their predictions. As previously mentioned, the prediction performances are ranged in the datasets. To integrate the ranged predictions and find a stable modeling method for genetic datasets, we used a voting protocol in this study to identify the datasets.

All of the weights were the same, except in the information theory methods such as *InfoGain*, *SymmetricalUncert*, and *GainRatio*. The weights in the information theory methods were modified to 1/3 when voting the predictions from the subdatasets of the OV-miRNA group according to the prediction distribution. More details may be found in "the coverage and reliability of selected genes" and "the distributions of the predictions."

Three voting schemes were used to arrive at a comprehensive conclusion. All of the methods were first used for voting. Then, some of the methods which performed poorly for all datasets were eliminated. Finally, mRNA and miRNA datasets were combined for voting.

2.5. Measurement Methods. Since there were many predictions, box plots were used to reflect the stability of the different classes. The quartiles Q1 and Q3, the interquartile range (IQR), and the whiskers (the lower whisker is Q1 −1.5 IQR, and upper whisker is Q3 +1.5 IQR) in the box plot are discussed. Two types of box plots were used to present the results in different angles, one based on the WEKA classes of modeling methods and another based on the scale of the subdatasets. Because the ratio of positive to negative samples was biased, the Matthews correlation coefficient (MCC) was used as the criterion for the plots. The MCC is one of the criteria used to evaluate the prediction performance, and the associated formula is

$$\text{MCC} = \frac{\text{TP} \times \text{TN} - \text{FP} \times \text{FN}}{\sqrt{(\text{TP} + \text{FN}) \times (\text{TP} + \text{FP}) \times (\text{TN} + \text{FP}) \times (\text{TN} + \text{FN})}}, \quad (2)$$

where TP, TN, FP, and FN are the number of true-positive predictions, true-negative predictions, false-positive predictions, and false-negative predictions, respectively.

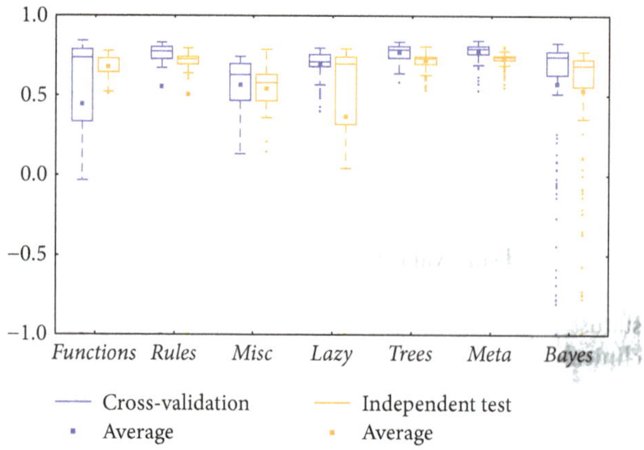

FIGURE 2: MCC of the BRCA-mRNA group by the *functions* class.

3. Results

Figure 1 is a flow chart that shows the modeling and integration of the preprocessed datasets. The associated results are listed below. Since there were 5880 modules in total (6 datasets × 980 predictive models per datasets), figures instead of tables were used to present the results (Figures 2–13). The individual cross-validation and independent test results were together listed in the Supplementary file "ModelingResults.xlsx" (available here).

3.1. Modeling Results. The prediction performance was ascertained in two ways: by the modeling method class defined by WEKA (see Table 2) and by the different subdatasets generated by the different variable selection methods. Therefore, a total of 3 cancers × 2 sequence methods × 2 kinds of plots = 12 figures were generated for the modeling results.

The *meta* class and *trees* class methods performed better than those in other classes for the two types of BRCA genomics data (mRNA in Figure 2 and miRNA in Figure 4), as evidenced by the best medians and averages. The box for the *trees* class had higher whiskers, but the box for the *meta* class had a smaller IQR.

The KIRC mRNA (Figure 6) and miRNA (Figure 8) datasets showed diverse prediction distributions. In the KIRC-mRNA group, the distributions were similar in the two BRCA groups but the *lazy* class performed in a similar manner in the *meta* and *trees* classes. The distributions were

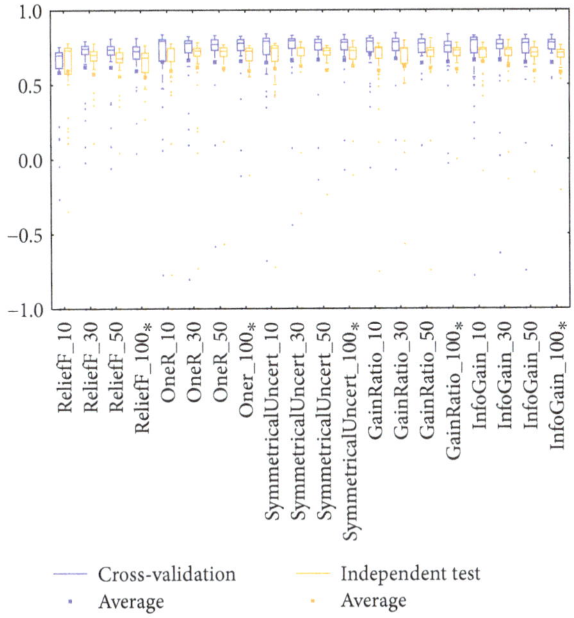

FIGURE 3: MCC of the BRCA-mRNA group by reduced datasets. *Note that the subdatasets from OV-miRNA have at most 83 micro-RNAs and thus the scale "100" of OV means 83.

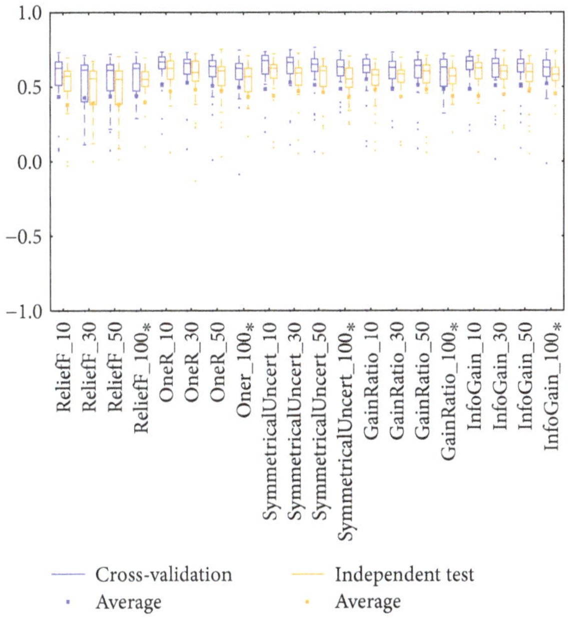

FIGURE 5: MCC of the BRCA-miRNA group by reduced datasets. *Note that the subdatasets from OV-miRNA have at most 83 micro-RNAs and thus the scale "100" of OV means 83.

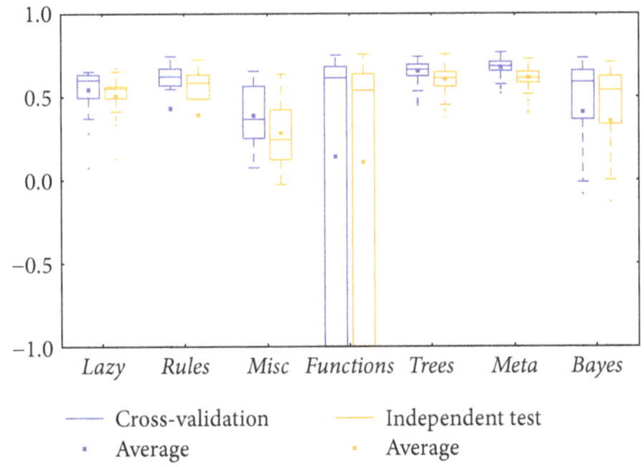

FIGURE 4: MCC of the BRCA-miRNA group by the *functions* class.

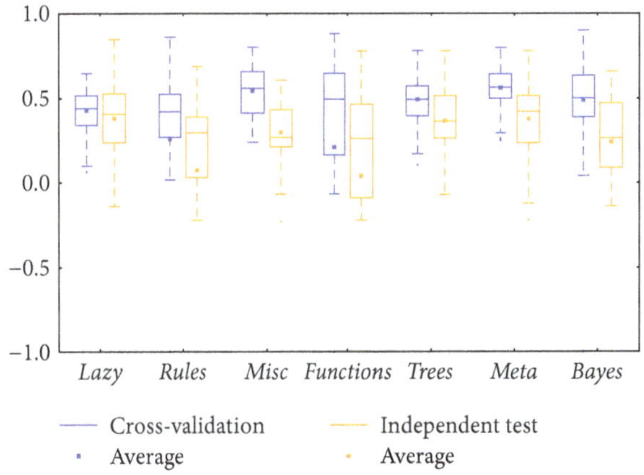

FIGURE 6: MCC of the KIRC-mRNA group by the *functions* class.

totally different in the KIRC-miRNA class. The *misc* class had the best distribution in the box plot, and the others were far worse. However, most of the medians and averages in the KIRC-miRNA group were less than those in the KIRC-mRNA group.

The situation was similar in the OV-mRNA group (Figure 10) and the OV-miRNA group (Figure 12). The *misc* class was best in the OV-mRNA group datasets, but the *meta* class had relatively better prediction performance for the OV-miRNA group. The *rules* class and *functions* class had very poor distributions in both groups.

Comparing between cross-validation and independent test, most of them were similar; the differences between the whiskers were not huge except for two classes: *functions* and *lazy* (could be found in Figures 2, 8, and 12). The *functions* class often has wider ranges between the two whiskers in cross-validation and is still larger than its independent test. On the contrary, the *lazy* class usually has narrower ranges in cross-validation than its independent test.

Distributions based on different datasets were clearer. In most cases, the dataset (mRNA or microRNA) that contained more attributes had a better distribution (i.e., a smaller IQR or a higher median). However, the peak was sometimes found in a smaller dataset. In more detail, the two BRCA groups (Figures 3 and 5) seemed to be insensitive to the attribute number and the distributions especially medians are very similar. The boxes were larger in the KIRC-miRNA group (Figure 9). The prediction results were sensitive to the attribute number, and the upper whisker increased as the attributes increased, but the lower whisker simultaneously decreased. In comparison, the distributions of the boxes in the KIRC-mRNA group (Figure 7) were better when the attribute number increased to 30 and were relatively stable afterwards. Differently in the OV-mRNA (Figure 11) and

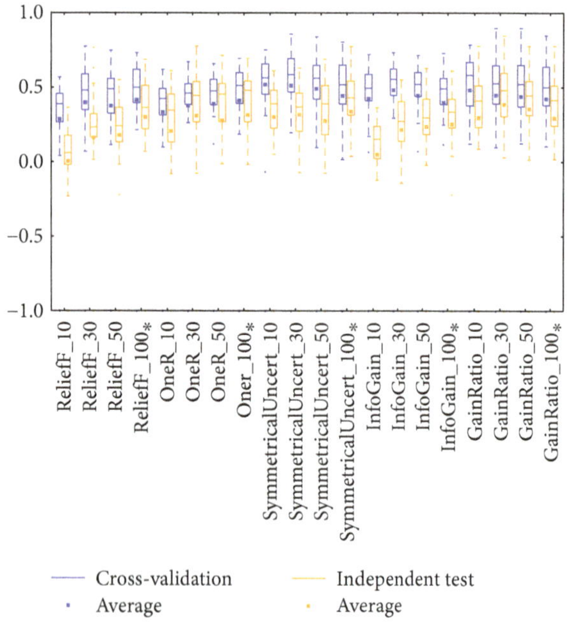

Figure 7: MCC of the KIRC-mRNA group by reduced datasets. *Note that the subdatasets from OV-miRNA have at most 83 micro-RNAs and thus the scale "100" of OV means 83.

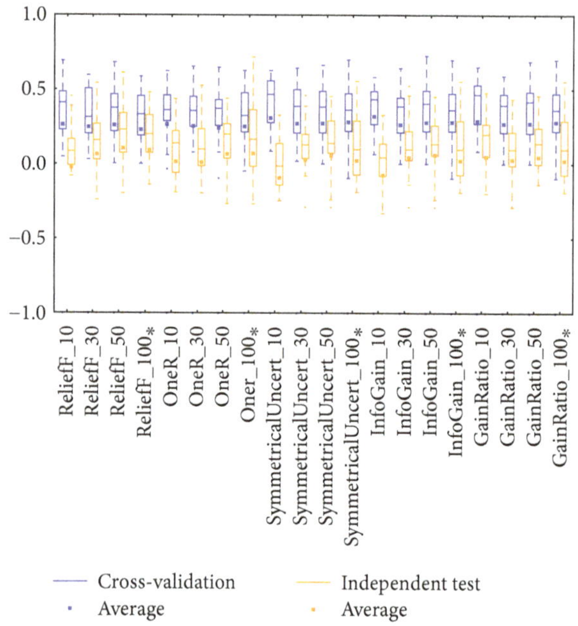

Figure 9: MCC of the KIRC-miRNA group by reduced datasets. *Note that the subdatasets from OV-miRNA have at most 83 micro-RNAs and thus the scale "100" of OV means 83.

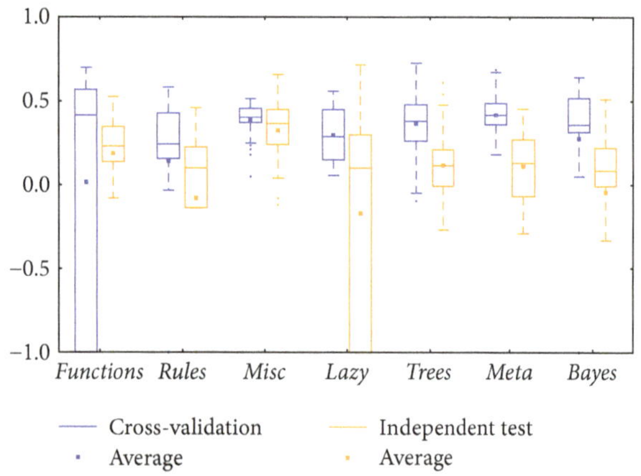

Figure 8: MCC of the KIRC-miRNA group by the *functions* class.

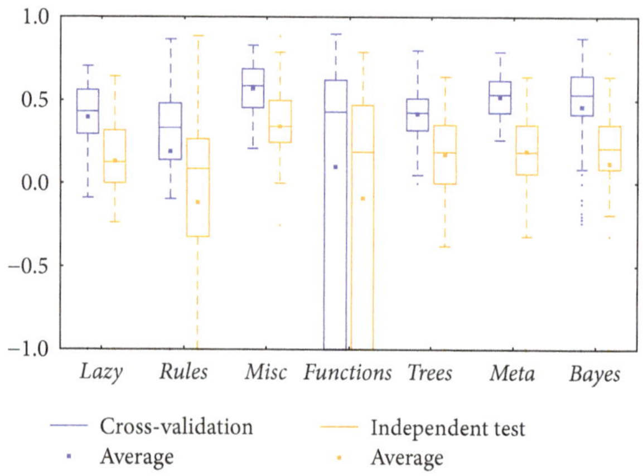

Figure 10: MCC of the OV-mRNA group by the *functions* class.

the OV-miRNA groups (Figure 13), the boxes contracted when attributes were added.

3.2. Voting Results from All Methods. The performance of all of the voted methods is listed in the column "Vote from all methods" of Tables 3, 4, and 5. There is no doubt that the overall voting performance would be lower than optimum, since not all of the voting methods are sufficiently good for the 6 datasets; nevertheless, most of the MCC achieved by voting are better than the average MCC. In addition, a part of the voting performance reached the upper bound (i.e., the maximum), for example, in the *bayes* class in the OV-miRNA group. In more detail, similar to the box plots, the voting performance based on the BRCA datasets was similar and near the upper whisker except for *misc*. The range was larger in the other datasets (e.g., KIRC-mRNA, KIRC-miRNA, OV-mRNA, and OV-miRNA). Different classes in turn, including *bayes*, *functions*, *meta*, *lazy*, *misc*, and *trees*, showed the top three best prediction performance. The *rules* class always ranked lower in the overall voting test, but the *bayes* class showed good voting performance even though its distribution, as indicated by the box plot, was not stable.

3.3. Voting Results from Eliminated Methods. The filtering rule was based on the distribution of the prediction results. Values that fell out of the range indicated by the whiskers in a box plot were considered to be outliers. Similarly, in our study, a method with an MCC below the lower whisker was considered as an outlier. There were 5 value selection methods, and each generated 4 subdatasets. If 6 datasets (3 cancers × 2 sequencing techniques) were considered, a modeling method was used: 5 variable selection methods × 4

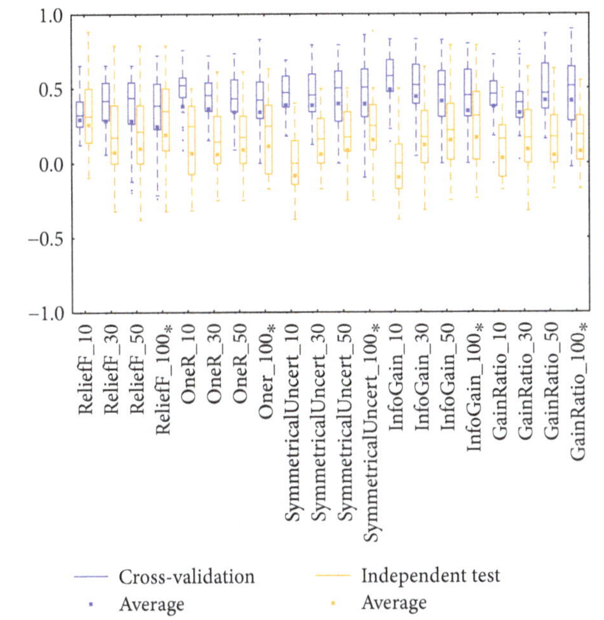

FIGURE 11: MCC of the OV-mRNA group by reduced datasets. *Note that the subdatasets from OV-miRNA have at most 83 micro-RNAs and thus the scale "100" of OV means 83.

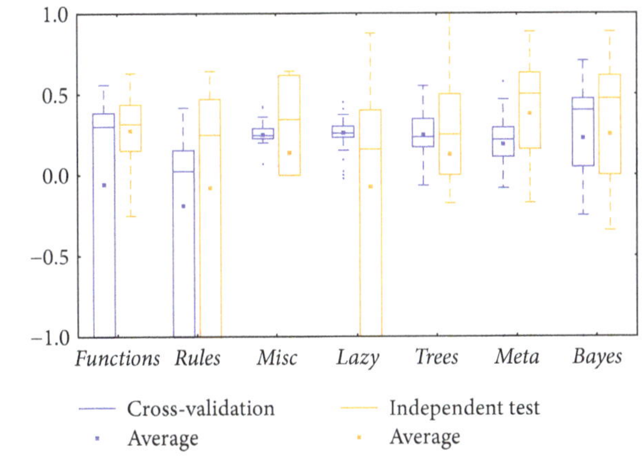

FIGURE 12: MCC of the OV-miRNA group by the *functions* class.

subdatasets × 6 datasets = 120 times. Therefore, if a method was designated as an outlier more than 6 times (5% of 120), it was not considered in this voting test.

The details of eliminated methods are listed in Table 6. The associated prediction performance was listed in the column "Vote after filtering" in Tables 3, 4, and 5. One or more methods were eliminated in all of the classes except the *meta*. It was evident that *ZeroR* had the most counts, and it was accordingly eliminated. Since only default parameters were used, *LibSVM* and *SPegasos*, which could be considered an optimized *SVM*, were designated as outliers with high counts. Similarly, the *BayesianLogisticRegression* method required parameter optimization and was thus eliminated by many counts. No methods in the *meta* class were eliminated, which indicated that methods based on resampling and an ensemble

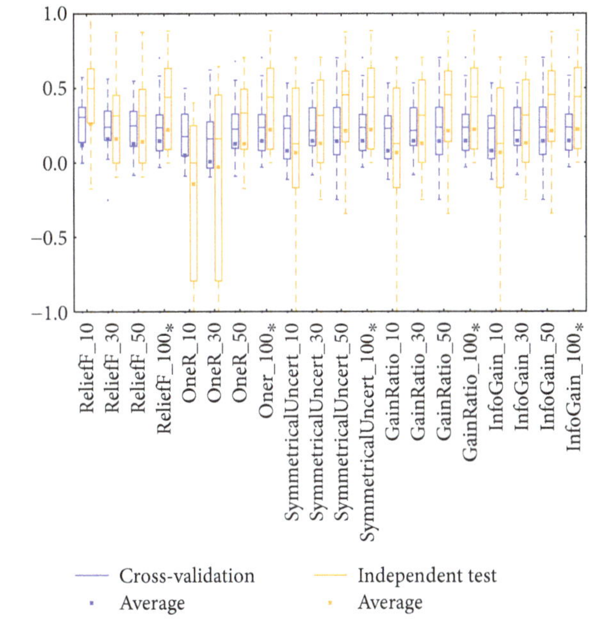

FIGURE 13: MCC of the OV-miRNA group by reduced datasets. In the 12 box plots, the line in the box is the median. The upper and lower boundaries of the box are Q1 and Q3. The boundaries of the dotted line are the whiskers. *The subdatasets from OV-miRNA have at most 83 microRNAs, and thus, the scale "100" of OV means 83. *Note that the subdatasets from OV-miRNA have at most 83 micro-RNAs and thus the scale "100" of OV means 83.

mechanism were stable and could address varied datasets even they were not always the best.

3.4. Voting Results from the Combined Datasets. As mentioned in the Preprocessing, the datasets for one cancer comprised the same samples. Therefore, voting from both mRNA and miRNA datasets is possible. The prediction performance is in the last row of Tables 3, 4, and 5. According to these three tables, the effects of the combination were different in the different classes. No class always benefited from the combination, but the MCCs determined by voting were not worse than the lowest MCC (Tables 4 and 5) and sometimes better than both (Table 3).

4. Discussion

In this section, the modeling results will be discussed. Therefore, the discussion is comprehensive, and the gene selection and the modeling results are discussed separately.

4.1. Coverage of Selected Genes. Five variable selection methods were used in this study, and different mRNA/miRNA datasets were separately generated. If the subdatasets generated are similar, combining multiple variable selection methods is worthless. Therefore, it is necessary to analyze the contents of the subdatasets and determine the similarity of the datasets. Moreover, the importance of the mRNA was evaluated by using a 3rd party database.

TABLE 3: Voting results of BRCA.

Platforms	Class	Vote from all methods	Vote after filtering	MCC avg	MCC max
BRCA-mRNA	*bayes*	0.7620	0.7620	0.5327	0.7766
	functions	0.7252	0.7766	0.3696	0.7938
	lazy	0.7620	0.7423	0.6797	0.7793
	meta	0.7274	0.7274	0.7295	0.8085
	misc	0.5821	0.6466	0.5449	0.7869
	rules	0.7274	0.7274	0.5069	0.7967
	trees	0.7274	0.7274	0.7207	0.8085
	overall	0.7274	0.7274	0.5985	0.8085
BRCA-miRNA	*bayes*	0.7237	0.6895	0.3544	0.7067
	functions	0.6733	0.6908	0.1080	0.7566
	lazy	0.6214	0.6214	0.5067	0.6736
	meta	0.6278	0.6278	0.6142	0.7269
	misc	0.3203	0.5773	0.2816	0.6383
	rules	0.5990	0.6278	0.3933	0.7234
	trees	0.6427	0.6602	0.6066	0.7566
	overall	0.6405	0.6908	0.4427	0.7566
BRCA-mRNA and BRCA-miRNA	*bayes*	0.7423	0.7252	0.4436	0.7766
	functions	0.6555	0.7915	0.2388	0.7938
	lazy	0.7595	0.7595	0.5932	0.7793
	meta	0.7595	0.7595	0.6719	0.8085
	misc	0.5624	0.6756	0.4133	0.7869
	rules	0.7080	0.7746	0.4501	0.7967
	trees	0.7595	0.7595	0.6636	0.8085
	overall	0.7407	0.7746	0.5206	0.8085

All of the measurements in the tables are MCCs, and the vote after filtering is the MCC based on the eliminated methods. The "avg" is the average of the MCCs.

The coverage fraction is the number of the genes which are used more than once of all genes, and accordingly, the formula is

$$\text{Coverage} = \frac{\text{NumOfSelectedGenes} - \text{NumOfIndependentGenes}}{\text{NumOfSelectedGenes}}. \quad (3)$$

According to Tables 7 and 8, the overall coverage was approximately 40% in the datasets from mRNA. However, the overall coverage of the microRNA datasets was much larger, due to dataset scale limitations. Especially in OV, the finally selected microRNAs were the same because there were only 83 microRNAs; therefore, all of them were selected and ranked in the top 100, and the OV-miRNA datasets must be carefully considered when the predictions are integrated. Conversely, the coverage of the two mRNA subdatasets was not large. According to the statistical results for the mRNA datasets, the frequency of shared mRNA increased as the coverage fraction increased. The shared mRNAs were usually ranked lower by the variable selection methods, which means the most commonly used mRNAs were not recognized as crucial genes that would be ranked higher. The reason might be an insufficient sample number to determine the relationships, since there are many mRNAs. Another probable reason is that the gene expression data could not be directly associated with a disease since cancer is a complex group of diseases, so not only one or a small number of gene are correlated, such as in coexpression [39].

4.2. Reliability of the Selected mRNAs. To determine if the shared mRNA is important, the Human Protein Atlas database [40] was used to evaluate its importance. The Human Protein Atlas contains a map from mRNA to tissue generated by an antibody-based approach. The gene reliability is recorded in the database, so that the mRNAs selected in this study could be evaluated by using the database. The evaluation had two steps. First, the fraction of Hits/Total in Table 9 was used to determine the number of selected mRNAs in the dataset. Next, the mRNAs' reliability was verified by using the associated record in the "reliability (IH)" table. An mRNA was considered reliable only if the record was designated as "approved" or "supported."

Table 9 shows that the fractions of hits in the database were approximately 60%, 55%, and 35% for BRCA, KIRC, and OV, respectively. However, the reliable hits ranged near 70% for the three datasets. The reliabilities based on more than 10 hits were always around 70% no matter which dataset was used. Few differences in reliability were found for the three datasets. The fraction of the reliable mRNAs from OV is relatively larger, and smaller for BRCA. Since the number of records in the Human Protein Atlas is still limited, a more

Table 4: Voting results of KIRC.

Platforms	Class	Vote from all methods	Vote after filtering	MCC avg	MCC max
KIRC-mRNA	*bayes*	0.4216	0.3672	0.2401	0.6590
	functions	0.6292	0.7162	0.0399	0.7785
	lazy	0.4682	0.6292	0.3801	0.8474
	meta	0.5261	0.5261	0.3780	0.7785
	misc	0.4176	0.4105	0.2991	0.6058
	rules	0.4385	0.6110	0.0723	0.6885
	trees	0.5421	0.5421	0.3649	0.7785
	overall	0.6110	0.5421	0.2535	0.8474
KIRC-miRNA	*bayes*	0.2368	0.1805	−0.0433	0.5131
	functions	0.0889	0.0530	−0.1698	0.7162
	lazy	0.4371	0.3410	0.1865	0.5249
	meta	0.1667	0.1667	0.1105	0.4542
	misc	0.4606	0.4795	0.3230	0.6590
	rules	0.0889	0.2689	−0.0795	0.4606
	trees	0.0530	0.1667	0.1165	0.6110
	overall	0.1667	0.1667	0.0323	0.7162
KIRC-mRNA and KIRC-miRNA	*bayes*	0.4795	0.4795	0.0984	0.6590
	functions	0.2605	0.6885	−0.0649	0.7785
	lazy	0.5514	0.4371	0.2833	0.8474
	meta	0.2300	0.2300	0.2442	0.7785
	misc	0.4105	0.5131	0.3110	0.6590
	rules	0.2605	0.5249	−0.0036	0.6885
	trees	0.4371	0.5249	0.2407	0.7785
	overall	0.4371	0.5249	0.1429	0.8474

All of the measurements in the tables are MCCs, and the vote after filtering is the MCC based on the eliminated methods. The "avg" is the average of the MCCs.

reliable discussion and conclusion must await further analysis based on more samples and records.

The records in the Human Protein Atlas will be updated, and more mRNAs will become available. Therefore, the fraction of hits and the reliability will change accordingly and the shared mRNAs that are currently not recorded in this database might be worthy of study. In addition, the reliabilities of the genes were similar whether a gene was independent or shared, and this indicates that the selected genes were representative for a prognosis but might have a redundant function; therefore, the shared genes are not significantly different from the independent genes in reliability.

4.3. Reliability of the Modules. The results from cross-validation could be used as the reference to evaluate the reliability by comparing the results with independent test. As an empirical conclusion, in most cases, the results from cross-validation could be better than an independent test due to various reasons such as the overfitting and batch effects, but if the difference is not too large, the modules could be identified as reliable. Reflected in the figures, except a few classes such as the *functions* class and *lazy* class, the MCC from cross-validation had a relatively better predicting performance (e.g., higher median and average or narrower IQR) than independent test. According to the whiskers, most of the classes had small differences, but there was still a lot of the modules that had a large difference which could be reflected by the outliers. Thus, it is still risky to get the unreliable prediction if we only use the modules which have good performance in cross-validation for predicting. However, since most of the methods were reliable, combining the methods together becomes useful and necessary to reduce the risk.

4.4. Distributions of the Predictions. A balanced ratio of positive samples to negative samples is an important factor for prediction. The BCRA datasets had the largest sample numbers; therefore, the IQRs were the smallest, which indicates that the MCCs were concentrated toward the median. The boxes become wider for the KIRC, OV-mRNA, and OV-miRNA datasets. The sample number should be guaranteed before modeling if a stable prediction is to be obtained. However, sometimes, many reasons such as cost, privacy, and difficulty limit the sample number, so it is insufficient to confirm the prediction stability. Such predictions should be considered very carefully because overfitting may have occurred.

A basic way to avoid overfitting is to reduce the attribute number for modeling, and that is why 4 subdatasets (i.e., datasets with 10, 30, 50, and 100 samples) were used for modeling. As shown in the Modeling Results, especially in Figures 9 and 11, more attributes relatively improved the

Table 5: Voting results of OV.

Platforms	Class	Vote from all methods	Vote after filtering	MCC avg	MCC max
OV-mRNA	*bayes*	0.4725	0.4725	0.1217	0.7906
	functions	0.1250	0.4725	−0.0890	0.7906
	lazy	0.3162	0.3162	0.1328	0.6447
	meta	0.1890	0.1890	0.1923	0.6447
	misc	0.6139	0.7500	0.3433	0.8864
	rules	0.1250	0.3162	−0.1160	0.8864
	trees	0.1890	0.1890	0.1734	0.6447
	overall	0.3162	0.3162	0.0890	0.8864
OV-miRNA	*bayes*	1.0000	0.8771	0.2508	0.8864
	functions	0.5000	0.6139	−0.0740	0.8771
	lazy	0.6447	0.6447	0.2756	0.6325
	meta	0.7559	0.7559	0.3786	0.8864
	misc	0.7559	0.6139	0.1384	0.6447
	rules	0.3430	0.7559	−0.0783	0.6447
	trees	0.5000	0.7559	0.1258	1.0000
	overall	0.7559	0.7559	0.1432	1.0000
OV-mRNA And OV-miRNA	*bayes*	0.4725	0.4725	0.1863	0.8864
	functions	−1.0000	0.7500	−0.0815	0.8771
	lazy	0.7500	0.8771	0.2042	0.6447
	meta	0.3162	0.3162	0.2854	0.8864
	misc	0.6139	0.7500	0.2408	0.8864
	rules	0.3430	0.3162	−0.0971	0.8864
	trees	0.3162	0.4725	0.1496	1.0000
	overall	0.3162	0.4725	0.1161	1.0000

All of the measurements in the tables are MCCs, and the vote after filtering is the MCC based on the eliminated methods. The "avg" is the average of the MCCs.

Table 6: Methods eliminated as outliers.

Class	Names	Counts
bayes	BayesianLogisticRegression	78
	DMNBtext	24
functions	SPegasos	113
	VotedPerceptron	19
	LibSVM	106
lazy	KStar	8
meta	*none*	/
misc	HyperPipes	12
rules	ConjunctiveRule	56
	OneR	9
	ZeroR	120
trees	BFTree	14
	DecisionStump	20

The counts are the number of methods whose MCC is lower than the lower whisker in the box plot.

overall prediction performance. However, the improvement was still limited by the sample number. All of the BRCA sub-datasets had the relatively smallest IQR compared to the others that had the same attribute number. This limitation might apply not only to modeling but also to many other studies which must use the samples as a template to measure the correlations among samples and genes. For example, in a gene set enrichment analysis, genes should be eliminated by statistical methods such as a *t*-test, fold-change, or FDR. In the analysis, the genes are independent from each other when calculating the correlations and thus the validity of the identified genes is only affected by the sample number. If the sample number is too small (less than 10), the *t*-test result is not reliable.

Limited sample and attribute numbers make a prediction sensitive to the datasets. The various distributions in Figures 2, 4, 6, 8, 10, and 12 indicate that the methods in the *meta* class are relatively stable for prediction, which is reasonable because the methods in the *meta* class use other methods for ensemble learning, so they are not sensitive to different dataset distributions as other types of methods are. The methods in the *trees* class were similar but had a more varied performance than the *meta* class methods because the classification and regression tree algorithms can be as simple as REPTree or as complex as random forest. Therefore, the boxes of the *trees* class usually had a larger IQR than the *meta* class. However, the *misc* class performed best for two datasets, but only two methods were contained in this class. Based on the algorithm, the methods in the *misc* class were much different and thus could have a much more variable performance for different datasets. Except for the *rules*

Table 7: Coverage of the selected genes from mRNA.

Disease	ShareNum	Subdata scale 10	30	50	100
BRCA	5	0	2	2	5
	4	4	5	10	25
	3	4	13	21	45
	2	2	16	26	42
	1	18	49	85	156
	Total	28	85	144	273
	Coverage fraction	35.7%	42.4%	40.1%	42.9%
KIRC	5	0	0	0	2
	4	0	3	3	8
	3	2	6	14	28
	2	6	22	34	71
	1	32	76	128	232
	Total	40	107	179	341
	Coverage fraction	20%	29%	28.5%	32%
OV	5	0	1	2	5
	4	1	2	7	16
	3	1	8	16	55
	2	9	27	34	43
	1	25	59	96	160
	Total	36	97	155	279
	Coverage fraction	30.6%	39.2%	38.1%	42.7%

The coverage fraction is the number of the genes which are used more than once in all of the genes, and accordingly, the formula is Coverage = NumOfSelectedGenes − NumOfIndependentGenes/NumOfSelectedGenes. The ShareNum is the number of a gene used in the subdatasets. For example, in Table 7, the value in the OV-miRNA group with ShareNum 3 and data scale 10 is 1; it means that there is one gene which is used by 3 subdatasets and each subdataset has 10 microRNAs as the attributes.

Table 8: Coverage of the datasets from miRNA.

Disease	ShareNum	Subdata scale 10	30	50	100*
BRCA	5	1	2	2	8
	4	3	8	17	47
	3	3	11	23	31
	2	3	9	12	26
	1	18	57	79	127
	Total	28	87	133	239
	Coverage fraction	35.7%	34.5%	40.6%	46.9%
KIRC	5	1	14	16	35
	4	3	6	13	53
	3	5	7	19	12
	2	2	7	11	9
	1	14	21	39	59
	Total	25	55	98	168
	Coverage fraction	44%	61.8%	60.2%	64.9%
OV	5	0	5	17	83
	4	3	15	27	0
	3	7	10	6	0
	2	3	5	12	0
	1	11	25	15	0
	Total	24	60	77	83
	Coverage fraction	54.2%	58.3%	80.5%	100%

The coverage fraction means the number of the genes which are used for more than 1 times in all of the genes, and accordingly, the formula is Coverage = NumOfSelectedGenes − NumOfIndependentGenes/NumOfSelectedGenes. The ShareNum is the number of a gene that is used for the subdatasets. For example, in Table 8, the value in the OV-miRNA group with ShareNum 3 and data scale 10 is 7; it means that there are 7 genes which are used by 3 subdatasets and each subdataset has 10 microRNAs as the attributes. *The subdatasets from the OV-miRNA group have at most 83 microRNAs, and thus, the scale "100" of OV-miRNA means 83.

class, the methods in the *bayes* class also had a relatively inferior performance, which might have been caused by the low sample number, because probability-based methods are sensitive on a modeling scale. The performance of methods in the *functions* class was more skewed; the boxes usually had a good upper whisker but a poor lower whisker. One reason for these might be that the algorithms in this class are also different from those in the *misc* class. There were 6 methods in the *functions* class, and thus the prediction performance varied widely. Another reason might be parameter optimization; less parameter optimization would affect all of the methods, but the methods in the *functions* class might be the most affected because they are much more sensitive to the parameters when only default parameters were used.

The high similarity of the gene data used would lead to a similar prediction performance, but when the similarity is lower than 50%, as reflected by the coverage in Table 7 or Table 8, the associated box plots were not significantly similar (as shown in Figures 3, 5, 7, 9, and 11).

According to the coverage of mRNA and microRNA used and shown in Tables 7 and 8, the coverage among the datasets was not large, but most of the medians from different subdatasets for the same cancer were similar, and this is reflected in the boxplots (Figures 3, 5, 7, 9, and 11), which indicates that some of the information contained in the genes was duplicated. In other words, the genes eliminated by the variable selection methods are representative for modeling, but not comprehensive. The duplicated genes could arise from similar genetics or pathology; for example, they could have the same genetic regulation pathway or simply be coexpressed, so that only one would be sufficient for modeling. Additionally, duplication could indicate that variable selection and machine learning methods are not sufficient to find out all of the disease-correlated genes. On the one hand, current machine learning methods can only determine some disease-associated genes, so further study might be necessary. On the other hand, the voting scheme provided in this study could be helpful for evaluating the relationship between cancer and genes.

As previously mentioned, the predictions using different datasets differed, meaning that we cannot determine which method is best for all datasets. The separate use of different

TABLE 9: Reliability of selected mRNAs.

Names	Data scale	Reliable/Hits/ShareNum 10	30	50	100*
BRCA	5	0	1/2/2	1/2/2	4/5/5
	4	4/4/4	4/5/5	8/9/10	13/17/25
	3	0/2/4	2/6/13	5/10/21	13/21/45
	2	0/0/2	5/6/16	8/12/26	18/29/42
	1	8/10/18	27/33/49	39/58/85	68/97/156
	Total	12/16/28	39/52/85	61/91/144	116/169/273
	Hits/Total	57.1%	61.2%	63.2%	61.9%
	Reliable/Hits	75%	75%	67%	68.6%
KIRC	5	0	0	0	0/0/2
	4	0	0/0/3	0/0/3	2/4/8
	3	2/2/2	2/2/6	4/6/14	12/17/28
	2	2/2/6	10/17/22	19/27/34	26/40/71
	1	15/18/32	31/42/76	45/63/128	98/129/232
	Total	19/22/40	43/61/107	68/96/179	138/190/341
	Hits/Total	55%	57%	53.6%	55.7%
	Reliable/Hits	86.4%	70.5%	70.8%	72.6%
OV	5	0	1/1/1	1/1/2	2/2/5
	4	1/1/1	0/1/2	2/3/7	2/4/16
	3	0/0/1	1/1/8	1/2/16	12/16/55
	2	0/2/9	5/8/27	6/10/34	14/18/43
	1	7/8/25	13/21/59	27/34/96	50/64/160
	Total	8/11/36	20/32/97	37/50/155	80/104/279
	Hits/Total	35%	33%	33.3%	37.3%
	Reliable/Hits	72.7	62.5%	74%	76.9%

The table that records the hits and reliability in the "Human Protein Atlas" database. The "ShareNum" is the same in Table 8, and the "Hits" is the number of mRNAs recorded in the "Human Protein Atlas" database. The "Reliable" is the number of reliable hits. The reliability is measured by using the associated record in the "reliability (IH)" table. An mRNA is considered reliable only if the record is "approved" or "supported." "Hits/Total" and "Reliable/Hits" are calculated simply by using the row "Total" for division. For example, the last two elements in the last column are 37.3% and 76.9%. They are calculated by the element in the associated row in "Total," such as 80/104/279, where 37.3% = 104/279 and 76.9% = 80/104. *The subdatasets from the OV-miRNA group have at most 83 microRNAs, and thus, the scale "100" of OV means 83.

modeling methods will result in a loss of information, so using ensemble methods to integrate the modeling results is necessary.

4.5. Effects of Voting. A sufficiently large fraction of coverage in the data space is necessary for good voting performance, and the average accuracy must not be too low. Therefore, if only the box plots are considered, the expected performance of the *bayes* class is much lower. However, the result is anomalous in that the *bayes* class showed a good ensemble classification performance (it was ranked in top three) for 5 datasets. On the other hand, the *meta* and *trees* classes were not as good as the *bayes* class even though they had relatively similar distributions. One reason is many ensemble algorithms are contained in those two classes, so that the voting had already been accomplished, so the prediction results were concentrated near the median or the average. It is not difficult to see that in most cases, the voting results of the two classes were near the average. Another reason is that only default parameters were used in the entire test. Many ensemble methods must optimize the submethods for ensemble learning, and the resampling methods also must be optimized.

As a comparison, the voting performance from the eliminated methods was similar. However, the voting results were not always better than the original results. The *bayes* class was negatively affected by the filter. The MCCs based on the *bayes* class were not larger than prior to voting. The *rules* and *functions* classes were benefited by the filter, and the MCCs were improved for most of the datasets. The *trees* class was slightly benefited in the BRCA-miRNA group but was not globally affected as the *meta* class was. The other classes, including the overall voting, were affected positively or negatively by different datasets. The biased effects could indicate that the methods were sensitive to the datasets; the prediction performance of a method changed greatly when the dataset changed. On the other hand, the overall performance was not affected too greatly. One reason was the *meta* class methods, which showed a stable prediction ability, were not eliminated, and thus, the overall results remained stable.

Another reason might be that the performance effects were polarized. For example, the *bayes* class that showed the best comprehensive performance was negatively affected but the *rules* class was benefited. Therefore, the overall differences from the two voting mechanisms could be less.

The biased effects mean that a filter might not be that necessary if there is no prior knowledge. However, the voting by combining the mRNA and miRNA datasets could produce better performance if the sample size is sufficiently large, as shown in Table 3. Moreover, as shown in Tables 4 and 5, even the sample size is insufficient, the results will not become worse. Since the methods are not weighted, the results support that most of the methods will produce the same prediction, so combining the two datasets will be beneficial.

5. Conclusions

The purpose of this study was to discover a reliable way to predict unknown data to reduce the risk of error prediction when not enough samples were used for modeling. The distribution of the modeling performance indicated that the best methods were different for different datasets; therefore, the methods were integrated using a voting protocol. Finally, we proposed a better way to model different gene expression datasets. In conclusion, no prior knowledge exists; a comparison of the prediction results for three cancers indicates that the methods in the *bayes* class show a good ensemble performance, even though the individual methods are not as stable as those in the *meta* or *trees* classes. The *meta* and *trees* classes already contain many ensemble methods; therefore, their performance is stable but, again, not good for ensemble twice. Therefore, using the methods in the *bayes* class as a group and one of the algorithms in the *meta* class might be a practical approach for a dataset without sufficient prior knowledge. If prior knowledge exists for a cancer, the methods and datasets used can be more specific. For example, this study indicates that using miRNA as an attribute for modeling the OV data could yield a better result than using mRNA, if we knew that at first, some of the negative effects could be avoided. We hope that the scheme can facilitate related studies of genetic data modeling and elucidate important genes to enhance the reliability of the final model.

Conflicts of Interest

The authors declare that they have no conflicts of interest.

Acknowledgments

This project was supported by the National Natural Science Foundation of China under Grant no. 21575094, Sichuan Province Key Laboratory of Special Waste Water Treatment under Grant no. SWWT2015-3, Fundamental Research Funds for Central Non-profit Scientific Institution under Grant no. 12017206030202209, the National Key Research and Development Program of China under Grant no. 2016YFB0201305, Guangdong Provincial Department of Science and Technology under Grant no. 2016B090918122, the Science Technology and Innovation Committee of Shenzhen Municipality under Grant nos. JCYJ20160331190 123578 and GJHZ20170314154722613, and Youth Innovation Promotion Association of the Chinese Academy of Sciences, to Yanjie Wei.

References

[1] A. Bashiri, M. Ghazisaeedi, R. Safdari, L. Shahmoradi, and H. Ehtesham, "Improving the prediction of survival in cancer patients by using machine learning techniques: experience of gene: a narrative review," *Iranian Journal of Public Health*, vol. 46, no. 2, pp. 165–172, 2017.

[2] K. Kourou, T. P. Exarchos, K. P. Exarchos, M. V. Karamouzis, and D. I. Fotiadis, "Machine learning applications in cancer prognosis and prediction," *Computational and Structural Biotechnology Journal*, vol. 13, pp. 8–17, 2015.

[3] S.-W. Chang, S. Abdul-Kareem, A. F. Merican, and R. B. Zain, "Oral cancer prognosis based on clinicopathologic and genomic markers using a hybrid of feature selection and machine learning methods," *BMC Bioinformatics*, vol. 14, no. 1, p. 170, 2013.

[4] Y. Ishibashi, N. Hanyu, K. Nakada et al., "Profiling gene expression ratios of paired cancerous and normal tissue predicts relapse of esophageal squamous cell carcinoma," *Cancer Research*, vol. 63, no. 16, pp. 5159–5164, 2003.

[5] X. Xu, Y. Zhang, L. Zou, M. Wang, and A. Li, "A gene signature for breast cancer prognosis using support vector machine," in *2012 5th International Conference on BioMedical Engineering and Informatics*, pp. 928–931, Chongqing, China, 2012.

[6] L. He, Y. Wang, Y. Yang, L. Huang, and Z. Wen, "Identifying the gene signatures from gene-pathway bipartite network guarantees the robust model performance on predicting the cancer prognosis," *BioMed Research International*, vol. 2014, Article ID 424509, 10 pages, 2014.

[7] L. Jiang, L. Huang, Q. Kuang et al., "Improving the prediction of chemotherapeutic sensitivity of tumors in breast cancer via optimizing the selection of candidate genes," *Computational Biology and Chemistry*, vol. 49, pp. 71–78, 2014.

[8] F. Xie, M. He, L. He et al., "Bipartite network analysis reveals metabolic gene expression profiles that are highly associated with the clinical outcomes of acute myeloid leukemia," *Computational Biology and Chemistry*, vol. 67, pp. 150–157, 2017.

[9] L. P. Petalidis, A. Oulas, M. Backlund et al., "Improved grading and survival prediction of human astrocytic brain tumors by artificial neural network analysis of gene expression microarray data," *Molecular Cancer Therapeutics*, vol. 7, no. 5, pp. 1013–1024, 2008.

[10] Y.-C. Chen, W.-C. Ke, and H.-W. Chiu, "Risk classification of cancer survival using ANN with gene expression data from multiple laboratories," *Computers in Biology and Medicine*, vol. 48, pp. 1–7, 2014.

[11] F. Sato, Y. Shimada, F. M. Selaru et al., "Prediction of survival in patients with esophageal carcinoma using artificial neural networks," *Cancer*, vol. 103, no. 8, pp. 1596–1605, 2005.

[12] O. Gevaert, F. D. Smet, D. Timmerman, Y. Moreau, and B. D. Moor, "Predicting the prognosis of breast cancer by integrating clinical and microarray data with Bayesian networks," *Bioinformatics*, vol. 22, no. 14, pp. e184–e190, 2006.

[13] A. Moslemi, H. Mahjub, M. Saidijam, J. Poorolajal, and A. R. Soltanian, "Bayesian survival analysis of high-dimensional microarray data for mantle cell lymphoma patients," *Asian Pacific Journal of Cancer Prevention*, vol. 17, no. 1, pp. 95–100, 2016.

[14] H. Wang, L. Huang, R. Jing et al., "Identifying oncogenes as features for clinical cancer prognosis by Bayesian nonparametric variable selection algorithm," *Chemometrics and Intelligent Laboratory Systems*, vol. 146, pp. 464–471, 2015.

[15] T. Ando, M. Suguro, T. Hanai, T. Kobayashi, H. Honda, and M. Seto, "Fuzzy neural network applied to gene expression profiling for predicting the prognosis of diffuse large B-cell lymphoma," *Japanese Journal of Cancer Research*, vol. 93, no. 11, pp. 1207–1212, 2002.

[16] L. Shi, G. Campbell, W. D. Jones et al., "The MicroArray Quality Control (MAQC)-II study of common practices for the development and validation of microarray-based predictive models," *Nature Biotechnology*, vol. 28, no. 8, pp. 827–838, 2010.

[17] C. Park, J. Ahn, H. Kim, and S. Park, "Integrative gene network construction to analyze cancer recurrence using semi-supervised learning," *PloS One*, vol. 9, no. 1, article e86309, 2014.

[18] J. Hou, J. Aerts, B. den Hamer et al., "Gene expression-based classification of non-small cell lung carcinomas and survival prediction," *PloS One*, vol. 5, no. 4, article e10312, 2010.

[19] J. Xu, R. Jing, Y. Liu, Y. Dong, Z. Wen, and M. Li, "A new strategy for exploring the hierarchical structure of cancers by adaptively partitioning functional modules from gene expression network," *Scientific Reports*, vol. 6, no. 1, 2016.

[20] F. M. Lopes, R. M. Cesar Jr., and L. D. F. Costa, "Gene expression complex networks: synthesis, identification, and analysis," *Journal of Computational Biology*, vol. 18, no. 10, pp. 1353–1367, 2011.

[21] O. Rozenblatt-Rosen, R. C. Deo, M. Padi et al., "Interpreting cancer genomes using systematic host network perturbations by tumour virus proteins," *Nature*, vol. 487, no. 7408, pp. 491–495, 2012.

[22] T. Wang, J. Gu, J. Yuan, R. Tao, Y. Li, and S. Li, "Inferring pathway crosstalk networks using gene set co-expression signatures," *Molecular BioSystems*, vol. 9, no. 7, pp. 1822–1828, 2013.

[23] R. L. Grossman, A. P. Heath, V. Ferretti et al., "Toward a shared vision for cancer genomic data," *The New England Journal of Medicine*, vol. 375, no. 12, pp. 1109–1112, 2016.

[24] M. Hall, E. Frank, G. Holmes, B. Pfahringer, P. Reutemann, and I. H. Witten, "The WEKA data mining software: an update," *ACM SIGKDD Explorations Newsletter*, vol. 11, no. 1, pp. 10–18, 2009.

[25] C.-C. Chang and C.-J. Lin, "LIBSVM: a library for support vector machines," *ACM Transactions on Intelligent Systems and Technology*, vol. 2, no. 3, pp. 1–27, 2011.

[26] S. Le Cessie and J. C. Van Houwelingen, "Ridge estimators in logistic regression," *Applied Statistics*, vol. 41, no. 1, pp. 191–201, 1992.

[27] Y. Freund and R. E. Schapire, "Experiments with a new boosting algorithm," in *Proceedings of the Thirteenth International Conference on Machine Learning*, vol. 96, pp. 148–156, Bari, Italy, 1996.

[28] L. Breiman, "Bagging predictors," *Machine Learning*, vol. 24, no. 2, pp. 123–140, 1996.

[29] G. H. John and P. Langley, "Estimating continuous distributions in Bayesian classifiers," in *UAI'95 Proceedings of the Eleventh conference on Uncertainty in artificial intelligence*, pp. 338–345, Montreal, Quebec, Canada, 1995.

[30] A. McCallum and K. Nigam, "A comparison of event models for naive bayes text classification," in *Learning for Text Categorization: Papers from the 1998 AAAI Workshop*, vol. 752, pp. 41–48, Stanford, CA, USA, 1998.

[31] A. Genkin, D. D. Lewis, and D. Madigan, "Large-scale Bayesian logistic regression for text categorization," *Technometrics*, vol. 49, no. 3, pp. 291–304, 2007.

[32] R. Kohavi, "The power of decision tables," in *ECML '95 Proceedings of the 8th European Conference on Machine Learning*, pp. 174–189, Heraclion, Crete, Greece, 1995.

[33] E. Frank, M. Hall, and B. Pfahringer, "Locally weighted naive bayes," in *UAI'03 Proceedings of the Nineteenth conference on Uncertainty in Artificial Intelligence*, pp. 249–256, Acapulco, Mexico, 2002.

[34] L. Breiman, "Random forests," *Machine Learning*, vol. 45, no. 1, pp. 5–32, 2001.

[35] G. Demiröz and H. Güvenir, "Classification by voting feature intervals," in *ECML '97 Proceedings of the 9th European Conference on Machine Learning*, pp. 85–92, Prague, Czech Republic, 1997.

[36] R. C. Holte, "Very simple classification rules perform well on most commonly used datasets," *Machine Learning*, vol. 11, no. 1, pp. 63–90, 1993.

[37] K. Kira and L. A. Rendell, "A practical approach to feature selection," in *ML92 Proceedings of the Ninth International Workshop on Machine Learning*, pp. 249–256, San Francisco, CA, USA, 1992.

[38] C. E. Shannon, "A mathematical theory of communication, part I, part II," *Bell System Technical Journal*, vol. 27, no. 4, pp. 623–656, 1948.

[39] Z. Wen, Z. Wang, S. Wang et al., "Discovery of molecular mechanisms of traditional Chinese medicinal formula Si-Wu-Tang using gene expression microarray and connectivity map," *PloS One*, vol. 6, no. 3, article e18278, 2011.

[40] M. Uhlén, L. Fagerberg, B. M. Hallström et al., "Tissue-based map of the human proteome," *Science*, vol. 347, no. 6220, article 1260419, 2015.

10

HuoXueJieDu Formula Alleviates Diabetic Retinopathy in Rats by Inhibiting SOCS3-STAT3 and TIMP1-A2M Pathways

Hongliang Wang,[1,2] Wei Xing,[1] Shijie Tang,[3] Zhenglin Wang,[1] Tiantian Lv,[1] Yan Wu,[4] Shuzhen Guo,[1] Chun Li,[5] Jing Han,[4] Ruixin Zhu,[6] and Wei Wang[1]

[1]*College of Basic Medicine, Key Laboratory of Ministry of Education (Syndromes and Formulas), Key Laboratory of Beijing (Syndromes and Formulas), Beijing University of Chinese medicine, Beijing 100029, China*
[2]*School of Public Health, Capital Medical University, Beijing 100069, China*
[3]*Key Laboratory of Systems Biology, CAS Center for Excellence in Molecular Cell Science, Innovation Center for Cell Signaling Network, Institute of Biochemistry and Cell Biology, Shanghai Institutes for Biological Sciences, Chinese Academy of Sciences, Shanghai 200032, China*
[4]*Institute of Chinese Medicine, Beijing University of Chinese medicine, Beijing 100029, China*
[5]*Modern Research Center for Traditional Chinese Medicine, Beijing University of Chinese Medicine, Beijing 100029, China*
[6]*Department of Bioinformatics, School of Life Sciences and Technology, Tongji University, Shanghai 200092, China*

Correspondence should be addressed to Jing Han; hanjing8585@163.com, Ruixin Zhu; rxzhu@tongji.edu.cn, and Wei Wang; wangwei26960@126.com

Hongliang Wang and Wei Xing contributed equally to this work.

Academic Editor: Guoxun Chen

HuoXueJieDu (HXJD) formula exerts protective effects against diabetic retinopathy (DR) in rats, but its underlying mechanism remains unknown. In the present study, the diabetic rats were established using streptozocin. The administration of HXJD was initiated at 20 weeks after diabetes induction and continued for 12 weeks. Whole genome expression profiles in rat retinas were examined using microarray technology. Differential gene expression and pathway enrichment analysis were conducted on the microarray data, with validation through real-time PCR and immunohistochemical staining. The results showed that 170 genes and several IPA canonical pathways related to inflammation, matrix metabolism, and phototransduction were regulated by HXJD. PCR validation of selected genes, including SOCS3, STAT3, TIMP1, and A2M, confirmed the gene expression changes influenced by HXJD. In addition, the immunohistochemical staining results suggested that critical members of the SOCS3-STAT3 pathway were also affected by HXJD. Taken together, these results indicated that SOCS3-STAT3 and TIMP1-A2M pathways might mediate the alleviation of HXJD activities in rats with diabetic retinopathy.

1. Introduction

Diabetic retinopathy (DR) is defined as a retinal disease resulting from diabetes, characterized by microvascular lesions in the retina, including vascular proliferation, vascular leakage, and retinal ischaemia [1]. DR has been attributed to 4.8% of the 37 million blind patients worldwide, and this disease is the most common cause of visual impairment in the working-age population [2]. The number of individuals suffering from DR has increased because diabetes has recently become a global epidemic disease.

Until recently, hyperglycaemia control, laser photocoagulation, and vitreoretinal surgery have been the main treatments for DR. However, the progress of DR in some patients has not been arrested or even reversed by these efforts. Thus, new therapies that could block the development of retinal neovascularization need further exploration.

HuoXueJieDu (HXJD) is a Chinese herbal formula comprising *Euonymus alatus* (Gui-Jian-Yu), *Radix trichosanthis* (Tian-Hua-Fen), *Panax notoginseng* (San-Qi), and *Coptis chinensis* (Huang-Lian). HXJD originates from clinical experience and could stabilize vision and reduce retinal oedema. In addition, previous studies have shown that in the central retinal artery of diabetic rats, HXJD improves peak systolic velocity, end diastolic velocity, and mean velocity and decreases the pulse index and resistance index [3]. In addition, HXJD increases the wave amplitudes (the maximal electric reaction for dark-adapted eyes) and the sum amplitude of the oscillatory potentials (OPs) in the electroretinogram [4]. Furthermore, HXJD reduces vascular density and the ratio of endothelial cells to pericytes in the retinas of diabetic rats [3]. This effect was consistent with the downregulation of apoptosis [5], VEGF (vascular endothelial growth factor) [6], and MMPs (matrix metalloproteinases) [7]. These results indicate that HXJD confers protection against retinal injuries in DR, but the mechanism underlying the effect of HXJD has not been elucidated.

In recent years, high-throughput omics technologies, particularly microarray transcriptomes, have been increasingly applied in traditional Chinese medicine research [8, 9]. Thus, to address the mechanism of HXJD, we assessed the protective effects of this compound on DR in STZ-induced rats, and microarray technology was subsequently used to analyze the influence of HXJD on the gene expression profile. Furthermore, some of the potential targets were validated using PCR. The results of the present study will reveal the biochemical pathways underlying the effect of HXJD and provide further insight into the therapeutic approaches for the prevention of DR.

2. Methods

2.1. Ethic Statement. All procedures involving the animals and their care were performed according to the governmental guidelines on animal experimentation and the "Principles of Laboratory Animal Care" of the National Institutes of Health. All experimental protocols were approved by the Institutional Animal Ethics Committee of Beijing University of Chinese Medicine, Beijing, China (permit number 26-1514).

2.2. Antibodies. The following antibodies were used in the present study: anti-JAK2 antibody (number 3230; Cell Signaling Technology, Boston, Massachusetts, USA); anti-p38 MAPK antibody (ab197348; Abcam, Cambridge, UK); anti-NF-κBp65 antibody (ab16502; Abcam); anti-phospho-stat3 antibody (number 9145; Cell Signaling Technology, Boston, Massachusetts, USA); anti-STAT3 (number 4904, Cell Signaling Technology, Boston, Massachusetts, USA); anti-TIMP1 (ab61224, Abcam, Cambridge, UK); anti-SOCS3 (ab16030, Abcam, Cambridge, UK); and anti-A2M (ab58703, Abcam, Cambridge, UK).

2.3. Plant Materials. E. alatus (Gui-Jian-Yu), *R. trichosanthis* (Tian-Hua-Fen), *P. notoginseng* (San-Qi), and *C. chinensis* (Huang-Lian) were obtained from Tongrentang (Beijing, China). Professor Peng Tan (College of Chinese Medicine, Beijing University of Chinese Medicine, Beijing) authenticated the processed Chinese medicinal materials. The plant specimens (Number TCM120305, 120306, 120307, and 120308) were stored at the YIFU Research Building of Beijing University of Chinese Medicine.

2.4. Preparation of Plant Extract. E. alatus, R. trichosanthis, P. notoginseng, and *C. chinensis* were extracted for 2 hours under reflux with water. The plants were extracted 2 times, and the extracts were pooled and vacuum concentrated at 60°C.

Ultraviolet spectrophotometry was applied to determine the major component contents in the extracts using rutin as a standard, with a content of 17%. Ginsenoside Rg1 was used to represent the content of saponins at 16%, while berberine represented the typical content of alkaloids at 20%, and glucose was used as the standard to assess the content of polysaccharides at 18%. The extracts were dissolved in water.

2.5. Animals. Healthy, male Sprague-Dawley rats (7 weeks of age, 200–250 g) were purchased from Vital River Laboratory Animal Technology Co. Ltd. (Beijing, China, certificate number SCXK (Beijing) 2007-0001). The animals were raised in a clean grade animal laboratory with a temperature of 22–24°C and 40%–60% humidity. The rats were provided commercial rat feed and water *ad libitum.* At the end of the experiments, the animals were sacrificed by intraperitoneal injection with pentobarbital (50 mg/kg). The blood glucose was examined every four weeks.

2.6. Induction of Diabetes. Streptozocin (STZ, Sigma, St. Louis, Mo USA, catalog number S0130) was dissolved in citrate buffer (PH4.5) with a concentration of 20 g/L. The dissolved STZ was preserved in the dark. The rats were administered an intraperitoneal injection of STZ at 65 mg/kg body weight [10]. The control rats received equal volumes of citrate buffer. Four weeks later, blood glucose level in a blood sample obtained from the tail vein was evaluated using a standard glucometer (One Touch Profile, LifeScan, San Francisco, California, USA). Rats with blood glucose levels higher than 16.7 mmol/L were selected for subsequent experiments.

2.7. Treatment Schedule. Twenty weeks after STZ injection, the diabetic rats were grouped into two groups according to the glucose concentration and body weight: diabetic ($n = 8$) and HXJD ($n = 8$, 15.4 g/kg). Subsequently, the diabetic rats in HXJD group were administered an HXJD decoction through intragastric gavage. The rats in the normal ($n = 8$) and diabetic groups were administered with water. After 12 weeks of treatment, the rats were anaesthetized, and the retinas were removed.

2.8. Haematoxylin-Eosin (HE) and Immunohistochemical Staining. The retinas were isolated, fixed in paraformaldehyde, and embedded in paraffin. The paraffin sections were deparaffinized and dehydrated. Subsequently, the sections were stained with haematoxylin-eosin. The thickness of retinas from the photoreceptor layer (PL) to the inner plexiform layer (IPL) in the midretina was determined, and the number of the retinal ganglion cells (RGC) were assessed. For immunohistochemistry, after dewaxing in benzene and dehydration in gradient alcohol, the sections were incubated with 0.3% hydrogen peroxidase to remove endogenous peroxidases. Subsequently, the sections were incubated with primary antibody (dilution of antibody: JAK2 1 : 50, p38 MAPK 1 : 100 and NF-κBp65 1 : 500) overnight at 4°C. After three washes with PBS, the sections were incubated with secondary antibody at room temperature. Finally, the sections were stained with 3,3′-diaminobenzidine (DAB) and haematoxylin. The sections were visualized using a microscope and analyzed using ImageJ software.

2.9. mRNA Microarray Analysis. Total RNA from the retinas of rats was extracted. The integrity of the RNA was assessed using the Agilent Bioanalyzer 2100 (Agilent Technologies, Palo Alto, California, USA). Only samples with RNA integrity values larger than 7 and 28S/18S ratios larger than 0.7 were considered qualified. Subsequently, 100 ng of RNA was amplified, labelled, and purified according to the manufacturer's instructions. After hybridization, the microarrays were washed and scanned with default settings according to the manufacturer's instructions (Agilent Technologies, USA). Background correction, quantile normalization, and differential expression analyses were performed using the limma package in the *R* (version 3.2.4) statistical computing environment. In the limma package, linear models were used to assess the differential expression in the designed microarray experiments. Genes with absolute fold change > 1.5, and P value <0.01 were considered differentially expressed genes (DEGs). Subsequently, pathway enrichment analysis was performed using the Ingenuity Pathway Analysis (IPA) tool. P values were calculated using Fisher's exact test, and P values <0.05 were considered significantly enriched in the canonical pathways [11].

2.10. Quantitative Real-Time PCR. Quantitative real-time PCR was performed to confirm the results of gene expression profiling. The RNA from retinas was extracted with the TRIzol reagent, and the RNA concentration of retinas was evaluated by Nanodrop2600 at 260 nm. Then, the RNA was reverse transcribed using a reverse transcription reagent kit (Roche, Basel, Switzerland) and 1 μg RNA was inputted for the reaction. Real-time PCR was performed with SYBR Green PCR Mix (Roche, Basel, Switzerland) on an ABI Step One Plus® instrument (ABI, Carlsbad, California, USA). Reactions were performed in a volume of 10 μL containing 2X SYBR Green PCR Mix, 5 μL; cDNA, 1 μL; forward and reverse primers, 0.4 μL; and ddH2O, 3.6 μL. β-Actin was served as the control. Differences in gene expression were quantified using the ΔΔCt method.

2.11. Western Blot. The retinas were added in ice-cold RIPA lysis buffer which contained a protease inhibitor mixture. The protein dissolved in supernatant was centrifuged and determined using the BCA protein assay reagent kit (Thermo Scientific, Waltham, MA, USA). SDS-PAGE gels containing 12% acrylamide were used to separate the samples (50 μg). The proteins were transferred onto membrane, blocked for 1.5 h with 5% milk at 37°C, and then incubated with primary antibodies at 4°C overnight. Subsequently, the PVDF membranes were incubated with secondary antibody for 1.5 h. In the end, the bands were visualized using an electrochemiluminescence (ECL) kit and quantified using image analysis software.

2.12. Addition Statistical Analysis. One-way ANOVA with LSD or Tamhane test was used for comparison between the three groups. The statistical tests were performed using SPSS statistical processing software and P values less than 0.05 were considered statistically significant. The Kolmogorov-Smirnov test was used to confirm normal distribution. The results were presented as the means ± SEM, indicating the variability reliability of the data.

3. Results

3.1. HXJD Protects Retinal Structure in Rats with DR. The HE staining was performed to observe the morphology of retinas. The retinas of the rats in normal group presented a clear structure, and the cells in every layer were uniformly distributed and densely packed. However, in the diabetic group, the retinas presented a reduction of cell density and oedema, and the cells were irregularly organized and loosely arranged. In addition, the retinal thickness and RGC number in the eyes of diabetic rats were decreased compared to those of normal rats (Figure 1). Expectedly, HXJD ameliorated the observed pathological changes. The retinal structure was improved, and the cells were organized with intact layers. Moreover, the retinal thickness and RGC number were evaluated in HXJD group (Figure 1).

3.2. HXJD Changes the Expression of Genes Associated with Cytokine Signalling and Matrix Metalloproteases. A comparative analysis between the normal, diabetic, and HXJD groups for genome-scale gene expression was performed. The mRNA array chip included nearly forty thousand genes. Microarray analysis revealed that compared with the normal group, 215 genes were upregulated in the diabetic group, among which 99 genes were reversed in the HXJD group. In addition, 539 genes were downregulated in the diabetic group compared with the normal group, among which 71 genes were reversed in the HXJD group (Figure 2(a)).

To examine the function of differentially expressed mRNAs, IPA pathway annotation was applied. The results revealed that 37 canonical pathways were enriched within the 99 genes upregulated in diabetic group and subsequently downregulated after HXJD treatment, including the role of macrophages, fibroblasts, and endothelial cells in rheumatoid arthritis, matrix metalloprotease inhibition, and IL-6 signalling. Moreover, 18 canonical pathways were enriched within the 71 genes downregulated in the diabetic group and subsequently upregulated after HXJD treatment, which have been implicated in phototransduction pathways, the visual cycle, and RAR activation. The top 10 enriched pathways are shown, respectively, in Tables 1 and 2.

To verify the results of the microarray analysis, real-time PCR was performed to examine the levels of differentially expressed genes. ATP12A, which functions as a

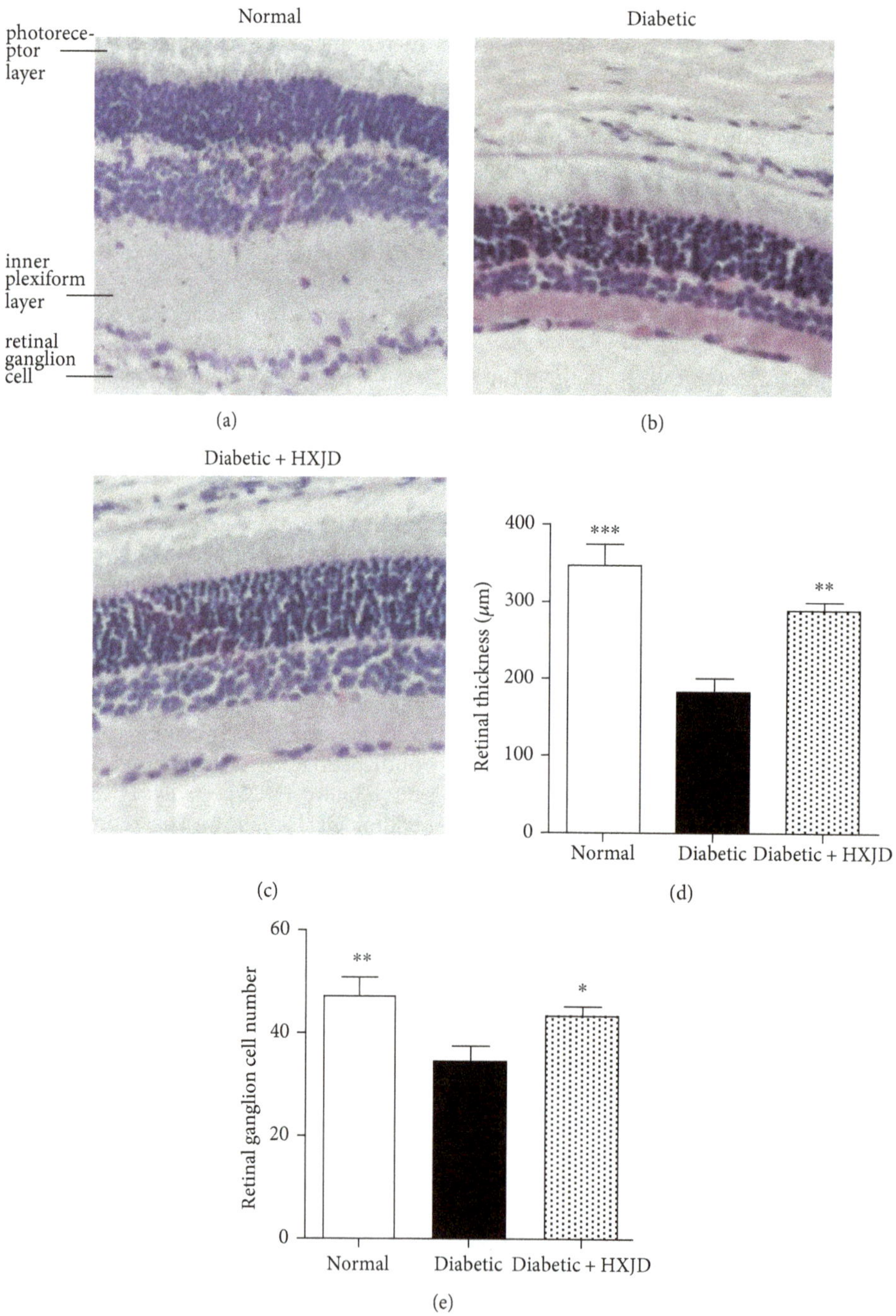

FIGURE 1: The effect of HXJD on retinal morphology. The retinas were stained with haematoxylin and eosin staining. (a) Retinas of normal rats. Retinal ganglion cell (RGC), inner plexiform layer (IPL), and photoreceptor layer (PL). (b) Retinas of diabetic rats. (c) Retinas of the rats in the HXJD group. Magnification: ×400. (d) The retinal thickness from IPL to PL in the midretina was measured. HXJD increased the retinal thickness. (e) The RGC number was counted. Retinas of the rats in the HXJD group showed an increased number of RGC. Data are represented as the means ± SEM. ($n = 3\sim5$). $^{*}P < 0.05$, $^{**}P < 0.01$, and $^{***}P < 0.001$ compared with the diabetic group.

P-type cation transport ATPase [12], was highlighted as remarkably decreased after HXJD treatment. Microarray data demonstrated that ATP12A expression in the diabetic group was 34 times higher than the HXJD group, and PCR experiments achieved similar results (Figure 2(b)).

In addition, several genes associated with DR were selected for further study. SOCS3 and STAT3 are signal transducers involved in the regulation of cytokines [13, 14], including IL-6, IL-10, IL-22, and IL-9 (Table 1). TIMP1 and A2M govern the degradation of the extracellular matrix [15, 16]. The gene levels of SOCS3, STAT3, TIMP1, and A2M were increased in the retinas of diabetic rats compared to normal rat retinas, whereas these genes were downregulated in the retinas of HXJD-treated diabetic rats (Figures 2(c)–2(f)).

Moreover, ADCY1 was selected for PCR verification. This gene encodes a member of the of adenylate cyclase gene family [17]. ADCY1 expression was reduced in diabetic rats

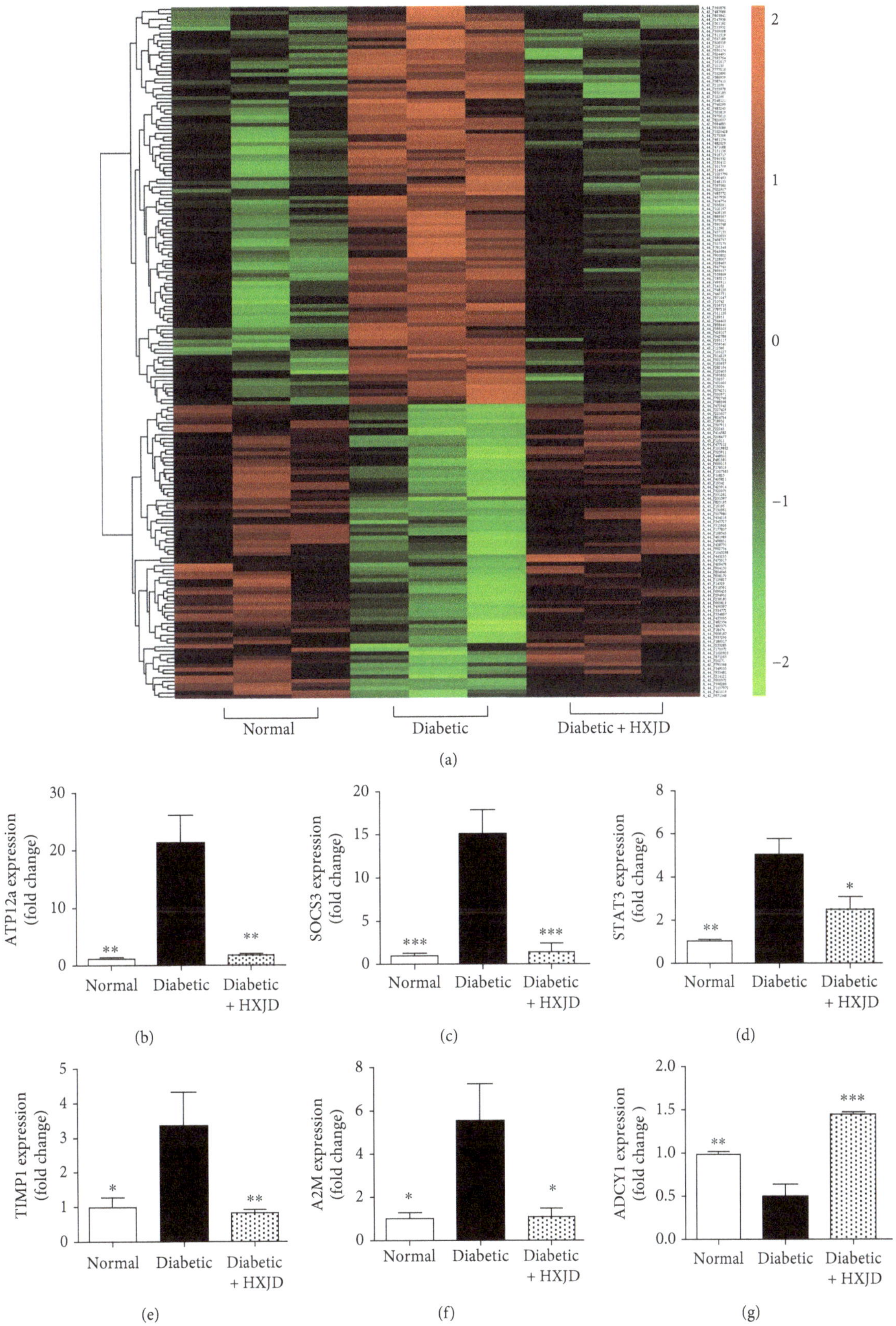

Figure 2: The effect of HXJD on the expression of genes associated with cytokine signalling and matrix metalloprotease expression. (a) Heat map of the regulated genes in normal, diabetic, and diabetic +HXJD rat groups. (b–g) The microarray data were validated using real-time PCR. Data are represented as the means ± SEM. ($n = 3\sim4$). $^{*}P < 0.05$, $^{**}P < 0.01$, and $^{***}P < 0.001$, compared with the diabetic group.

TABLE 1: Top 10 canonical pathways identified using IPA (downregulated by HXJD).

Ingenuity canonical pathways	P value	Affected molecules
Role of macrophages, fibroblasts, and endothelial cells in rheumatoid arthritis	1.15E−04	TLR2, IL1R2, SOCS3, WNT10A, STAT3, IL17RA, and IRAK2
Inhibition of matrix metalloproteases	4.07E−04	MMP7, TIMP1, and A2M
IL-6 signalling	9.12E−04	IL1R2, SOCS3, STAT3, and A2M
IL-10 signalling	2.04E−03	IL1R2, SOCS3, and STAT3
Growth hormone signalling	2.14E−03	SOCS3, STAT3, and A2M
IL-22 signalling	3.55E−03	SOCS3, STAT3
Role of JAK family kinases in IL-6-type cytokine signalling	3.89E−03	SOCS3, STAT3
IL-9 signalling	6.31E−03	SOCS3, STAT3
Role of JAK2 in hormone-like cytokine signalling	7.41E−03	SOCS3, STAT3
Tetrahydrobiopterin biosynthesis I	1.10E−02	GCH1

TABLE 2: Top 10 canonical pathways identified using IPA (upregulated by HXJD).

Ingenuity canonical pathways	P value	Affected molecules
Phototransduction pathway	2.57E−04	ARR3, GRK1, and PDE6D
The visual cycle	5.50E−04	RDH11, RDH12
RAR activation	1.00E−03	RDH11, ADCY1, RDH12, and CITED2
Retinoate biosynthesis I	2.69E−03	RDH11, RDH12
Retinol biosynthesis	2.69E−03	RDH11, RDH12
Creatine-phosphate biosynthesis	1.17E−02	CKMT1A/CKMT1B
Protein kinase A signalling	1.26E−02	H3F3A/H3F3B, ADCY1, GRK1, and PDE6D
NAD biosynthesis III	1.26E−02	Nmnat3
NAD biosynthesis from 2-amino-3-carboxymuconate semialdehyde	1.62E−02	Nmnat3
NAD salvage pathway III	1.62E−02	Nmnat3

compared to normal rats, and the expression levels of this gene were reversed after treatment with HXJD (Figure 2(g)).

The PCR results demonstrated consistency between the microarray and real-time PCR data. This consistency indicated the reliability of the microarray results and confirmed the influence of HXJD on the expression of these genes.

3.3. HXJD Reduces the Protein Expression of p-STAT3, STAT3, SOCS3, A2M, and TIMP1. Western blot was carried out to examine the protein expression of p-STAT3, STAT3, SOCS3, A2M, and TIMP1 in the retinas of rats. As shown in Figure 3, in the diabetic rats, the protein levels of STAT3 and SOCS3 remarkably increased compared with the normal rats (Figures 3(c) and 3(d)). Meanwhile, HXJD reduced the protein expression of STAT3 and SOCS3 in the rat retinas (Figures 3(c) and 3(d)). Moreover, the protein levels of p-STAT3, A2M, and TIMP1 showed an increasing tendency in the diabetic group, compared with the normal group. And HXJD presented a decreasing tendency of p-STAT3, A2M, and TIMP1 (Figures 3(b), 3(e), and 3(f)).

3.4. HXJD Reduces the Expression of JAK2, p38 MAPK, and NF-κB in the Retinas of Diabetic Rats. Immunohistochemistry was performed to evaluate the expression levels of JAK2, p38 MAPK, and NF-κB in rat retinas. As shown in Figure 4, in the diabetic group, the expression of JAK2, p38 MAPK, and NF-κB significantly increased in the retinas compared with that in the normal group (Figure 4). However, rats treated with HXJD presented the decreased expression of JAK2, p38 MAPK, and NF-κB in the retinas (Figure 4).

4. Discussion

The aim of the present study was to clarify the mechanism of HXJD against DR. We used microarray technology to examine the retinal mRNA expression changes affected by HXJD. The results showed that 170 genes were altered after HXJD treatment, and these genes played critical roles in inflammation, proliferation, cytokine signalling, and retinal function. Among these processes, two mechanisms were verified: reduced SOCS3-STAT3 signalling pathway and inhibited TIMP1-A2M expression.

4.1. Reduced SOCS3-STAT3 Signalling Pathway. SOCS3 and STAT3 have been implicated in most of the pathways regulated by HXJD (Table 1). These proteins are signal transducers that govern essential cellular processes, such as proliferation, apoptosis, and inflammatory gene transcription. STAT3

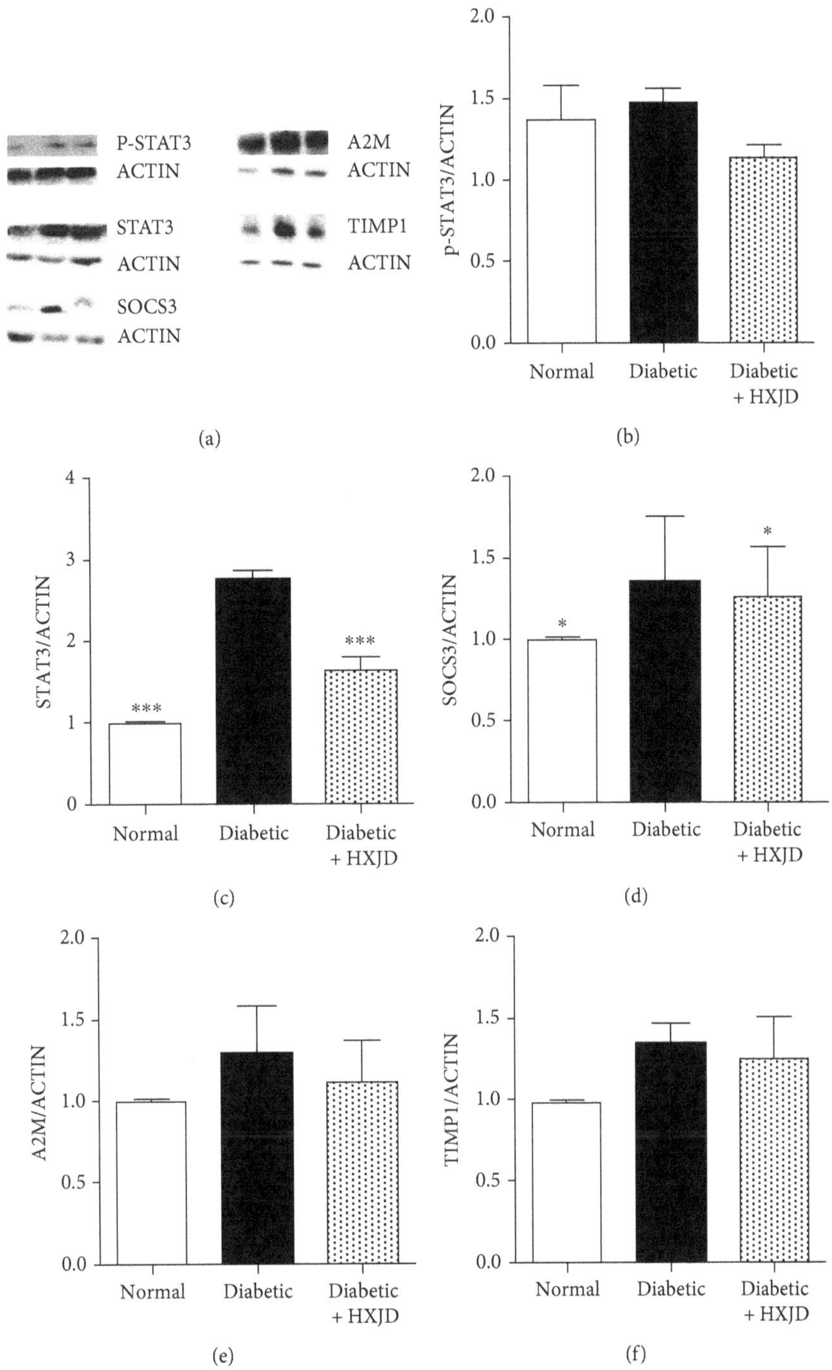

FIGURE 3: HXJD reduces the protein expression of p-STAT3, STAT3, SOCS3, A2M, and TIMP1 in the retinas of diabetic rats. The protein expression of p-STAT3, STAT3, SOCS3, A2M, and TIMP1 were measured using western blot. (a) Western blot images. (b–f) Image analysis of western blot. Data are represented as the means ± SEM. ($n = 3$). $^{*}P < 0.05$ and $^{***}P < 0.001$, compared with the diabetic group.

mediates the signals from several receptor tyrosine kinases, including EGF, PDGF, IL-6, IL-10, and IL-12 [13, 18]. SOCS3 negatively regulates the STAT3 pathway [14]. In recent years, an increasing number of studies have demonstrated that SOCS3 and STAT3 mediate inflammation, apoptosis, and insulin signalling in hyperglycaemia-stimulated retinal endothelial cells [19]. Moreover, STAT3 is reported to regulate the expression of VEGF in various cells in different kinds of circumstances [20, 21]. The results of the present study showed that HXJD reduced SOCS3 and STAT3 levels in the retinas of diabetic rats. In addition, previous studies have shown that HXJD attenuated retinal apoptosis and VEGF expression in diabetic rats [5]. Thus, we speculated that HXJD might ameliorate hyperglycaemia-induced apoptosis and VEGF in the retinas of diabetic rats through controlling SOCS3 or STAT3 expression.

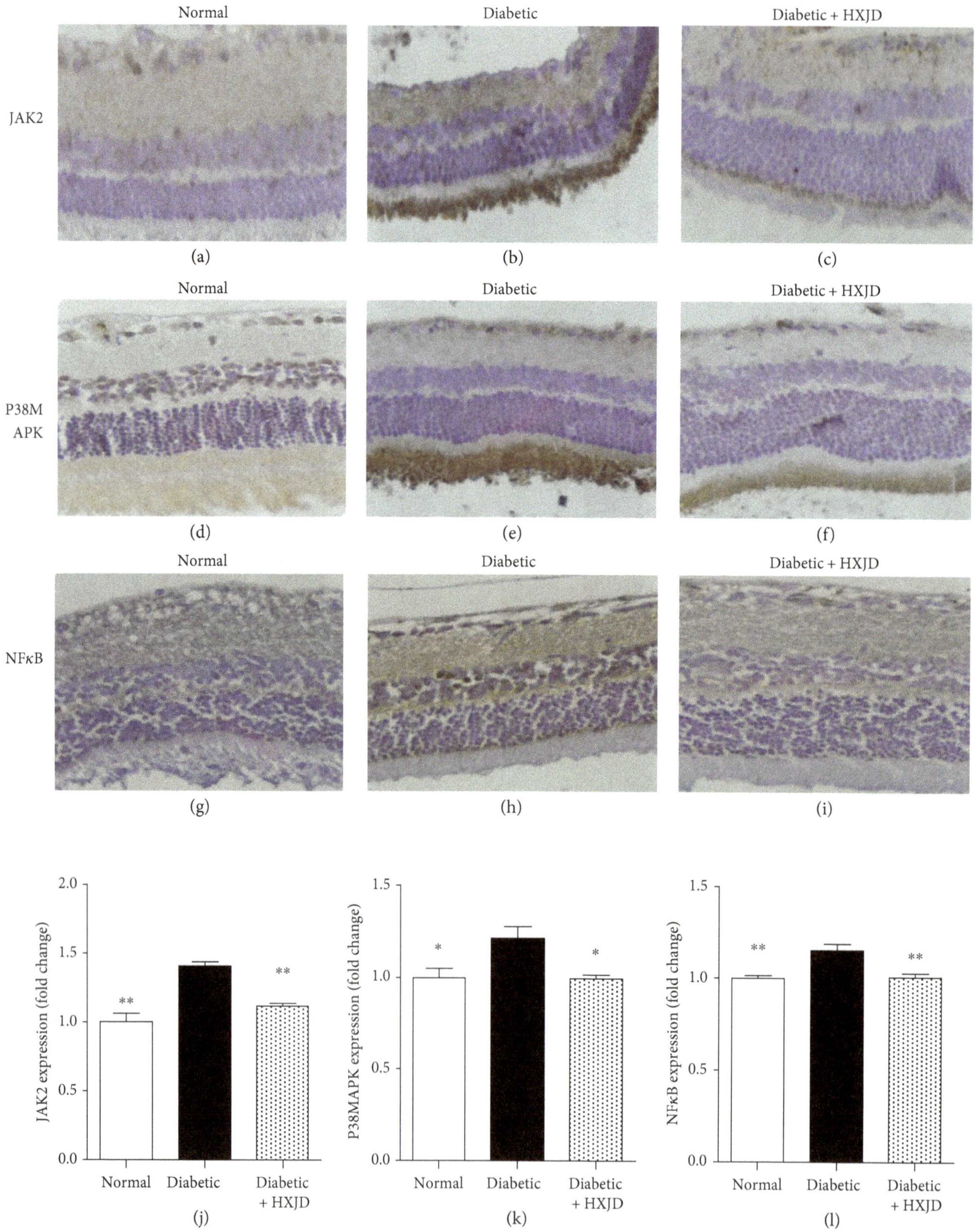

FIGURE 4: HXJD reduces the expression of JAK2, p38 MAPK, and NF-κB in the retinas of diabetic rats. Distribution and expression of JAK2, p38 MAPK, and NF-κB were measured using IHC. (a–c) IHC staining of JAK2. (d–f) IHC staining of p38 MAPK. (g–i) IHC staining of NF-κB. Magnification: ×400. (j–l) Image analysis of JAK2, p38 MAPK, and NF-κB. A reduction of JAK2, p38 MAPK, and NF-κB was observed in the HXJD group. Data are represented as the means ± SEM. ($n = 3$~6). $^{*}P < 0.05$ and $^{**}P < 0.01$, compared with the diabetic group.

Since SOCS3-STAT3 signal transduction is fundamental for a wide variety of physiological processes and pathological status, we further evaluated the proteins involved in SOCS3-STAT3 pathway. JAK2 activates and phosphorylates STAT3 [22]. In addition, NF-κB [23, 24] and p38 MAPK [25] promote SOCS3-STAT3 signalling. HXJD decreased the levels of JAK2, p38 MAPK, and NF-κB in the retinas of diabetic rats, suggesting that HXJD may arrest the progression of DR through the inhibition of the SOCS3-STAT3 pathway.

In addition to clarifying the pharmacological mechanisms of HXJD in preventing DR, the present study enhances the current understanding of the factors underlying the development of DR. In the early-middle stages of diabetes, the levels of STAT3 and SOCS3 in the retinas are increased [26–28]. Here, we provided new evidence that SOCS3 and STAT3 gene expression were also elevated in the late period of diabetes, as the present study was conducted for 32 weeks. Moreover, the results revealed that other key members of the SOCS3-STAT3 pathway were also disturbed in the diabetic retina. Taken together, these results suggest that the SOCS3-STAT3 signalling pathway participates in the development of DR. Furthermore, these results emphasize the necessity to assess the levels of SOCS3 and STAT3 in patients with DR. Thus, SOCS3 and STAT3 may represent new markers in the process of DR.

Since the SOCS3-STAT3 pathway plays a fundamental role in the pathogenesis of DR, it is necessary to explore new type of drugs for the treatment of DR by targeting SOCS3 or STAT3. Some drugs used to treat diabetes or DR, such as statins [27], enalapril [29], and tazones [30], decrease SOCS3 and STAT3 expression in diabetic animals. However, until recently, there are no drugs in clinical use that specifically target SOCS3-STAT3. We propose the exploration of a new active compound, HXJD, which targets SOCS3 or STAT3. In previous studies, we identified seven compounds in the active fractions of HXJD, including notoginsenoside R4, ginsenoside Ra3, ginsenoside Rb1, 20S-ginsenoside Rg2, ginsenoside Re, ginsenoside Rd, and notoginsenoside R1 [31]. And this time, we examined the content of rutin, berberine and ginsenoside Rg1 in HXJD. It has been reported berberine from *C. chinensis* notably attenuates the excessive expression of p-STAT3 in the rats with autoimmune myocarditis [32]. Ginsenoside Rb1 [33], from *P. notoginseng*, has been reported to decrease the levels of p-STAT3 and SOCS3 in ovarian cancer cells, microglial cells, and obese mice. Berberine and ginsenoside Rb1 may be among the active compounds of HXJD. However, until recently, studies supporting a role for the SOCS3-STAT3 pathway in the effect of HXJD are only correlative, as studies to determine the action of HXJD by directly blocking SOCS3-STAT3 signalling molecules are more rigorous.

4.2. Inhibited TIMP1-A2M Expression. The inhibition of matrix metalloproteases (MMPs) showed the second highest score among the HXJD-downregulated pathways (Table 1). TIMP1 belongs to a family of proteinase inhibitors that regulate matrix metalloproteinase activity by forming one-to-one stoichiometric complexes [15]. TIMP1 can irreversibly inactivate substrates, such as MMP2 and MMP9 [34], but the final response is dependent on the ratio of TIMPs/MMPs

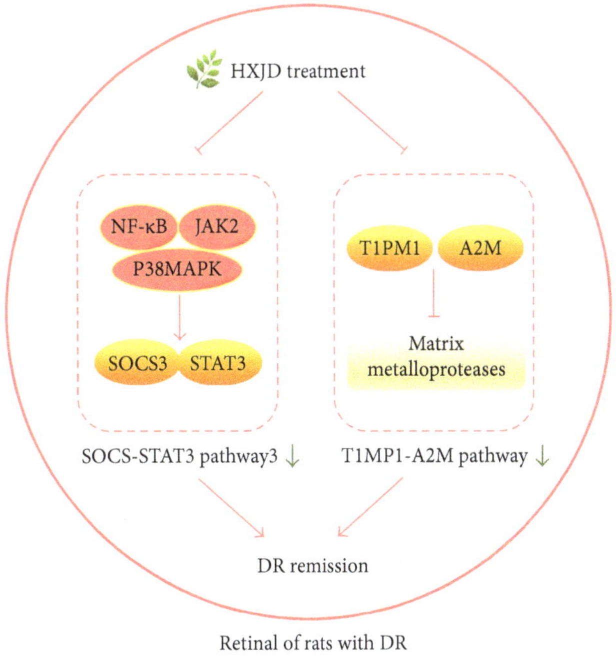

FIGURE 5: The molecular mechanism of HDJX in attenuating diabetic retinopathy in rats via the inhibition of SOCS3-STAT3 and T1MP1-A2M pathways. JAK2, NF-κB, and p38 MAPK promote SOCS3-STAT3 signalling, then SOCS3-STAT3 is involved in the progression of DR. TIMP1 and A2M play roles in matrix degradation which also take part in the process of DR. HXJD has the ability of inhibiting these pathological changes and thereafter it will ameliorate DR.

rather than the separate levels of TIMPs and MMPs [35]. In addition, TIMP1 shows protease-independent regulatory activities [36]. A2M is a broad-spectrum protease inhibitor that regulates the activities of MMPs [16]. Studies have demonstrated that TIMP1 and A2M expression is increased in the eyes of patients or animals with diabetic retinopathy [37, 38]. Thus, the TIMP1-A2M pathway likely plays important roles in the pathological processes of DR. In the present study, HXJD inhibited the dysregulated expression of TIMP1 and A2M genes. However, the gene levels of MMP2/9 showed no significant differences between the three groups analyzed. In a previous experiment, we observed that HXJD reduced the protein content of MMP2/9 in the retinas of diabetic rats [7], likely reflecting the fact that HXJD effects the posttranslation of MMP2/9. Taken together, these results suggested that HXJD might impede the progress of DR through the inhibition of the TIMP1-A2M pathway.

5. Conclusions

HXJD ameliorated retinal structure and pathological changes in DR. Moreover, the microarray data showed that HXJD reduced the expression of SOCS3 and STAT3 and decreased the levels of TIMP1 and A2M expression. In conclusion, the protective role of HXJD against DR is likely achieved by affecting SOCS3-STAT3 and TIMP1-A2M pathways (Figure 5). These findings lay the foundation for the active research and therapeutic application of HXJD.

Abbreviations

DR: Diabetic retinopathy
SOCS: Suppressor of cytokine signalling
STAT: Signal transducer and activator of transcription
TIMP: Tissue inhibitor matrix metalloproteinase
A2M: a2-macroglobulin
VEGF: Vascular endothelial growth factor
MMP: Matrix metalloproteinase
STZ: Streptozocin
ADCY1: Type 1 adenylyl cyclase
JAK: Janus kinase
NF-κB: Nuclear factor-kappa B
MAPK: Mitogen-activated protein kinase
ONL: Outer nuclear layer
INL: Inner nuclear layer
RGC: Retinal ganglion cells.

Conflicts of Interest

The authors declare that they have no competing interests.

Authors' Contributions

Wei Wang, Ruixin Zhu, and Jing Han conceived and designed the experiments. Hongliang Wang, Wei Xing, Tiantian Lv, and Zhenglin Wang performed the experiments. Yan Wu, Chun Li, Shuzhen Guo, and Shijie Tang analyzed the data. Jing Han and Ruixin Zhu contributed to the drafting of the manuscript. Hongliang Wang and Wei Xing contributed equally to this work. All authors read and approved the final manuscript.

Acknowledgments

This work was financially supported by grants from the National Nature Science Foundation of China (nos. 81673705, 31200986, and 41530105), the Natural Science Foundation of the Shanghai Committee of Science and Technology (no. 16ZR1449800), and the Fundamental Research Funds for the Central Universities (nos. 10247201546 and 2000219083).

References

[1] J. H. Wang, D. Ling, L. Tu, P. van Wijngaarden, G. J. Dusting, and G. S. Liu, "Gene therapy for diabetic retinopathy: are we ready to make the leap from bench to bedside?," *Pharmacology & Therapeutics*, 2017.

[2] M. S. Vasant, I. S. Kim, and D. K. Choi, "Recent update on the role of Chinese material medica and formulations in diabetic retinopathy," *Molecules*, vol. 22, p. 76, 2017.

[3] Q. Yao, J. Han, J. Yu, L. Huang, L. Pan, and W. Wang, "Effects of Huoxue Jiedu recipe on retinopathy in diabetic rats," *Zhong Guo Zhong Xi Yi Jie He Za Zhi*, vol. 32, no. 3, pp. 362–366, 2012.

[4] Q. Yao, J. Han, L. Huang, J. Yu, and W. Wang, "Effects of Huoxue Jiedu recipe on retinopathy in early diabetic rats," *Zhong Guo Zhong Xi Yi Jie He Za Zhi*, vol. 32, no. 9, pp. 1271–1274, 2012.

[5] Q. YAO, *Diabetic Retinopathy Intervened by Huo Xue JieDu Formula*, Beijing University of Chinese Medicine, Bei Jing, 2012.

[6] Q. Yao, J. Han, L. Huang, J. Yu, and W. Wang, "Effect of Huoxue Jiedu formula on vascular endothelial growth factor in retina of diabetic rats," *Zhong Hua Zhong Yi Yao Za Zhi*, vol. 27, 2012.

[7] H. Jie, H. Gai-mei, W. Zhi-qian et al., "Effects of Huoxue Jiedu recipe on retinopathy in diabetic rats," *Zhong Guo Quan Ke Yi Xue*, vol. 18, pp. 507–512, 2015.

[8] Q. Feng, W. Liu, S. S. Baker et al., "Multi-targeting therapeutic mechanisms of the Chinese herbal medicine QHD in the treatment of non-alcoholic fatty liver disease," *Oncotarget*, vol. 8, no. 17, pp. 27820–27838, 2017.

[9] Y. Wang, W. Lin, C. Li et al., "Multipronged therapeutic effects of Chinese herbal medicine Qishenyiqi in the treatment of acute myocardial infarction," *Frontiers in Pharmacology*, vol. 8, p. 98, 2017.

[10] N. Nagai, S. Deguchi, H. Otake, N. Hiramatsu, and N. Yamamoto, "Therapeutic effect of cilostazol ophthalmic nanodispersions on retinal dysfunction in streptozotocin-induced diabetic rats," *International Journal of Molecular Sciences*, vol. 18, 2017.

[11] A. Baufeld, D. Koczan, and J. Vanselow, "Induction of altered gene expression profiles in cultured bovine granulosa cells at high cell density," *Reproductive Biology and Endocrinology*, vol. 15, p. 3, 2017.

[12] D. Streif, E. Iglseder, C. Hauser-Kronberger, K. G. Fink, M. Jakab, and M. Ritter, "Expression of the non-gastric H^+/K^+ ATPase ATP12A in normal and pathological human prostate tissue," *Cellular Physiology and Biochemistry*, vol. 28, pp. 1287–1294, 2011.

[13] E. J. Hillmer, H. Zhang, H. S. Li, and S. S. Watowich, "STAT3 signaling in immunity," *Cytokine & Growth Factor Reviews*, vol. 31, pp. 1–15, 2016.

[14] A. R. Hackett, D. H. Lee, A. Dawood et al., "STAT3 and SOCS3 regulate NG2 cell proliferation and differentiation after contusive spinal cord injury," *Neurobiology of Disease*, vol. 89, pp. 10–22, 2016.

[15] G. Song, S. Xu, H. Zhang et al., "TIMP1 is a prognostic marker for the progression and metastasis of colon cancer through FAK-PI3K/AKT and MAPK pathway," *Journal of Experimental & Clinical Cancer Research*, vol. 35, p. 148, 2016.

[16] L. Bian, J. D. Yang, T. W. Guo et al., "Association study of the *A2M* and *LRP1* genes with Alzheimer disease in the Han Chinese," *Biological Psychiatry*, vol. 58, pp. 731–737, 2005.

[17] F. Sethna, W. Feng, Q. Ding, A. J. Robison, Y. Feng, and H. Wang, "Enhanced expression of ADCY1 underlies aberrant neuronal signalling and behaviour in a syndromic autism model," *Nature Communications*, vol. 8, article 14359, 2017.

[18] C. Garbers, S. Aparicio-Siegmund, and S. Rose-John, "The IL-6/gp130/STAT3 signaling axis: recent advances towards specific inhibition," *Current Opinion in Immunology*, vol. 34, pp. 75–82, 2015.

[19] Y. Jiang, Q. Zhang, C. Soderland, and J. J. Steinle, "TNFα and SOCS3 regulate IRS-1 to increase retinal endothelial cell apoptosis," *Cellular Signalling*, vol. 24, pp. 1086–1092, 2012.

[20] J. Lv, B. Sun, Z. Mai, M. Jiang, and J. Du, "STAT3 potentiates the ability of airway smooth muscle cells to promote angiogenesis by regulating VEGF signalling," *Experimental Physiology*, vol. 102, pp. 598–606, 2017.

[21] H. Okazaki, S. Tokumaru, Y. Hanakawa et al., "Nuclear translocation of phosphorylated STAT3 regulates VEGF-A-induced lymphatic endothelial cell migration and tube formation," *Biochemical and Biophysical Research Communications*, vol. 412, pp. 441–445, 2011.

[22] M. Hedvat, D. Huszar, A. Herrmann et al., "The JAK2 inhibitor AZD1480 potently blocks Stat3 signaling and oncogenesis in solid tumors," *Cancer Cell*, vol. 16, pp. 484–497, 2009.

[23] G. He and M. Karin, "NF-κB and STAT3 - key players in liver inflammation and cancer," *Cell Research*, vol. 21, pp. 159–168, 2011.

[24] V. Mishra, V. Baranwal, R. K. Mishra, S. Sharma, B. Paul, and A. C. Pandey, "Titanium dioxide nanoparticles augment allergic airway inflammation and Socs3 expression via NF-κB pathway in murine model of asthma," *Biomaterials*, vol. 92, pp. 90–102, 2016.

[25] C. Ehlting, W. S. Lai, F. Schaper et al., "Regulation of suppressor of cytokine signaling 3 (SOCS3) mRNA stability by TNF-α involves activation of the MKK6/p38^{MAPK}/MK2 cascade," *Journal of Immunology*, vol. 178, pp. 2813–2826, 2007.

[26] Y. Jiang, S. Thakran, R. Bheemreddy, W. Coppess, R. J. Walker, and J. J. Steinle, "Sodium salicylate reduced insulin resistance in the retina of a type 2 diabetic rat model," *PLoS One*, vol. 10, article e0125505, 2015.

[27] M. Al-Shabrawey, M. Bartoli, A. B. El-Remessy et al., "Role of NADPH oxidase and Stat3 in statin-mediated protection against diabetic retinopathy," *Investigative Ophthalmology & Visual Science*, vol. 49, pp. 3231–3238, 2008.

[28] J. H. Yun, S. W. Park, K. J. Kim et al., "Endothelial STAT3 activation increases vascular leakage through downregulating tight junction proteins: implications for diabetic retinopathy," *Journal of Cellular Physiology*, vol. 232, pp. 1123–1134, 2017.

[29] H. W. Kim, J. L. Kim, H. K. Lee, D. Y. Hur, I. H. Yun, and S. D. Kim, "Enalapril alters expression of key growth factors in experimental diabetic retinopathy," *Current Eye Research*, vol. 34, pp. 976–987, 2009.

[30] Y. Jiang, S. Thakran, R. Bheemreddy et al., "Pioglitazone normalizes insulin signaling in the diabetic rat retina through reduction in tumor necrosis factor α and suppressor of cytokine signaling 3," *The Journal of Biological Chemistry*, vol. 289, pp. 26395–26405, 2014.

[31] G. M. Hao, Y. G. Liu, Y. Wu et al., "The protective effect of the active components of ERPC on diabetic peripheral neuropathy in rats," *Journal of Ethnopharmacology*, vol. 202, pp. 162–171, 2017.

[32] X. Liu, X. Zhang, L. Ye, and H. Yuan, "Protective mechanisms of berberine against experimental autoimmune myocarditis in a rat model," *Biomedicine & Pharmacotherapy*, vol. 79, pp. 222–230, 2016.

[33] Y. Wu, Y. Yu, A. Szabo, M. Han, and X. F. Huang, "Central inflammation and leptin resistance are attenuated by ginsenoside Rb1 treatment in obese mice fed a high-fat diet," *PLoS One*, vol. 9, article e92618, 2014.

[34] G. P. Botta, M. Reichert, M. J. Reginato, S. Heeg, A. K. Rustgi, and P. I. Lelkes, "ERK2-regulated TIMP1 induces hyperproliferation of K-*Ras*^{G12D}-transformed pancreatic ductal cells," *Neoplasia*, vol. 15, pp. 359–372, 2013.

[35] F. J. Descamps, E. Martens, D. Kangave et al., "The activated form of gelatinase B/matrix metalloproteinase-9 is associated with diabetic vitreous hemorrhage," *Experimental Eye Research*, vol. 83, pp. 401–407, 2006.

[36] H. W. Jackson, V. Defamie, P. Waterhouse, and R. Khokha, "TIMPs: versatile extracellular regulators in cancer," *Nature Reviews Cancer*, vol. 17, pp. 38–53, 2017.

[37] M. C. Sanchez, J. D. Luna, P. F. Barcelona et al., "Effect of retinal laser photocoagulation on the activity of metalloproteinases and the α_2-macroglobulin proteolytic state in the vitreous of eyes with proliferative diabetic retinopathy," *Experimental Eye Research*, vol. 85, pp. 644–650, 2007.

[38] W. M. Freeman, G. V. Bixler, R. M. Brucklacher et al., "A multistep validation process of biomarkers for preclinical drug development," *The Pharmacogenomics Journal*, vol. 10, pp. 385–395, 2010.

11

Urinary Metabolomic Study of Chlorogenic Acid in a Rat Model of Chronic Sleep Deprivation using Gas Chromatography-Mass Spectrometry

Wei-ni Ma,[1] Ming-mei Zhou,[1] Xiao-jun Gou,[2] Le Zhao,[1] Fang Cen,[1,3] Ying Xu,[4] and Hong-yi Shen[5]

[1]*Center for Chinese Medicine Therapy and Systems Biology, Shanghai University of Traditional Chinese Medicine, Shanghai 201203, China*
[2]*Central Laboratory, Baoshan District Hospital of Integrated Traditional Chinese and Western Medicine of Shanghai, Shanghai University of Traditional Chinese Medicine, Shanghai 201999, China*
[3]*Experiment Center of Teaching & Learning, Shanghai University of Traditional Chinese Medicine, Shanghai 201203, China*
[4]*Department of Physiology, Shanghai University of Traditional Chinese Medicine, Shanghai 201203, China*
[5]*Research Center for Health and Nutrition, Shanghai University of Traditional Chinese Medicine, Shanghai 201203, China*

Correspondence should be addressed to Ming-mei Zhou; zhoumm368@163.com and Xiao-jun Gou; gouxiaojun1975@163.com

Academic Editor: Yong Wang

The urinary metabolomic study based on gas chromatography-mass spectrometry (GC-MS) had been developed to investigate the possible antidepressant mechanism of chlorogenic acid (CGA) in a rat model of sleep deprivation (SD). According to pattern recognition analysis, there was a clear separation among big platform group (BP), sleep deprivation group (SD), and the CGA (model + CGA), and CGA group was much closer to the BP group by showing a tendency of recovering towards BP group. Thirty-six significantly changed metabolites related to antidepressant by CGA were identified and used to explore the potential mechanism. Combined with the result of the classic behavioral tests and biochemical indices, CGA has significant antidepressant effects in a rat model of SD, suggesting that the mechanism of action of CGA might be involved in regulating the abnormal pathway of nicotinate and nicotinamide metabolism; glyoxylate and dicarboxylate metabolism; glycine, serine, and threonine metabolism; and arginine and proline metabolism. Our results also show that metabolomics analysis based on GC-MS is a useful tool for exploring biomarkers involved in depression and elucidating the potential therapeutic mechanisms of Chinese medicine.

1. Introduction

With the accelerated pace of modern society life and the rapid increase of the working pressure, more and more people are sleep deprived for various reasons, such as extending working hours into the night and curtailing or delaying sleep. Accumulating studies have demonstrated that sleep loss is closely coupled with an increased risk of immune system imbalance [1], neuroendocrine dysfunction [2], and depression [3, 4]. Depression and sleep disturbance are closely linked and have a mutual cause-and-effect relationship. Depressive patients usually suffer from poor sleep quality, while sleep disorder has also become a diagnostic criterion for major depressive disorder (MDD) [5]. However, the mechanism by which sleep loss causes depression is still unclear. At present, most of the drugs for the treatment of depression are mainly western medicine in clinical work, but, there are many problems such as recurrence and toxic side effects after drug withdrawal, which greatly affects the clinical efficacy. Chinese medicine has the characteristics of stable curative effect, lasting effect, and little adverse reaction. It has a long history in the prevention and treatment of depression and has become a hot topic both at home and abroad [6]. Recent studies demonstrate that polyphenols from natural herbs have the feasibility to be potential therapeutic agents in promoting resilience against SD-induced dysfunction [7, 8]. Chlorogenic acid (CGA), one of the major polyphenols of many fruits, has been reported in a number of previous studies as an antioxidant and neuroprotectant compound [9, 10]. CGA is also enriched in many traditional Chinese herbs, such as the bark of *Eucommia ulmoides* Oliver, leaves of *Hypericum origanifolium*, flowers of *Lonicera japonica*, fruits of *Crataegus oxyacantha* Linn., and whole herb of *Artemisia capillaris* Thunb. [11]. Studies showed that CGA could exhibit an antidepressive effect as shown in animal behavioral tests, accompanied by neuron protection and promotion of serotonin release in the brain [11, 12]. In addition, the evidence that decaffeinated coffee enriched in CGA elevated mood in 39 healthy volunteers also demonstrated CGA's character as an antidepressant/mood-elevator agent. However, the studies to address CGA's efficacy as well as the mechanism in treating depression are not sufficient [13].

As one of the hypothesis-free approaches, metabolomics is a powerful tool in discovering novel molecules involved in the pathophysiological process of various diseases [14]. Our previous metabolomic studies have revealed significant changes of systemic metabolites in novel depression rat model [15, 16]. Therefore, in order to develop pathophysiological understanding of the underlying relationship between SD stress and depression [17, 18], a metabolic method based on GC-MS with multivariate statistical techniques was used to assess the efficacy of CGA in a rat model of sleep deprivation in this study. In addition, potential biomarkers involved in antidepressant effects were identified and explored the underlying mechanism.

2. Methods

2.1. Rats and Treatments. Eighteen healthy male Wistar rats (purchased from Shanghai Sippr-BK Laboratory Animal Co. Ltd.) weighing 230–250 g were kept at a density of 6 per cage. All rats were maintained in a standard laboratory environment (20 ± 5°C and 55 ± 15% humidity) under a 12 h dark/light cycle (07:00–19:00 at 40 W light condition) with free access to a semipurified diet [19] and water ad libitum for 1 week prior to the experiment. The rats were randomly divided into three groups ($n = 6$): (1) big platform group (BP), (2) small platform group also known as sleep deprivation group (SD), and (3) CGA-treated group (SD + CGA); animals received CGA (purity ≥ 98%, HPLC method, Sichuan Weikeqi Bio-Tech Co. Ltd., China) extracted from *Eucommia ulmoides* Oliver bark and dissolved in anhydrous ethanol as a published paper performed [12] at a daily dose of 50 mg/kg of body weight, orally, while BP and SD group received normal saline solution from day 1 to day 7. Sleep deprivation procedure was carried out as previously reported [18]. During the study period, the experimental room was maintained at controlled conditions (20 ± 5°C and 55 ± 15% humidity) under a 12 h dark/light cycle (07:00–19:00 at 40 W light condition). The BP group was kept in the same environment with wide platforms (16 cm in diameter). Food and water were provided ad libitum by hanging on top of the tank. The research was approved by the Ethics Committee of the Shanghai University of TCM. Animal welfare and experimental protocols were strictly in accordance with the Guide for the Care and Use of Laboratory Animals and the ethics and regulations of the Shanghai University of TCM.

2.2. Behavioral Test. At the end of the SD paradigm, rats were subjected to FST and TST as previously performed [12] with little modification. Briefly, in the FST, rats were placed in a glass cylinder (19 cm in diameter and 50 cm in height) filled with 23–25°C water (30 cm in depth) for 6 min, individually. Immobility was defined as rats hopelessly kept heads above the water without other motions. In the TST, rats were suspended 50 cm above the floor by fixing the tip of the tail (1 cm). The total test procedure was counted during a test period of 5 min. Rats were considered immobile only when they hung passively in the absence of all movements.

2.3. Sample Collection. Each rat was put into a metabolic cage for 24 h urine sample collection after SD paradigm. Urine was centrifuged at 4°C, 13,000 rpm for 15 min, and the supernatant was stored at −80°C prior to GC-MS analysis. Subsequently, rats were sacrificed with chloral hydrate (300 mg/kg body weight, i.p.) anaesthesia for collection of abdominal aorta blood, which was centrifuged at 13,000 rpm and 4°C for 15 min, and the serum was collected and stored at −80°C.

2.4. Biochemical Analysis. Plasma concentrations of IL-6, TNF-α, CORT, and NE were determined by an automatic biochemistry analyzer using the commercial ELISA kits according to the protocols provided by the manufacturer.

2.5. Urine Sample Preparation, Derivation, and GC-MS Analysis. Prior to analysis, urine samples were thawed at room temperature and vortex-mixed for 30 s. Pretreatment procedure was in accordance with our previous published method [20] with little modification. Briefly, each 200 μL aliquot of urine samples was added into a 1.5 mL tube for centrifugation (13,000 rpm, 4°C for 10 min). Later, a 50 μL supernatant was transferred to a new tube containing 10 μL urease (30 U) and incubated at 37°C for 15 min. Then the metabolite extraction procedure was carried out after adding two internal standard solutions (10 μL of L-2-chlorophenylalanine in water, 0.3 mg/mL; 10 μL of heptadecanoic acid in methanol, 1 mg/mL) and 170 μL of methanol. After vortexing for 30 s, the mixture was centrifuged at 13,000 rpm (4°C) for 5 min. A 200 μL supernatant was transferred into GC vial and dried under in a stream of pure nitrogen gas at 30°C. The two-step derivatization process was the same as Qiu et al.'s method [21].

Each 1 μL derivatized sample was injected into the GC-MS instrument Agilent 6890/5975B GC/MSD system (Agilent Technologies, California, USA) in splitless injection mode. A programmed column temperature was optimized for successful separation (Table 1). Ion separation was achieved on a capillary column (Agilent J&W DB-5ms Ultra Inert 30 m × 250 μm, i.d., 0.25 μm film thickness) with high-purity carrier gas (helium) at a constant flow rate of

Table 1: Temperature program of column incubator in GC-MS.

Rate (°C·min^{-1})	Temperature (°C)	Hold time (min)
/	70	2
2.5	160	0
5	240	16

1.0 mL·min^{-1}. The temperatures of the injection port, transfer interface, and ion source were set to 280°C, 260°C, and 230°C, respectively. The measurements were collected using electron impact ionization (70 eV) in full scan mode (*m*/*z* 50–600).

2.6. Data Analysis. Data from the behavioral test and biochemical detection were analyzed using SPSS statistical package (SPSS program, version 21.0). Statistical analysis was performed using one-way ANOVA followed by LSD *t*-test. The results were expressed as means ± standard errors of the mean (SEM). Significant differences were considered at a level of $P < 0.05$ or $P < 0.01$. The graphs were generated using Prism 6.0 (GraphPad Software Inc., USA).

All the GC-MS raw files (.D) of urine samples were converted into easily identified NetCDF format via Agilent MSD workstation and subsequently preprocessed using the XCMS toolbox script with R 2.13.2 package with default settings for baseline correction, peak recognition, and calibration. The resulting data were exported into Microsoft Excel, and the peaks were normalized to the total sum of spectrum. The resulting three-dimensional matrix involving peak index (RT-*m*/*z* pair), sample numbers (observations), and normalized peak area percent were introduced into SIMCA-P 11.0 software package (Umetrics, Umeå, Sweden) for multivariate analyses including principal components analysis (PCA) and orthogonal partial least squares discriminant analysis (OPLS-DA). The quality of the constructed model was evaluated by R2X, R2Y, and Q2 parameters. The former two were used to assess the goodness of fit, and the last one was used to assess the predictive reliability of the model [22].

2.7. Metabolite Identification and Pathway Analysis. Metabolite identification was performed firstly with an already constructed standard library including retention time and mass spectra. The peaks which could not match with standard library were imported to the NIST MS 2.0 (NIST, Gaithersburg, MD) software to search compound information from the NIST 11 library. Compounds with a similarity of more than 70% were finally verified as available reference compounds. Online resource HMDB (http://www.hmdb.ca/) was applied to confirm endogenous metabolites from the above-identified compounds. The MetaboAnalyst 3.0 was used to forecast metabolic pathways significantly affected by those metabolites.

3. Results

3.1. CGA Intervention Reversed Depressive Behaviors Induced by SD. As shown in Figure 1, the immobility time of rats in the FST and TST was increased after SD paradigm ($P < 0.01$ in the FST, $P < 0.001$ in the TST) compared with BP group, meaning SD treatment could induce depressive symptoms. However, CGA intervention reversed the phenomenon ($P < 0.05$ in the FST and TST) relative to SD group.

3.2. Biochemical Index Determination. After SD treatment, the result of biochemical index (Figure 2) showed that the levels of serum IL-6 ($P < 0.001$), TNF-α ($P < 0.01$), CORT ($P < 0.001$), and NE ($P < 0.001$) were higher than those of control. However, after intervention of CGA for one week, these indexes were significantly reduced ($P < 0.05$) except for the level of TNF-α, which showed a tendency to decrease, yet.

3.3. Urine Metabolomics

3.3.1. Multivariate Statistical Analysis of the Urine Metabolic Data. PCA and OPLS-DA pattern were built to observe general clustering and trend among groups. Score plot analysis including Hotelling's T^2 plot revealed no outliers. In the PCA score plot, the SD group was separated from the BP and SD + CGA group (R2X = 0.776, $Q^2 = 0.426$) with little overlap, but no discernible clustering was observed between the BP and SD + CGA group (Figure 3(a)). However, in the OPLS-DA score plot, a clear separation was seen among three groups (Figure 3(b)). We found that SD samples were obviously discriminated from the BP group, and SD + CGA group showed a tendency to approach BP group (R2X = 0.758, R2Y = 0.975, $Q^2 = 0.853$). The result indicated that the model was constructed successfully and it seemed that CGA treatment could improve urine metabolic alternation induced by SD model.

The OPLS-DA pattern was also applied to for pairwise comparison and differential metabolites recognition. The BP and SD groups displayed significant deviation (R2X = 0.811, R2Y = 0.967, $Q^2 = 0.718$) in the OPLS-DA plot (Figure 4(a)). OPLS-DA score plots (Figure 4(b)) also showed that the CGA-treated groups had distinctive metabolic profiles from the SD group (R2X = 0.776, R2Y = 0.988, $Q^2 = 0.797$). It is worth noting that main automatic modeling parameters R^2 and Q^2 in pairwise groups were larger than 0.5, implying that patterns were robust and had good fitness as well as predictive ability.

3.3.2. Identification of Differential Metabolites for SD and Interventional Effect of CGA. On the base of OPLS-DA analysis, the V-plot (Figures 4(c) and 4(d)) and variable importance for projection (VIP) statistics were used for selecting discriminating ions responsible for group separation [23, 24]. Ions were firstly selected with a VIP value threshold set to 1.0. Then, a two-tailed Student's *t*-test was carried out to further validate discriminating variables with a *P* value less than 0.05, for the purpose of decreasing false-positive ions. These discriminating metabolites were identified and listed in Table 2.

Compared with BP group, 23 metabolites were significantly altered in SD group: creatinine, azelaic acid, protocatechuic acid, and 3,4-dihydroxyhydrocinnamic acid were upregulated, yet lactate, glycolic acid, alanine, nicotinic

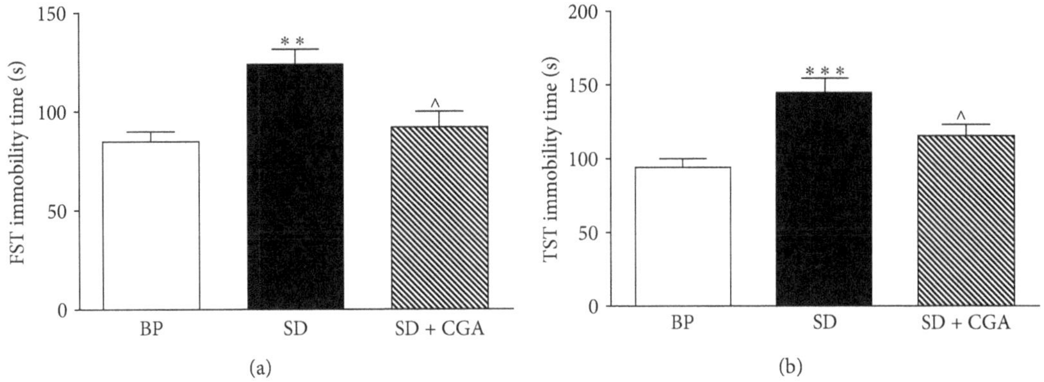

FIGURE 1: Behavioral test results of three groups (BP: big platform group, SD: sleep deprivation group, SD + CGA: SD group with CGA intervention). Data are expressed as mean ± SEM ($n = 6$). Compared with the BP group, $^{***}P < 0.001$; compared with the control group, $^{**}P < 0.01$; and compared with SD group, $^{\wedge}P < 0.05$.

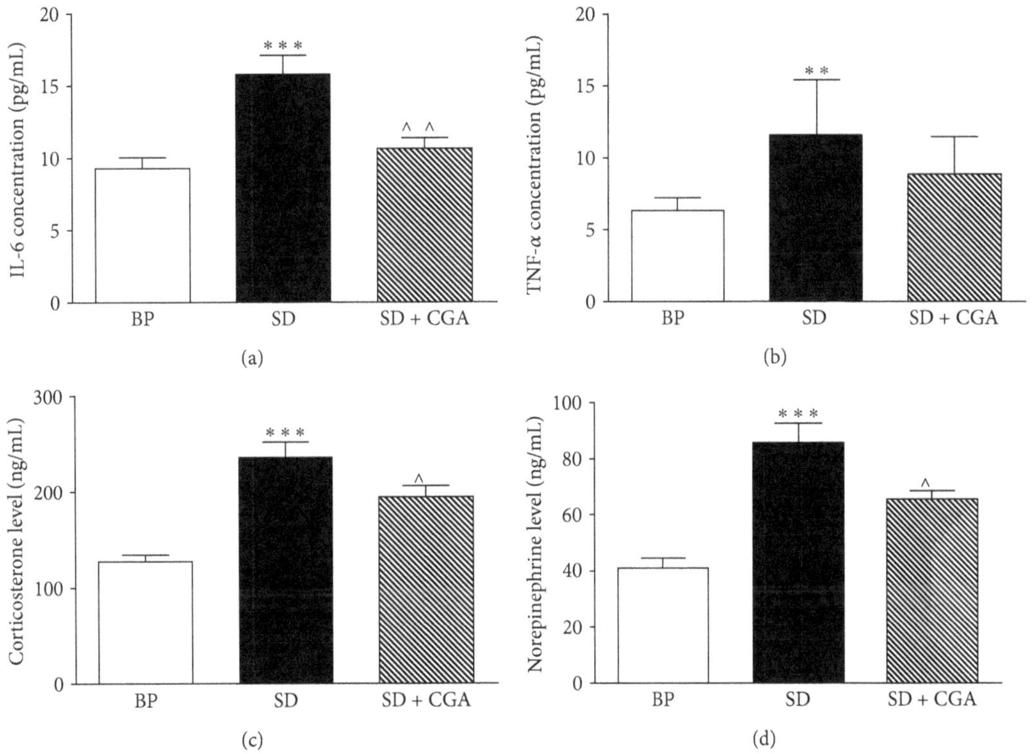

FIGURE 2: Biochemical index of the three rat group (BP: big platform group, SD: sleep deprivation group, SD + CGA: SD group with CGA intervention). Data are expressed as mean ± SEM ($n = 6$). Compared with the BP group, $^{***}P < 0.001$; compared with the control group, $^{**}P < 0.01$; compared with model group, $^{\wedge\wedge}P < 0.01$; and compared with SD group, $^{\wedge}P < 0.05$.

acid, phenylacetic acid, glycine, 4-deoxyerythronic acid, pipecolic acid, 2,4-dihydroxybutanoic acid, nicotinamide, uracil, threonic acid, suberic acid, putrescine, citric acid, hippurate, lysine, urate, and ascorbic acid were downregulated in SD group as shown in the heat map (Figure 5(a)). Intervention of SD rats with CGA moderated some metabolic alternations concomitantly with new metabolic changes as exhibited in the heat map (Figure 5(b)). Compared with SD rats, CGA raised glycolic acid, phenylacetic acid, putrescine, and urate apart from CGA conversion products (*trans*-ferulic acid, m-coumaric acid, 3,4-dihydroxyhydrocinnamic acid, and caffeic acid) [25]. In addition, CGA intervention solely increased 2-ethylhydracrylic acid, pyrocatechol, and 3-hydroxybenzoic acid, accompanied by decreased 4-deoxyerythronic acid, nicotinamide, asparagine, D-arabitol, fructose, D-mannose, 5-hydroxyindoleacetic acid, and 2-hydroxyglutarate. CGA modulated the above metabolites to display an antidepressant effect. These variations indicated that CGA not only exerted an antidepressant effect through modulating metabolic changes in SD rats but may also upregulate certain related compounds for a synergistic action.

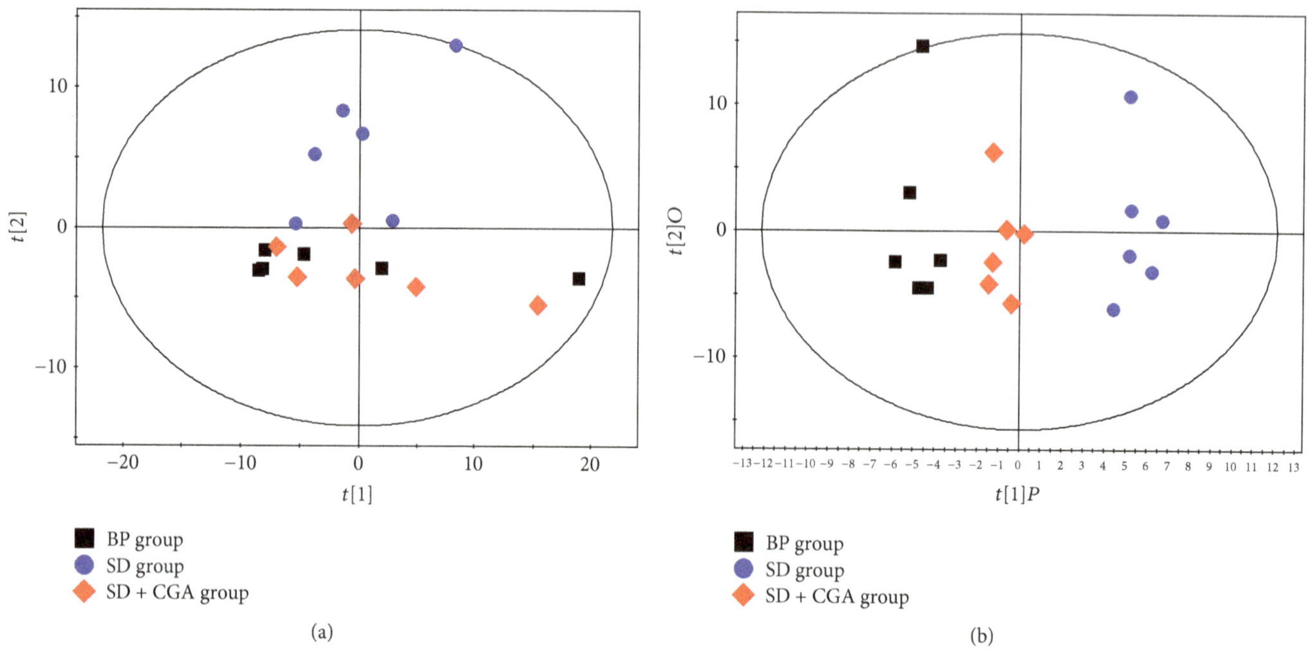

FIGURE 3: PCA and OPLS-DA score plot derived from the GC-MS analysis of urine from BP, SD, and SD + CGA groups.

3.3.3. Metabolic Pathway Analysis for SD and Interventional Effect of CGA. In order to further explore the underlying mechanism about perturbed metabolites induced by SD model and interventional effect of CGA, a comprehensive metabolic network was mapped by means of MetaboAnalyst 3.0 (http://www.metaboanalyst.ca/) through integration of all potential biomarkers identified in present research [26]. As a result, four disturbed metabolic pathways were found to be the most relevant pathways involved in SD-induced metabolic dysbiosis (impact > 0.1) [27]. They were nicotinate and nicotinamide metabolism; glyoxylate and dicarboxylate metabolism; glycine, serine, and threonine metabolism; and arginine and proline metabolism (Figure 6(a)). After CGA intervention, we observed that glyoxylate and dicarboxylate metabolism and glycine, serine, and threonine metabolism had little impact value, suggesting that CGA may play an intervening role through these two pathways (Figure 6(b)). The remaining two pathways which did not change could be potential targets for future drug design.

4. Discussion

4.1. CGA Intervention Reversed Depression-Like Behavior and Biochemical Alternation Induced by SD. In many studies related to SD, a close correlation between SD and depression has been established. Some researches demonstrated a positive connection, saying that SD could weaken depressive symptoms [28, 29], while others are on the opposite [30, 31]. Here, we applied the FST and TST to evaluate depressive symptoms of the rats. The results strengthened the point that SD could lead to a depression-like phenotype. And that CGA really could exert an antidepressant effect as mentioned earlier [12]. A great deal of research has shown that proinflammatory cytokines, such as IL-6, are possibly contributed to the emergence of depression-like symptoms [32] and sleep perturbations [33, 34]. After 7 days of experiment, the accumulation of proinflammatory cytokines (IL-6, TNF-α) was consistent with previous research [35], which validated the point that disturbance of the circadian clock by SD was involved in the regulation of inflammation [36]. As a consequence of SD intervention, changes in the concentration of the stress response hormones (CORT and NE) were observed in accord with published paper [37]. Serum CORT is a vital central nervous excitation enhancer, which is responsible for stress in rat by modulating hypothalamic-pituitary-adrenocortical (HPA) axis. Here, growing serum CORT was regarded as a biological indicator of SD stress [38]. Based on highly expressed brain adrenergic receptor mRNA levels, research demonstrated that norepinephrine system partly participated in chronic sleep restriction [39]. Therefore, higher levels of serum NE in SD group may confirm a possible neurochemical mechanism underlying SD. It is gratifying that CGA intervention reversed almost all stress-induced biochemical changes. All results above proved that depression could not only be caused by SD but also be attenuated or reversed by CGA to some extent.

4.2. Biochemical Interpretation. By applying the rat model of SD and the GC-MS metabolomics coupled with multivariate statistical analysis, the urine metabolic characteristics of CON, SD, and CGA-intervened rats were described. Thirty-six differential metabolites were identified as displayed in Table 2. Most interestingly, some of the metabolites such as critic acid, azelaic acid, alanine, and glycine have been identified as biomarkers for depression [40]. Through pathway analysis, we found four pathways to be disturbed after SD paradigm. In particular, two pathways were reversed after CGA intervention.

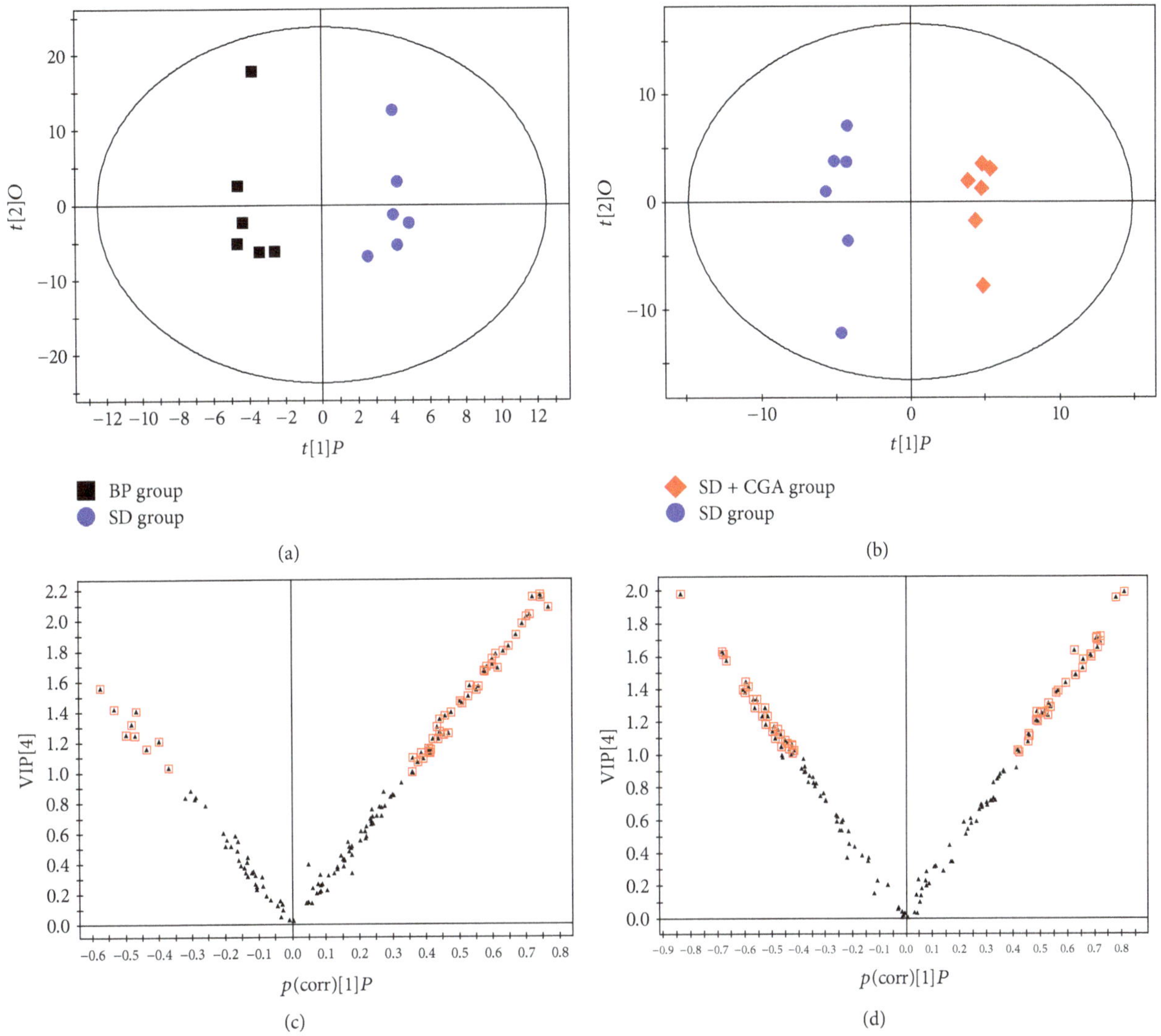

FIGURE 4: OPLS-DA score plots and V-plots between pairwise groups.

4.2.1. Disturbed Nicotinate and Nicotinamide Metabolism. In this altered pathway, nicotinic acid and nicotinamide were significantly reduced ($P < 0.01$) in SD group relative to BP group. Nicotinic acid, also known as vitamin B3, one of the thirteen essential vitamins for the human body, could be transformed into nicotinamide *in vivo*. As the major precursor of the coenzyme nicotinamide adenine dinucleotide (NADH/NAD^+), nicotinamide is crucial to life. In cells, it participates in the synthesis of NAD^+ and $NADP^+$, which are coenzymes involved in a wide variety of enzymatic oxidation-reduction reactions for energy production, and glycolysis, tricarboxylic acid cycle (TCA cycle), and the electron transport chain are the most notable events of these [41]. Here, reduction of nicotinamide in the urinary metabolites may be owing to oversynthesis of NAD^+ and $NADP^+$. To a certain degree, the symptom implied energy metabolism imbalance in rat organism. In the meantime, intervention with CGA did not have a regulatory effect on nicotinate and nicotinamide metabolism perturbation. But it could be potential targets for future drug design.

4.2.2. Disturbed Glyoxylate and Dicarboxylate Metabolism. The most significant alterations were attributed to changes in glyoxylate and dicarboxylate metabolism which include a variety of reactions involving glyoxylate or dicarboxylates. Glyoxylate is the conjugate base of glyoxylic acid. Likewise, dicarboxylates are the conjugate bases of dicarboxylic acids, a general class of organic compounds containing two carboxylic acid groups, such as oxalic acid or succinic acid. The glyoxylate cycle describes an important subset of these reactions involved in biosynthesis of carbohydrates from fatty acids or two-carbon precursors which enter the system as acetyl-coenzyme A. Our study found that two metabolites glycolic acid and citrate were involved in this metabolic process with marked reduction ($P < 0.01$) in SD rats. Interestingly, the alternation of glycolic acid was also found in the peripheral blood mononuclear cells from a rodent model of depression [42]. It is known to us all that citric acid plays a vital role in physiology which transforms fats, proteins, and sugars into carbon dioxide and provide the main energy for organisms. The alternations of metabolites strengthened the

Table 2: Discriminating urinary metabolites of three groups.

RT (min)	Metabolite	SD versus BP		SD + CGA versus SD	
		Fold change[a]	*t*-test (*P*)	Fold change[a]	*t*-test (*P*)
5.39	Lactate	0.42	0.02	—	—
5.62	Glycolic acid	0.55	0.00	1.68	0.01
6.01	Alanine	0.59	0.03	—	—
7.83	2-Ethylhydracrylic acid	—	—	3.21	0.00
8.87	Nicotinic acid	0.31	0.04	—	—
8.93	Phenylacetic acid	0.33	0.00	4.43	0.02
9.01	Glycine	0.58	0.01	—	—
9.19	Pyrocatechol	—	—	2.71	0.05
9.32	Methylsuccinic acid	—	—	0.57	0.01
9.56	4-Deoxyerythronic acid	0.36	0.05	—	—
9.98	Pipecolic acid	0.40	0.01	—	—
10.79	2,4-Dihydroxybutanoic acid	0.58	0.05	—	—
12.14	Nicotinamide	0.44	0.00	0.48	0.05
12.25	Uracil	0.62	0.01	—	—
12.56	Asparagine	—	—	0.34	0.00
13.66	Creatinine	2.28	0.01	—	—
13.79	Threonic acid	0.53	0.01	—	—
13.95	3-Hydroxybenzoic acid	—	—	3.14	0.00
14.32	Suberic acid	0.50	0.00	—	—
17.46	D-Arabitol	—	—	0.57	0.00
17.81	Putrescine	0.38	0.00	2.00	0.04
19.33	Azelaic acid	3.45	0.05	—	—
19.81	Protocatechuic acid	1.63	0.03	—	—
19.82	Citrate	0.42	0.00	—	—
20.28	Hippurate	0.45	0.04	—	—
20.97	Fructose	—	—	0.35	0.03
21.14	m-Coumaric acid	—	—	14.21	0.00
22.24	Lysine	0.46	0.03	—	—
22.73	3,4-Dihydroxyhydrocinnamic acid	9.48	0.02	87.95	0.02
26.08	*trans*-Ferulic acid	—	—	30.35	0.01
26.30	Urate	0.42	0.00	1.69	0.03
26.82	D-Mannose	—	—	0.51	0.03
26.96	Caffeic acid	—	—	59.50	0.01
27.43	5-Hydroxyindoleacetic acid	—	—	0.26	0.01
28.13	2-Hydroxyglutarate	—	—	0.43	0.05
33.58	Ascorbic acid	0.53	0.02	—	—

Note: [a]fold change = SD (SD + CGA)/BP (SD); —: unable to be detected.

former point for energy metabolism disorder. The CGA in the current study obviously restored glyoxylate and dicarboxylate metabolism by upregulated glycolic acid.

4.2.3. Disturbed Glycine, Serine, and Threonine Metabolism. The glycine, serine, and threonine metabolic pathway has been thought to provide a major energy metabolism precursor substance for TCA cycle [43]. Here, the core compound glycine in the pathway was decreased ($P < 0.01$) dramatically with a similar variation trend in the peripheral blood mononuclear cell of rat with chronic unpredictable mild stress rat model of depression [42], in the brain of depressive mice with chronic imipramine treatment [44] and in the serum of MDD patients [45]. Glycine is a simple, nonessential amino acid, which is involved in the production of phospholipids and collagen and releases energy, as well [46]. Literature documented that the concentration of plasma glycine may be responsible for regulating the mobilization of amino acids from peripheral tissues [47]. Glycine was also known as an inhibitory neurotransmitter, which possessed the ability to combine correlative NMDA receptor antagonist so as to exert an antidepressant effect [48]. Although after

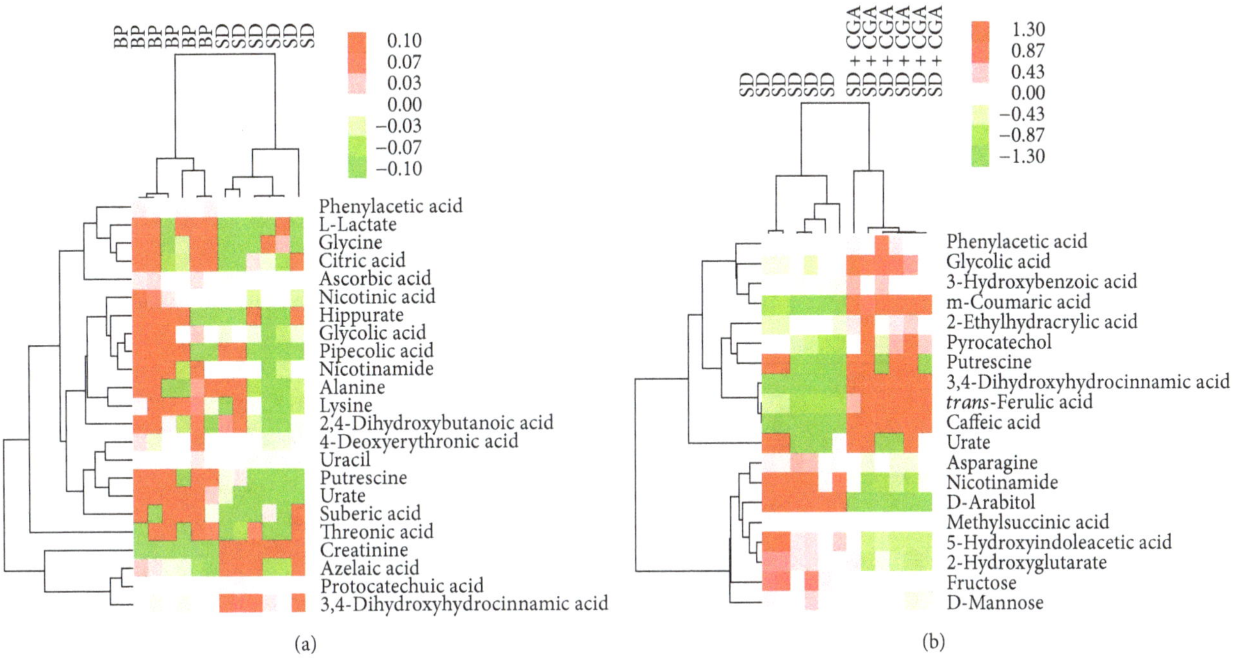

FIGURE 5: Heat maps described metabolite difference between pairwise groups.

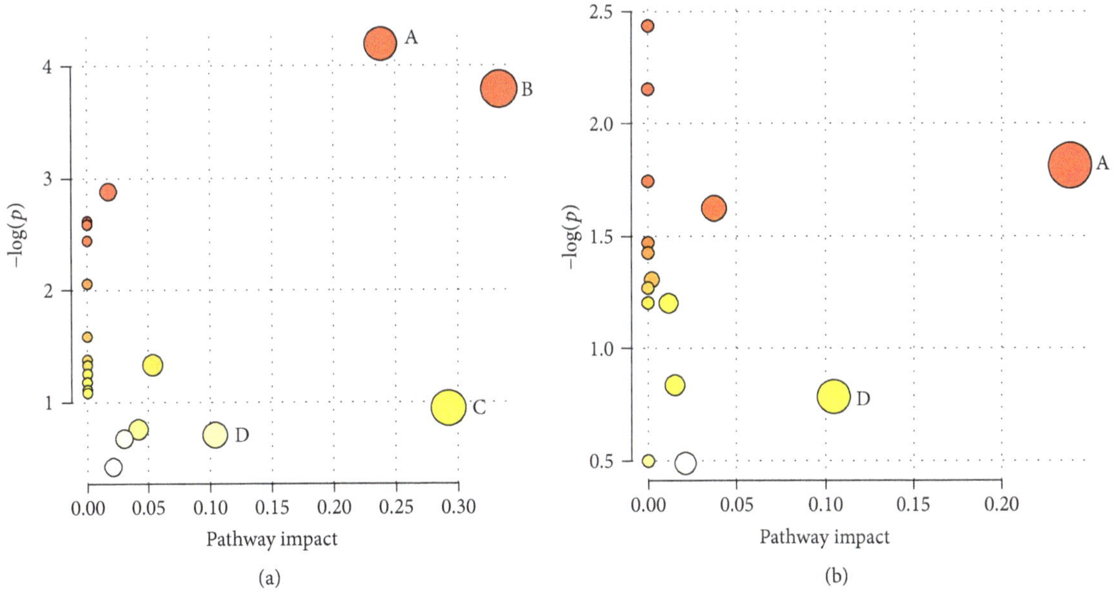

FIGURE 6: Summary of pathway analysis. (A) Nicotinate and nicotinamide metabolism, (B) glyoxylate and dicarboxylate metabolism, (C) glycine, serine, and threonine metabolism, and (D) arginine and proline metabolism.

CGA treatment, the impact value of glycine, serine, and threonine metabolism decreased, we did not find any relative metabolites got modulated. The question was warranted to be solved in the future.

4.2.4. Disturbed Arginine and Proline Metabolism. Arginine and proline metabolism is one of the central pathways for the biosynthesis of the amino acids. In addition to our discovery about disturbed arginine and proline metabolism after SD treatment, other researchers also found this significantly perturbed pathway in the prefrontal cortex of depressive rat model [24, 49]. We found that putrescine in the core position was the only disturbed metabolite ($P < 0.01$) in this pathway and had a consistent variation trend with published literature [50]. Actually, the small polyamine putrescine (1,4-diaminobutane) is ubiquitously and readily found in all three domains of life. It is a precursor, through N-aminopropylation or N-aminobutylation, to

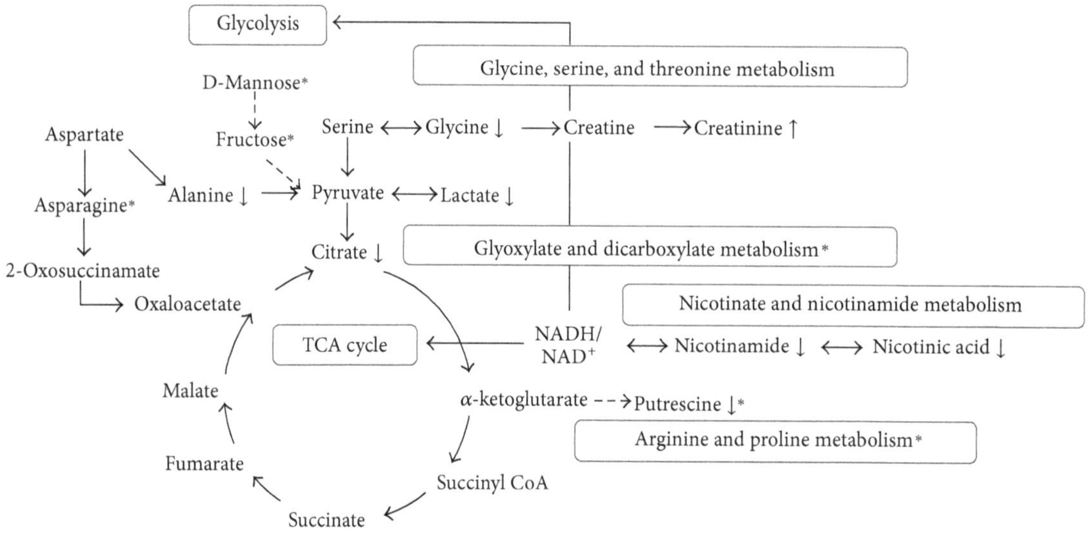

Figure 7: The energy perturbed process in response to SD exposure and CGA intervention. The levels of changed metabolites in SD group compared to BP group were labeled with (↓) downregulated or (↑) upregulated; *metabolites or metabolic pathway could be regulated by CGA. Solid arrows represent direct connections, and dashed arrows represent multiple and indirect connections between two compounds. The contents of the blue box represent energy processes or metabolic pathways.

biosynthesize longer polyamines spermidine, sym-homospermidine, spermine, and thermospermine or even longer and branched chain polyamines, which enhance phosphorylation processes of functional proteins in neurons involved in the therapeutic mechanisms of antidepressants [51]. Putrescine is also biochemically modified for purposes of metabolic regulation and catabolism. Previous studies proved that polyamine levels are also possible to be regulated by stress and might play a role in depression. For instance, the levels of putrescine in the nucleus accumbens, spermidine, and spermine in the hippocampus were rescored after antidepressant treatment [50]. A research by administering spermidine intracerebroventricularly further revealed the antidepressant-like activity of the spermidine [52]. In light of existing studies, we hypothesized that the metabolic change of putrescine not only is too simple but also may be related to the development of depression; further research is warranted.

4.3. Metabolite-Energy Network. Depending on metabolites discussed above and KEGG Pathway Database (http://www.genome.jp/kegg/), we drew a diagram concerning energy metabolism process (Figure 7). All in all, these findings corroborated previously published works which indicated that depression may be a kind of energy metabolic disorder.

Limitations to this study must be declared. First of all, the sample numbers in this study were relatively small; thus, future studies with larger cohorts should be performed to validate our findings. Regrettably, nearly all experimental rats were involved in stress. Actually, the CON group rats in normal cage-housing should be taken into consideration. In terms of value, our studies displayed how SD affected rats' organism alterations. Finally, owing to the diverse biochemical properties and wide concentration range of metabolites, further metabolomic methods or multiple metabolomic platforms (i.e., targeted metabolomics and nuclear magnetic resonance) should be employed in future studies.

5. Conclusion

It is the first time for us to demonstrate that SD indeed caused depressive symptoms in rat experiments. Depression-like behavioral phenotypes are reflected by alterations in serum inflammatory cytokines and hormone levels, and alternation of urinary metabolites. We further suggest that the development of depression is involved in alternations of energy-related pathways. CGA displays an antidepressant role by regulating inflammatory cytokines and hormone levels and metabolite related pathways. This research opens up new sights for a functional link between sleep disturbance and depression as well as the molecular mechanism of CGA in SD paradigm of depression.

Abbreviations

SD:	Sleep deprivation
GC-MS:	Gas chromatography coupled to mass spectrometry
CGA:	Chlorogenic acid
MDD:	Major depressive disorder
BP:	Big platform group
SD + CGA:	CGA-treated group
FST:	Forced swimming test
TST:	Tail suspension test
IL-6:	Interleukin- 6
TNF-α:	Tumor necrosis factor alpha
NE:	Norepinephrine
CORT:	Corticosterone
PCA:	Principal component analysis
OPLS-DA:	Orthogonal projections to latent structures discriminant analysis.

Conflicts of Interest

The authors declare that they have no competing interests.

Authors' Contributions

All authors participated in the writing and discussed the final text.

Acknowledgments

This study was financially supported by the National Natural Science Foundation of China (no. 81473475), Baoshan District Hospital of Integrated Traditional Chinese and Western Medicine of Shanghai combined with the National Natural Education fund (no. GZRPYJJ-201605), the innovation project of Shanghai University of TCM (no. ZYX-CXYJ-018), and Shanghai Municipal Education Committee (no. 2013JW17), further accelerated by the development of Chinese medicine three-year action plan of Shanghai (2014–2016) (no. ZY3-RCPY-3-1013), New Emerging Interdisciplinary of Traditional Chinese Medicine in Shanghai Municipality: Nutrition of Traditional Chinese Medicine, and Key Laboratory of Medicinal Animal and Plant Resources in Qinghai-Tibet Plateau, Xining, Qinghai (no. 2017-z-y25).

References

[1] C. M. Almeida and A. Malheiro, "Sleep, immunity and shift workers: a review," *Sleep Science*, vol. 9, no. 3, pp. 164–168, 2016.

[2] L. Y. Chen, T. Cheng, C. H. Tsai et al., "Early-life sleep deprivation persistently depresses melatonin production and bio-energetics of the pineal gland: potential implications for the development of metabolic deficiency," *Brain Structure & Function*, vol. 220, no. 2, pp. 663–676, 2015.

[3] C. Baglioni, G. Battagliese, B. Feige et al., "Insomnia as a predictor of depression: a meta-analytic evaluation of longitudinal epidemiological studies," *Journal of Affective Disorders*, vol. 135, no. 1–3, pp. 10–19, 2011.

[4] R. E. Roberts and H. T. Duong, "The prospective association between sleep deprivation and depression among adolescents," *Sleep*, vol. 37, no. 2, pp. 239–244, 2014.

[5] American Psychiatric Association, *Diagnostic and Statistical Manual of Mental Disorders*, American Psychiatric Association, Washington, 5 edition, 2013.

[6] S. L. Liu, Y. X. Xu, Y. J. Huang, and C. Q. Zeng, "Antidepressant effect of total alkaloids from Uncaria in chronic restraint stress mice," *Journal of Guangdong Pharmaceutical University*, vol. 33, no. 1, pp. 72–76, 2017.

[7] W. Zhao, J. Wang, W. Bi et al., "Novel application of brain-targeting polyphenol compounds in sleep deprivation-induced cognitive dysfunction," *Neurochemistry International*, vol. 89, pp. 191–197, 2015.

[8] S. Cho, M. Yoon, A. N. Pae et al., "Marine polyphenol phlorotannins promote non-rapid eye movement sleep in mice via the benzodiazepine site of the GABAA receptor," *Psychopharmacology*, vol. 231, no. 14, pp. 2825–2837, 2014.

[9] Y. F. Chu, P. H. Brown, B. J. Lyle et al., "Roasted coffees high in lipophilic antioxidants and chlorogenic acid lactones are more neuroprotective than green coffees," *Journal of Agricultural and Food Chemistry*, vol. 57, no. 20, pp. 9801–9808, 2009.

[10] J. Bouayed, H. Rammal, A. Dicko, C. Younos, and R. Soulimani, "Chlorogenic acid, a polyphenol from *Prunus domestica* (Mirabelle), with coupled anxiolytic and antioxidant effects," *Journal of the Neurological Sciences*, vol. 262, no. 1-2, pp. 77–84, 2007.

[11] J. Wu, H. Chen, H. Li et al., "Antidepressant potential of chlorogenic acid-enriched extract from eucommia ulmoides oliver bark with neuron protection and promotion of serotonin release through enhancing synapsin i expression," *Molecules*, vol. 21, no. 3, p. 260, 2016.

[12] S. H. Park, Y. B. Sim, P. L. Han, J. K. Lee, and H. W. Suh, "Antidepressant-like effect of chlorogenic acid isolated from *Artemisia capillaris* Thunb," *Animal Cells and Systems*, vol. 14, no. 4, pp. 253–259, 2010.

[13] V. Cropley, R. Croft, B. Silber et al., "Does coffee enriched with chlorogenic acids improve mood and cognition after acute administration in healthy elderly? A pilot study," *Psychopharmacology*, vol. 219, no. 3, pp. 737–749, 2012.

[14] H. Kitano, "Systems biology: a brief overview," *Science*, vol. 295, no. 5560, pp. 1662–1664, 2002.

[15] X. Wang, T. Zhao, Y. Qiu et al., "Metabonomics approach to understanding acute and chronic stress in rat models," *Journal of Proteome Research*, vol. 8, no. 5, pp. 2511–2518, 2009.

[16] X. Y. Wang, C. Y. Zeng, J. C. Lin et al., "Metabonomics approach to assessing the modulatory effects of St John's wort, ginsenosides, and clomipramine in experimental depression," *Journal of Proteome Research*, vol. 11, no. 12, pp. 6223–6230, 2012.

[17] V. Hajali, V. Sheibani, H. Ghazvini et al., "Effect of castration on the susceptibility of male rats to the sleep deprivation-induced impairment of behavioral and synaptic plasticity," *Neurobiology of Learning and Memory*, vol. 123, pp. 140–148, 2015.

[18] M. J. Maturana, C. Pudell, A. D. Targa et al., "REM sleep deprivation reverses neurochemical and other depressive-like alterations induced by olfactory bulbectomy," *Molecular Neurobiology*, vol. 51, no. 1, pp. 349–360, 2015.

[19] M. P. Gonthier, M. A. Verny, C. Besson, C. Rémésy, and A. Scalbert, "Chlorogenic acid bioavailability largely depends on its metabolism by the gut microflora in rats," *The Journal of Nutrition*, vol. 133, no. 6, pp. 1853–1859, 2003.

[20] W. Liao, H. Wei, X. Y. Wang et al., "Metabonomic variations associated with AOM-induced precancerous colorectal lesions and resveratrol treatment," *Journal of Proteome Research*, vol. 11, no. 6, pp. 3436–3448, 2012.

[21] Y. P. Qiu, G. X. Cai, M. M. Su et al., "Serum metabolite profiling of human colorectal cancer using GC-TOFMS and UPLC-QTOFMS," *Journal of Proteome Research*, vol. 8, no. 10, pp. 4844–4850, 2009.

[22] S. Mahadevan, S. L. Shah, T. J. Marrie, and C. M. Slupsky, "Analysis of metabolomic data using support vector machines," *Analytical Chemistry*, vol. 80, no. 19, pp. 7562–7570, 2008.

[23] J. Zhao, Y.-H. Jung, C. G. J. Ung, K. H. Chun, S. W. Kwon, and J. Lee, "Metabolomic identification of biochemical changes induced by fluoxetine and imipramine in a chronic mild stress mouse model of depression," *Scientific Reports*, vol. 5, no. 1, p. 8890, 2015.

[24] Y. Wu, Y. Fu, C. Rao et al., "Metabolomic analysis reveals metabolic disturbances in the prefrontal cortex of the

lipopolysaccharide-induced mouse model of depression," *Behavioural Brain Research*, vol. 308, pp. 115–127, 2016.

[25] G. Baeza, E. M. Bachmair, S. Wood, R. Mateos, L. Bravo, and B. de Roos, "The colonic metabolites dihydrocaffeic acid and dihydroferulic acid are more effective inhibitors of *in vitro* platelet activation than their phenolic precursors," *Food & Function*, vol. 8, no. 3, pp. 1333–1342, 2017.

[26] M. Yu, H. M. Jia, C. Zhou, Y. Yang, L. L. Sun, and Z. M. Zou, "Urinary and fecal metabonomics study of the protective effect of Chaihu-Shu-Gan-San on antibiotic-induced gut microbiota dysbiosis in rats," *Scientific Reports*, vol. 7, article 46551, 2017.

[27] X. Wang, B. Yang, A. Zhang, H. Sun, and G. Yan, "Potential drug targets on insomnia and intervention effects of Jujuboside A through metabolic pathway analysis as revealed by UPLC/ESI-SYNAPT-HDMS coupled with pattern recognition approach," *Journal of Proteomics*, vol. 75, no. 4, pp. 1411–1427, 2012.

[28] H. Azizi, J. Hwang, V. Suen et al., "Sleep deprivation induces changes in 5-HT actions and 5-HT 1A, receptor expression in the rat hippocampus," *Neuroscience Letters*, vol. 655, pp. 151–155, 2017.

[29] L. Muzio, V. Brambilla, L. Calcaterra, P. D'Adamo, G. Martino, and F. Benedetti, "Increased neuroplasticity and hippocampal microglia activation in a mice model of rapid antidepressant treatment," *Behavioural Brain Research*, vol. 311, pp. 392–402, 2016.

[30] Z. Wang, L. Chen, L. Zhang, and X. Wang, "Paradoxical sleep deprivation modulates depressive-like behaviors by regulating the MAOA levels in the amygdala and hippocampus," *Brain Research*, vol. 1664, pp. 17–24, 2017.

[31] T. M. D. C. Daniele, P. F. C. de Bruin, E. R. V. Rios, and V. M. S. de Bruin, "Effects of exercise on depressive behavior and striatal levels of norepinephrine, serotonin and their metabolites in sleep-deprived mice," *Behavioural Brain Research*, vol. 332, pp. 16–22, 2017.

[32] J. Hannestad, N. DellaGioia, and M. Bloch, "The effect of antidepressant medication treatment on serum levels of inflammatory cytokines: a meta-analysis," *Neuropsychopharmacology*, vol. 36, no. 12, pp. 2452–2459, 2011.

[33] A. A. Prather, M. Rabinovitz, B. G. Pollock, and F. E. Lotrich, "Cytokine-induced depression during IFN-α treatment: the role of IL-6 and sleep quality," *Brain, Behavior, and Immunity*, vol. 23, no. 8, pp. 1109–1116, 2009.

[34] C. H. Chan, X. W. Ji, J. S. Chan et al., "Effects of the integrative mind-body intervention on depression, sleep disturbances and plasma IL-6," *Psychotherapy and Psychosomatics*, vol. 86, no. 1, pp. 54–56, 2016.

[35] M. Chennaoui, D. Gomez-merino, C. Drogou et al., "Effects of exercise on brain and peripheral inflammatory biomarkers induced by total sleep deprivation in rats," *Journal of Inflammation*, vol. 12, no. 1, pp. 1–10, 2015.

[36] S. Norah and D. F. Dinges, "Sleep and inflammation," *Nutrition Reviews*, vol. 65, no. 12, pp. 244–252, 2007.

[37] P. J. Martins, M. S. Marques, S. Tufik, and V. D'Almeida, "Orexin activation precedes increased NPY expression, hyperphagia, and metabolic changes in response to sleep deprivation," *American Journal of Physiology-Endocrinology and Metabolism*, vol. 298, no. 3, pp. E726–E734, 2010.

[38] I. S. Hairston, N. F. Ruby, S. Brooke et al., "Sleep deprivation elevates plasma corticosterone levels in neonatal rats," *Neuroscience Letters*, vol. 315, no. 1-2, pp. 29–32, 2001.

[39] Y. Kim, L. Chen, R. W. Mccarley, and R. E. Strecker, "Sleep allostasis in chronic sleep restriction: the role of the norepinephrine system," *Brain Research*, vol. 1531, no. 39, pp. 9–16, 2013.

[40] J. J. Chen, C. J. Zhou, P. Zheng et al., "Differential urinary metabolites related with the severity of major depressive disorder," *Behavioural Brain Research*, vol. 332, pp. 280–287, 2017.

[41] P. Belenky, K. L. Bogan, and C. Brenner, "NAD metabolism in health and disease," *Trends in Biochemical Sciences*, vol. 32, no. 1, pp. 12–19, 2007.

[42] J. Li, G. Tang, K. Cheng et al., "Peripheral blood mononuclear cell-based metabolomic profiling of a chronic unpredictable mild stress rat model of depression," *Molecular BioSystems*, vol. 10, no. 11, pp. 2994–3001, 2014.

[43] R. G. Schwartz, E. J. Barrett, C. K. Francis, R. Jacob, and B. L. Zaret, "Regulation of myocardial amino acid balance in the conscious dog," *Journal of Clinical Investigation*, vol. 75, no. 4, pp. 1204–1211, 1985.

[44] M. Nagasawa, T. Murakami, S. Tomonaga, and M. Furuse, "The impact of chronic imipramine treatment on amino acid concentrations in the hippocampus of mice," *Nutritional Neuroscience*, vol. 15, no. 5, pp. 26–33, 2012.

[45] P. Zheng, H. C. Gao, Q. Li et al., "Plasma metabonomics as a novel diagnostic approach for major depressive disorder," *Journal of Proteome Research*, vol. 11, no. 3, pp. 1741–1748, 2012.

[46] W. Gillota, "Energy Drink Formula and Method," 2003, US, US20030104107.

[47] R. L. Landau and K. Lugibihl, "Effect of glucagon on concentration of several free amino acids in plasma," *Metabolism*, vol. 18, no. 4, pp. 265–276, 1969.

[48] W. L. Zhu, S. J. Wang, M. M. Liu et al., "Glycine site N-methyl-D-aspartate receptor antagonist 7-CTKA produces rapid antidepressant-like effects in male rats," *Journal of Psychiatry & Neuroscience*, vol. 38, no. 5, pp. 306–316, 2013.

[49] X. Zhou, L. Liu, Y. Zhang et al., "Metabolomics identifies perturbations in amino acid metabolism in the prefrontal cortex of the learned helplessness rat model of depression," *Neuroscience*, vol. 343, pp. 1–9, 2017.

[50] S. Genedani, S. Saltini, A. Benelli, M. Filaferro, and A. Bertolini, "Influence of SAMe on the modifications of brain polyamine levels in an animal model of depression," *Neuroreport*, vol. 12, no. 18, pp. 3939–3942, 2001.

[51] A. Benelli, M. Filaferro, A. Bertolini, and S. Genedani, "Influence of S-adenosyl-L-methionine on chronic mild stress-induced anhedonia in castrated rats," *British Journal of Pharmacology*, vol. 127, no. 3, pp. 645–654, 1999.

[52] S. Wośko, A. Serefko, K. Socała et al., "An anti-immobility effect of spermine in the forced swim test in mice," *Pharmacological Reports*, vol. 66, no. 2, pp. 223–227, 2014.

12

The Genetic Diversity and Geographic Differentiation of the Wild Soybean in Northeast China based on Nuclear Microsatellite Variation

Hongkun Zhao,[1,2] Yumin Wang,[3] Fu Xing,[1] Xiaodong Liu,[3] Cuiping Yuan,[3] Guangxun Qi,[2] Jixun Guo,[1] and Yingshan Dong[2,3]

[1]*Key Laboratory of Vegetation Ecology, Ministry of Education, Institute of Grassland Sciences, Northeast Normal University, Changchun 130024, China*
[2]*Institute of Crop Germplasms, Jilin Academy of Agricultural Sciences, Gongzhuling 136100, China*
[3]*Institute of Soybean Research, Jilin Academy of Agricultural Sciences, Changchun 130033, China*

Correspondence should be addressed to Jixun Guo; gjixun666@163.com and Yingshan Dong; ysdong@cjaas.com

Academic Editor: Marco Gerdol

In this study, the genetic diversity and population structure of 205 wild soybean core collections in Northeast China from nine latitude populations and nine longitude populations were evaluated using SSR markers. A total of 973 alleles were detected by 43 SSR loci, and the average number of alleles per locus was 22.628. The mean Shannon information index (I) and the mean expected heterozygosity were 2.528 and 0.879, respectively. At the population level, the regions of 42°N and 124°E had the highest genetic diversity among all latitudes and longitudes. The greater the difference in latitude was, the greater the genetic distance was, whereas a similar trend was not found in longitude populations. Three main clusters (1N, <41°N-42°N; 2N, 43°N-44°N; and 3N, 45°N–>49°N) were assigned to populations. AMOVA analysis showed that the genetic differentiation among latitude and longitude populations was 0.088 and 0.058, respectively, and the majority of genetic variation occurred within populations. The Mantel test revealed that genetic distance was significantly correlated with geographical distance ($r = 0.207$, $p < 0.05$). Furthermore, spatial autocorrelation analysis showed that there was a spatial structure ($\omega = 119.58$, $p < 0.01$) and the correlation coefficient (r) decreased as distance increased within a radius of 250 km.

1. Introduction

The annual wild soybean (*Glycine soja* Sieb. and Zucc.), the direct progenitor of the cultivated soybean (*Glycine max* (Linn.) Merr.), is a predominantly self-pollinated annual plant species [1, 2]. It is widely distributed across most provinces of China, with the exceptions of Qinghai, Xinjiang, and Hainan [3]. In Northeast China, the wild soybean is well known for its abundant populations, high population density, and rich phenotypic types [4]. A total of 48 wild soybean in situ reserves have been established in China, 14 of which are located in Northeast China. A total of 8518 wild soybean accessions have been ex situ conserved in the National Gene Bank, and nearly half were collected from Northeast China [5]. Some reports have suggested that Northeast China was probably a very important diversity center for wild soybeans [6–8].

Genetic diversity is essential to population stability and is the basis of the evolution of species [9]. The genetic diversity and population structure of wild soybeans have been described in many reports [10–13]. The wild soybean in Northeast China, which is a very important ecotype, has been used in many previous studies [14–18]. The genetic diversity of wild soybeans in Northeast China was often compared with that in other areas using morphological traits and various molecular markers. Alternatively, wild soybeans from some specific areas have been analyzed for their distribution patterns, origin, evolution, classification, and so on. Accordingly, results have not been consistent because different accessions (or populations), numbers of samples, and

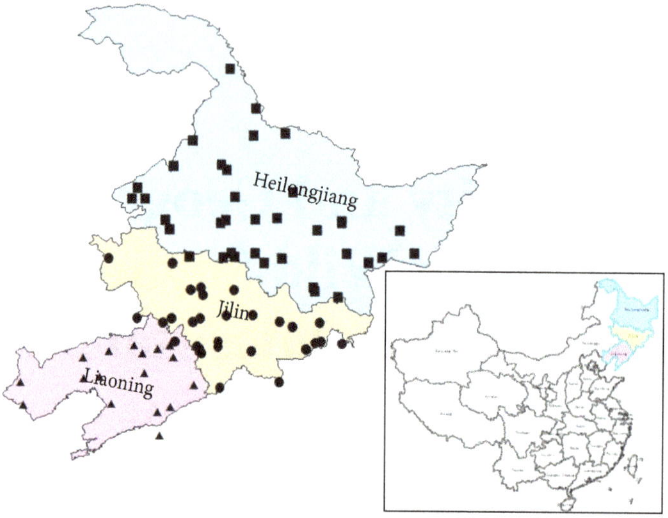

Figure 1: Geographic distribution map of wild soybean accessions in Northeast China. ■, ●, and ▲ represent the wild soybean core collections from Heilongjiang (HLJ), Jilin (JL), and Liaoning (LN), respectively.

analysis methods have been used. SSR markers have proven to be a reliable tool for determining the diversity of the wild soybean [19]. However, regarding wild soybeans in Northeast China as a single population, their genetic diversity and population structure have not been fully understood using molecular markers; in particular, the genetic differentiation of geographical populations of this region has not been well investigated.

The objectives of this work were to determine the extent of genetic variation in the wild soybean in Northeast China, to elucidate the geographical structure and genetic differentiation within and between latitude or longitude populations, and, ultimately, to provide valuable information for the scientific protection and efficient utilization of wild soybean resources in Northeast China.

2. Materials and Methods

2.1. Plant Material. All the wild soybean accessions in this study were preserved in the Gene Bank of Jilin Academy of Agricultural Sciences. A total of 205 wild soybean accessions from Northeast China were selected from the core collection established by Zhao et al. [20]. The collection sites of these samples were located within the region of 39–52°N and 119–133°E, covering 84 counties or districts. Nine latitude populations and nine longitude populations were divided based on a degree interval when the sample size was higher than 10; otherwise, the samples were combined into adjacent latitude or longitude population (see Figure 1 and Supplementary Table 1).

2.2. SSR Analysis. One single fresh leaf was used to extract genomic DNA for each accession using a modified CTAB method [21]; 43 SSR markers (see Supplementary Table 2), developed from 60 core loci [22] on 20 genetic linkage groups, were used to detect nuclear DNA variation. The primer sequences, with their linkage group locations, are available at https://www.soybase.org/dlpages/#soybasedata.

The 25 μL PCR reaction buffer consisted of 2.5 μL 10x PCR buffer (100 mmol/L), Tris-HCl (pH 8.3), 500 mmol/L KCl, 200 mmol/L $MgCl_2$ (0.001% gelatin, 0.1% Np-40), 0.4 μL dNTP (2.5 mmol/L), 2.0 μL each of forward and reverse primers (10 pmol/L), 2.0 μL total DNA (20 ng/μL), 1.0 μL Taq DNA polymerase (2 U/μL), and 15.1 μL ddH_2O. PCR was performed using the T100™ thermal cycler (Bio-Rad, USA) with the following cycle conditions: an initial denaturing at 95°C for 5 min, followed by 35 cycles of 95°C denaturing for 45 s, 52–57°C annealing for 45 s, 68°C extension for 45 s, and a final extension at 72°C for 10 min.

Amplified products were fractionated by electrophoresis through 6% denaturing polyacrylamide gels and stained with silver staining, which could detect 60 bp to 500 bp fragments and had a high resolution of 2 bp. The size of the stained band was analyzed based on its migration distance relative to the 100 bp DNA ladder (MBI Fermentas) using AlphaView software (version 1.3.0.7).

2.3. Data Analysis. The amplification fragments of genomic DNA by each SSR marker were scored based on the migration difference. The data format was converted accordingly in Microsoft Excel. The number of alleles (N_a), Shannon-Weaver index (I), expected heterozygosity (H_e), observed heterozygosity (H_0), fixation index (F_{is}), genetic differentiation coefficient (F_{st}), genetic identity and genetic distance, molecular variation analysis of variance (AMOVA), Mantel tests, and spatial autocorrelation coefficients were computed by GenAIEx v6.5 [23]. The Shannon-Weaver index (I) and expected heterozygosity (H_e) for evaluating the diversity were measured according to the formulas $I = -1 \times \text{sum}\,(p_i \times \text{Ln}\,(p_i))$ and $H_e = 1 - \text{sum}\,p_i^2$, where p_i is the frequency of the ith allele. H_0 is generally lower than H_e due to inbreeding, and $F_{is} = (1 - H_0)/H_e$. Outcrossing rate (t) was calculated using the equation $t = (1 - F_{is})/(1 + F_{is})$ [24]. Based on the matrix of Nei's genetic distance [25], the dendrogram was constructed using NTSYSpc21 [26]. Spatial autocorrelation analysis was performed using "spatial-single

TABLE 1: Genetic diversity of 205 wild soybean accessions by 43 nSSRs.

Number	Primer	LG	N_a	I	H_0	H_e	F_{is}	t
1	satt005	Dlb + W	28	2.862	0.025	0.924	0.973	0.014
2	satt022	N	23	2.593	0.010	0.890	0.989	0.006
3	satt099	L	17	2.257	0.020	0.862	0.976	0.012
4	satt112	E	19	2.571	0.010	0.909	0.989	0.006
5	satt146	F	21	2.582	0.034	0.902	0.962	0.019
6	satt168	B2	21	2.743	0.000	0.915	1.000	0.000
7	satt180	C1	18	2.145	0.005	0.813	0.994	0.003
8	satt184	Dla + Q	24	2.658	0.005	0.900	0.995	0.003
9	satt197	B1	31	2.633	0.005	0.876	0.994	0.003
10	satt216	Dlb + W	26	2.295	0.005	0.835	0.994	0.003
11	satt226	D2	22	2.477	0.005	0.888	0.994	0.003
12	satt236	A1	17	2.495	0.015	0.903	0.984	0.008
13	satt239	I	22	2.295	0.005	0.822	0.994	0.003
14	satt242	K	22	2.580	0.005	0.882	0.994	0.003
15	satt243	O	26	2.951	0.015	0.937	0.984	0.008
16	satt267	Dla + Q	18	2.440	0.059	0.890	0.934	0.034
17	satt268	E	18	2.284	0.005	0.859	0.994	0.003
18	satt279	H	28	2.815	0.005	0.916	0.995	0.003
19	satt281	C2	29	2.869	0.005	0.920	0.994	0.003
20	satt286	C2	36	3.124	0.020	0.941	0.979	0.011
21	satt300	A1	20	2.540	0.005	0.903	0.995	0.003
22	satt307	C2	26	2.877	0.061	0.923	0.934	0.034
23	satt308	M	27	2.718	0.015	0.893	0.983	0.009
24	satt309	G	13	1.424	0.005	0.586	0.992	0.004
25	satt334	F	20	2.384	0.010	0.862	0.988	0.006
26	satt345	O	26	2.827	0.005	0.921	0.995	0.003
27	satt346	M	16	2.075	0.005	0.787	0.994	0.003
28	satt352	G	24	2.675	0.010	0.902	0.989	0.006
29	satt373	L	22	2.311	0.005	0.847	0.994	0.003
30	satt386	D2	21	2.489	0.005	0.891	0.994	0.003
31	satt390	A2	25	2.593	0.000	0.894	1.000	0.000
32	satt429	A2	26	2.786	0.005	0.916	0.995	0.003
33	satt431	J	23	2.628	0.005	0.893	0.995	0.003
34	satt434	H	19	2.394	0.000	0.878	1.000	0.000
35	satt453	B1	20	2.301	0.005	0.847	0.994	0.003
36	satt462	L	23	2.618	0.010	0.895	0.989	0.006
37	satt487	O	19	2.279	0.005	0.866	0.994	0.003
38	satt530	N	19	2.330	0.005	0.868	0.994	0.003
39	satt571	B2	21	2.520	0.026	0.897	0.971	0.015
40	satt586	F	33	2.994	0.005	0.932	0.995	0.003
41	satt588	K	26	2.619	0.000	0.896	1.000	0.000
42	satt590	M	22	2.570	0.029	0.899	0.967	0.017
43	satt596	J	16	2.071	0.005	0.835	0.994	0.003
	Mean		22.628	2.528	0.011	0.879	0.987	0.007

LG = linkage group.

pop"; spatial distance was set as 100 km, and the number of permutations and the bootstraps of the selection mode were set as 999 times.

The population genetic structure was predicted by STRUCTURE 2.3.4 [27]. The default k value was set to 1 to 12, and ten runs were performed for each value of k to test stability of the results. The MCMC (Markov chain Monte Carlo) value was set as 100,000 burn-in with 200,000 iterations. The correct number of genetic clusters was inferred according to a value of Δk ($\Delta k = \text{mean}|\ln P(k+1) - 2\ln P(k) + \ln P(k-1)|/\text{Sd}|\ln P(k)|$) [28].

3. Results

3.1. SSR Polymorphism and Geographic Variation. Two hundred five representative wild soybean accessions across Northeast China were used in this study. A total of 973 alleles were detected from 43 SSR loci, and the percentage of polymorphic loci was 100%. The number of alleles per SSR marker varied from 13 (Satt309) to 36 (Satt286), with an average of 22.628. The mean Shannon information index (I) and expected heterozygosity (H_e) were 2.528 and 0.879, respectively (see Table 1).

At the population level (see Table 2), the Shannon information index (I) of nine latitude populations ranged from 1.300 to 2.419, with an average of 1.716; the expected heterozygosity (H_e) ranged from 0.661 to 0.884, with an average of 0.756. The region of 42°N had the highest genetic diversity among all latitudes. The Shannon information index (I) of nine longitudes varied from 1.659 to 2.368, averaging 1.919; the expected heterozygosity (H_e) varied from 0.751 to 0.880, averaging 0.805. The region of 124°E had the highest genetic diversity among all longitudes. As shown in Tables 3 and 4, the smaller the latitude difference was, the higher the genetic identity was; however, similar trends were not found in longitude populations. Two major geographical clustering groups (N and S) can be seen in the UPGMA dendrogram. Group N consists of four northern latitude groups (46°N–>49°N), and group S consists of five southern latitude groups (<41°N–45°N) (see Figure 2). The results indicate that the genetic diversity of wild soybean accessions in Northeast China is related to their latitudinal origin.

3.2. Population Structure and Genetic Differentiation. The STRUCTURE procedure was run to predict genetic structure for each predefined latitude and longitude population. When k was 3, Δk was the highest, which indicated that 3 main clusters had been identified (see Table 5). For the latitude population, 83.8% of individuals from the <41°N region and 50.2% of individuals from the 42°N region were assigned to Cluster1N, most individuals from the 43°N to 44°N region to Cluster2N, and individuals from the 45°N to >49°N regions to Cluster3N. These results were roughly consistent with those of hierarchical cluster analysis (see Figure 2). Some admixtures were found among the 41°N–45°N regions, which probably were important transitional areas. For the longitude population, most of the individuals from the <122°E region were separated from all others and formed a cluster (Cluster1E); three regions of 123°E, 125°E, and 129°E were assigned to Cluster2E, and three regions of 127°E, 128°E, and 130°E to Cluster3L. Admixtures were found widely among 9 predefined longitude populations. The regions of 42°N, 124°E, and 126°E were special, as they were not dominated (<60%) by the three clusters.

Table 2: Genetic diversity of different geographical populations in Northeast China.

Longitude Pop.	I	H_e	Longitude Pop.	I	H_e
<41°N	1.915	0.808	<122°E	1.959	0.820
42°N	2.419	0.884	123°E	1.808	0.789
43°N	2.063	0.816	124°E	2.368	0.880
44°N	1.621	0.751	125°E	2.067	0.833
45°N	1.882	0.790	126°E	2.094	0.827
46°N	1.302	0.661	127°E	1.771	0.780
47°N	1.521	0.733	128°E	1.661	0.769
48°N	1.300	0.662	129°E	1.880	0.797
>49°N	1.419	0.699	>130°E	1.659	0.751
Total	1.716	0.756	Total	1.919	0.805

Pop. = population.

The F_{st} value was used to evaluate the genetic differentiation of wild soybean populations at the scales of latitude and longitude in Northeast China (see Tables 3 and 4). Pairwise F_{st} values for latitude populations ranged from 0.040 to 0.183, and pairwise F_{st} values for longitude populations ranged from 0.034 to 0.132. In general, moderate differentiation ($0.05 < F_{st} < 0.15$) [29] was observed between most latitude and longitude populations, and the genetic differentiation among adjacent groups was relatively low. These results were confirmed by AMOVA; most of the genetic variations were found within latitude and longitude populations (see Table 6).

The Mantel test indicated that there was a positive correlation between geographic and genetic distance ($r = 0.207$, $p < 0.05$), which suggests that geographic distance limits gene flow among populations and influences the genetic structure. For further analyses, spatial autocorrelation analysis was performed by distance classes of 100 km, and a general decline was found in the correlation coefficient (r) with distance. The correlation values were negative and significant up to 250 km. This revealed that there is a clinal spatial structure in wild soybeans in Northeast China ($\omega = 119.58$, $p < 0.01$) (see Figure 3).

4. Discussion

The wild soybean in Northeast China is an important ecotype, and its genetic diversity has been widely studied by using phenotypic traits [7, 30, 31] and molecular markers [17]. The common view has been that the wild soybean in this region possesses a high genetic variation. In this study, the average allele number, Shannon index (I), and expected heterozygosity (H_e) were 22.628, 2.528, and 0.879, respectively, which were significantly higher than previously reported results [32–34]. The rich genetic variation could be attributed to the fact that the samples in this study were selected from the core collection, which has been defined as a subset of a crop species preserved with the most abundant repetitiveness [35, 36]. On the other hand, the large sample size and various geographical origins might also result in high diversity; 205 samples used in this study, accounting for

TABLE 3: Nei's genetic distance and genetic differentiation among latitude populations.

Group	<41°N	42°N	43°N	44°N	45°N	46°N	47°N	48°N	>49°N
<41°N	—	0.042	0.079	0.101	0.111	0.183	0.150	0.178	0.162
42°N	0.447	—	0.040	0.061	0.062	0.120	0.091	0.119	0.105
43°N	0.674	0.346	—	0.049	0.066	0.139	0.111	0.140	0.119
44°N	0.790	0.489	0.334	—	0.064	0.156	0.120	0.142	0.126
45°N	0.941	0.499	0.435	0.402	—	0.096	0.080	0.118	0.112
46°N	1.428	0.801	0.818	0.825	0.446	—	0.093	0.155	0.163
47°N	1.441	0.788	0.800	0.784	0.486	0.417	—	0.086	0.098
48°N	1.432	0.858	0.886	0.762	0.615	0.640	0.414	—	0.074
>49°N	1.491	0.896	0.823	0.77	0.674	0.780	0.550	0.349	—

Pop. = population; genetic differentiation coefficient (F_{st}) (above diagonal); Nei's genetic identity (below diagonal).

TABLE 4: Nei's genetic distance and genetic differentiation among longitude populations.

Group	<122°E	123°E	124°E	125°E	126°E	127°E	128°E	129°E	>130°E
<122°E	—	0.103	0.049	0.082	0.096	0.124	0.106	0.095	0.132
123°E	0.974	—	0.055	0.052	0.044	0.066	0.075	0.065	0.083
124°E	0.535	0.535	—	0.036	0.039	0.060	0.055	0.049	0.069
125°E	0.818	0.438	0.404	—	0.036	0.052	0.065	0.042	0.076
126°E	0.929	0.345	0.374	0.306	—	0.035	0.035	0.037	0.034
127°E	1.304	0.504	0.563	0.426	0.286	—	0.060	0.061	0.072
128°E	0.980	0.572	0.520	0.537	0.294	0.460	—	0.050	0.056
129°E	0.897	0.500	0.479	0.364	0.298	0.468	0.397	—	0.068
>130°E	1.168	0.545	0.530	0.533	0.227	0.463	0.377	0.443	—

Pop. = population; genetic differentiation coefficient (F_{st}) (above diagonal); Nei's genetic identity (below diagonal).

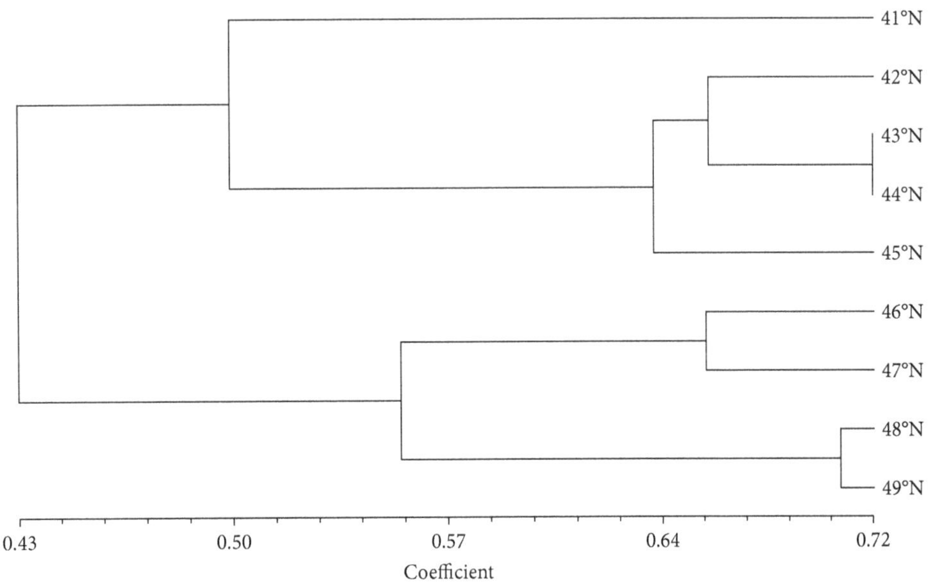

FIGURE 2: UPGMA dendrogram based on Nei's genetic identity among the latitudes.

85% of core collections in Northeast China, were selected from 242 core collections developed by Zhao et al. [20]. These accessions may have a continuous distribution in Northeast China (see Supplementary Table 1). Furthermore, 43 SSR primers covering 20 linkage groups might also be important causes for the detection of richer genetic variation [37].

The wild soybean is widely distributed in Northeast Asia. The higher its genetic diversity is, the greater its habitat-expansion capacity and environmental adaptation are [38]. Therefore, it also forms a specific natural distribution pattern [39]. In our study, the results indicate that the region of 42°N and 124°E has the highest level of genetic diversity

TABLE 5: Inferred population structure based on latitude populations and longitude populations.

Pop.	Inferred clusters Cluster1N	Cluster2N	Cluster3N	Pop.	Inferred clusters Cluster1E	Cluster2E	Cluster3E
<41°N	0.158	0.004	0.838	<122°E	0.891	0.106	0.003
42°N	0.431	0.066	0.502	123°E	0.078	0.690	0.232
43°N	0.916	0.019	0.065	124°E	0.462	0.360	0.179
44°N	0.943	0.054	0.002	125°E	0.122	0.714	0.164
45°N	0.246	0.741	0.013	126°E	0.029	0.433	0.539
46°N	0.034	0.962	0.003	127°E	0.002	0.303	0.695
47°N	0.021	0.977	0.003	128°E	0.053	0.305	0.642
48°N	0.052	0.946	0.002	129°E	0.110	0.614	0.276
>49°N	0.151	0.847	0.002	>130°E	0.003	0.203	0.795

Pop. = population.

TABLE 6: AMOVA analysis of different geographical populations.

Group	Source of variation	SS	MS	Est. Var.	Percentage of variation	F_{st}	p
Latitude	Among pops	723.151	90.394	1.695	9%	0.088	0.001
	Within pops	7058.358	17.602	17.602	91%		
Longitude	Among pops	539.012	67.376	1.111	6%	0.058	0.001
	Within pops	7242.498	18.061	18.061	94%		

Probability, p (rand ≥ data), for F_{st} is based on standard permutation across the full data set. F_{st} = Est. Var. among pops/(Est. Var. among pops + Est. Var. within pops); SS = the sums of squares; MS = the mean sums of squares; Est. Var. = the estimated variance.

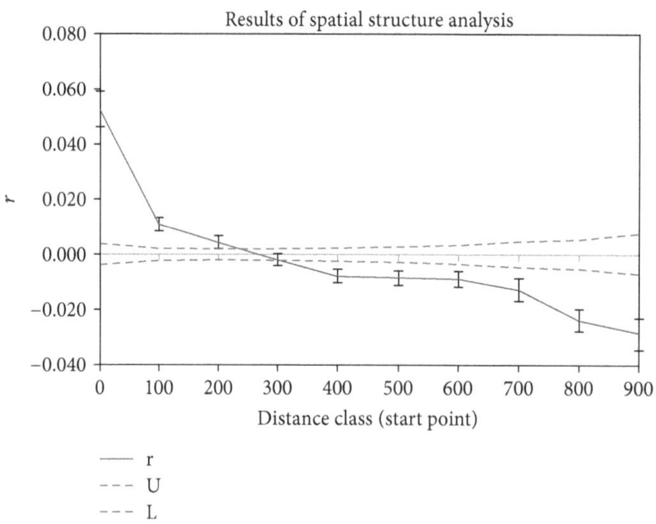

FIGURE 3: Results of spatial structure analysis. r: solid lines represent spatial autocorrelation coefficients; U and L: dashed lines represent 95% confidence interval.

(see Table 2), and the results roughly agree with those of previous studies based on morphological traits [7]. The results support the view that the genetic diversity of wild soybeans in Northeast China is related to latitude but not to longitude; three evolutionarily significant units were distinguished by latitude, corresponding to regions of <41°N-42°N, 43°N-44°N, and >45°N (see Tables 3–5). Moderate differentiation among the latitude populations (the mean F_{st} value was 0.088) and longitude populations (the mean F_{st} value was 0.058) occurred (see Table 6). This implies that natural selection might be the main cause of genetic structure [8, 29]. Previous studies have revealed that wild soybean genotypes exhibit regional distributions at different geographical scales [15, 40, 41], which are especially associated with latitudinal origin. However, some studies have also reported that the genetic differentiations were associated with longitude origins; for example, Leamy et al.'s results showed that the four genetic groups (Central China, Northern China, Korea, and Japan) differed more in longitude than in latitude [13]. Possible explanations for those results may include small sample size, large geographic span, strait isolation, and diverse ecosystems.

Genetic structure was mainly determined by the breeding system, gene flow, distance isolation, and so on [42]. Wild soybean is a strictly self-pollinating plant, with limited pollen flow. In general, for self-pollination-dominated plants, with an average $G_{st} = 0.51$, the total genetic variation among the populations accounts for more than half of the genetic structure; for out-crossing-dominated species $G_{st} = 0.10$, 90% of genetic variation occurs within populations [43]. In the present study, most of the genetic variation was found between individuals within populations, with less than 10% among populations (see Table 6). This suggests that the genetic differentiation among latitude or longitude populations in Northeast China is similar to that of out-crossing-dominated species [17, 40]. Zhao speculated that this phenomenon could be explained by out-crossing rates and long-distance gene flow [44]. Our results show that the out-crossing rate of Northeast China wild soybeans is only 0.7% ($F_{is} = 0.987$) (see Table 1), which confirms its selfing mating reproductive system and plays an important role in keeping a strong genetic structure.

The Mantel test revealed that there was a positive relationship between geographic and genetic distance ($r = 0.207$, $p < 0.05$), which indicates that geographical isolation has also been an important factor in forming the current genetic structure of wild soybeans in Northeast China. Spatial autocorrelation analysis revealed that the correlation between geographical distance and genetic distance is limited.

The wild soybean in Northeast China has become one of the most severely endangered wild plant species due to the interference of human activities [45]. The genetic diversity of wild soybeans is high, indicating that the wild soybean in this region has great potential for evolution, and in situ conservation is preferable. Although wild soybean accessions ex situ conserved in the National Gene Bank are more substantial than those from other areas, the collection in this area is still very limited, so further investigation and collection works in this region are necessary. According to distribution patterns of the genetic diversity of wild soybeans in Northeast China, the conservation strategy should emphasize individual protection, and protection in areas with high genetic diversity should be prioritized.

5. Conclusions

In summary, the genetic diversity and geographic population structure of the wild soybean in Northeast China were fully investigated as a single population, or at different latitude or longitude populations for the first time. The distribution pattern of genetic variation is related to latitude, and the highest level of genetic diversity was found at 42°N, and protection in areas with higher genetic diversity should be prioritized. This study disclosed that natural selection to adapt temperature and photoperiod, selfing mating reproductive system, and distance isolation resulted in the current population structure of wild soybean in Northeast China.

Conflicts of Interest

All authors have reported no financial interests or potential conflicts of interest regarding the publication of this article.

Acknowledgments

This work was supported by the National Science Foundation of China (31300282), the Agricultural Science and Technology Innovation Project (CXGC2017JQ018, CXGC2017ZD014), and the National Key Research and Development Program (2016YFD0100201-19).

References

[1] S. L. Broich and R. G. Palmer, "A cluster analysis of wild and domesticated soybean phenotypes," *Euphytica*, vol. 29, no. 1, pp. 23–32, 1980.

[2] J. Abe, A. Hasegawa, H. Fukushi, T. Mikami, M. Ohara, and Y. Shimamoto, "Introgression between wild and cultivated soybeans of Japan revealed by RFLP analysis for chloroplast DNAs," *Economic Botany*, vol. 53, no. 3, pp. 285–291, 1999.

[3] B. C. Zhuang, *Biological Studies on Wild Soybean in China*, Science press, Beijing, China, 1999.

[4] F. S. Li, "Study on origin and evolution of soybean," *Soybean Science*, vol. 13, pp. 61–66, 1994.

[5] K. J. Wang and X. H. Li, "Exploration and studies of wild soybean germplasm resources in the China Genebank during recent decade," *Journal of Plant Genetic Resources*, vol. 13, no. 508, pp. 507–514, 2012.

[6] Y. Fukuda, "Cytogenetical studies on the wild and cultivated Manchurian soybeans (*Glycine* L)," *Japanese Journal of Botany*, vol. 6, pp. 489–506, 1933.

[7] Y. S. Dong, B. C. Zhuang, L. M. Zhao, H. Sun, and M. Y. He, "The genetic diversity of annual wild soybeans grown in China," *TAG Theoretical and Applied Genetics*, vol. 103, no. 1, pp. 98–103, 2001.

[8] Z. Wen, Y. Ding, T. Zhao, and J. Gai, "Genetic diversity and peculiarity of annual wild soybean (*G. soja* Sieb. et Zucc.) from various eco-regions in China," *Theoretical and Applied Genetics*, vol. 119, no. 2, pp. 371–381, 2009.

[9] P. Taberlet, L. Fumagalli, A.-G. Wust-Saucy, and J.-F. Cosson, "Comparative phylogeography and postglacial colonization routes in Europe," *Molecular Ecology*, vol. 7, no. 4, pp. 453–464, 1998.

[10] Y. H. Li, W. Li, C. Zhang et al., "Genetic diversity in domesticated soybean (*Glycine max*) and its wild progenitor (*Glycine soja*) for simple sequence repeat and single-nucleotide polymorphism loci," *New Phytologist*, vol. 188, no. 1, pp. 242–253, 2010.

[11] S. He, Y. Wang, S. Volis, D. Li, and T. Yi, "Genetic diversity and population structure: implications for conservation of wild soybean (*Glycine soja* Sieb. et Zucc) based on nuclear and chloroplast microsatellite variation," *International Journal of Molecular Sciences*, vol. 13, no. 12, pp. 12608–12628, 2012.

[12] J. Guo, Y. Liu, Y. Wang et al., "Population structure of the wild soybean (*Glycine soja*) in China: implications from microsatellite analyses," *Annals of Botany*, vol. 110, no. 4, pp. 777–785, 2012.

[13] L. J. Leamy, C.-R. Lee, Q. Song et al., "Environmental versus geographical effects on genomic variation in wild soybean (*Glycine soja*) across its native range in northeast Asia," *Ecology and Evolution*, vol. 6, no. 17, pp. 6332–6344, 2016.

[14] J. J. Doyle, "5S ribosomal gene variation in the soybean and its progenitor," *Theoretical and Applied Genetics*, vol. 75, no. 4, pp. 621–624, 1988.

[15] D. Xu, J. Abe, J. Gai, and Y. Shimamoto, "Diversity of chloroplast DNA SSRs in wild and cultivated soybeans: evidence for multiple origins of cultivated soybean," *TAG Theoretical and Applied Genetics*, vol. 105, no. 5, pp. 645-653, 2002.

[16] Z. Li and R. L. Nelson, "RAPD marker diversity among cultivated and wild soybean accessions from four Chinese provinces," *Crop Science*, vol. 42, no. 5, pp. 1737–1744, 2002.

[17] X. H. Li, K. J. Wang, and J. Z. Jia, "Genetic diversity and differentiation of Chinese wild soybean germplasm (*G. soja* Sieb. & Zucc.) in geographical scale revealed by SSR markers," *Plant Breeding*, vol. 128, no. 6, pp. 658–664, 2009.

[18] Y. Wang, F. Ghouri, M. Q. Shahid, M. Naeem, and F. S. Baloch, "The genetic diversity and population structure of wild soybean evaluated by chloroplast and nuclear gene sequences," *Biochemical Systematics and Ecology*, vol. 71, pp. 170–178, 2017.

[19] Y. Kuroda, N. Tomooka, A. Kaga, S. M. S. W. Wanigadeva, and D. A. Vaughan, "Genetic diversity of wild soybean (*Glycine soja* Sieb. et Zucc.) and Japanese cultivated soybeans [*G. max* (L.) Merr.] based on microsatellite (SSR) analysis and the selection of a core collection," *Genetic Resources and Crop Evolution*, vol. 56, no. 8, pp. 1045–1055, 2009.

[20] Z. Limei, D. Yingshan, L. Bao, H. Shui, W. Kejing, and L. Xianghua, "Establishment of a core collection for the Chinese annual wild soybean (*Glycine soja*)," *Chinese Science Bulletin*, vol. 50, no. 10, p. 989, 2005.

[21] S. O. Rogers and A. J. Bendich, "Extraction of DNA from milligram amounts of fresh, herbarium and mummified plant tissues," *Plant Molecular Biology*, vol. 5, no. 2, pp. 69–76, 1985.

[22] H. Xie, R. Z. Chang, Y. S. Cao, M. H. Zhang, Z. F. Feng, and L. J. Qiu, "Selection of core SSR loci by using Chinese autumn soybean," *Scientia Agricultura Sinica*, vol. 36, pp. 360–366, 2003.

[23] R. Peakall and P. E. Smouse, "GenALEx6.5: genetic analysis in Excel. Population genetic software for teaching and research-an update," *Bioinformatics*, vol. 28, no. 19, pp. 2537–2539, 2012.

[24] B. S. Weir, *Genetic Data Analysis II*, Sinauer Associates, Sunderland, MA, 1996.

[25] M. Nei, "Genetic distance between populations," *The American Naturalist*, vol. 106, no. 949, pp. 283–292, 1972.

[26] F. J. Rohlf, *NTSYSps: Numerical Taxonomy System, ver.2.20. Setauket*, Exeter publishing Ltd., NY, USA, 2008.

[27] J. K. Pritchard, M. Stephens, and P. Donnelly, "Inference of population structure using multilocus genotype data," *Genetics*, vol. 155, no. 2, pp. 945–959, 2000.

[28] G. Evanno, S. Regnaut, and J. Goudet, "Detecting the number of clusters of individuals using the software STRUCTURE: a simulation study," *Molecular Ecology*, vol. 14, no. 8, pp. 2611–2620, 2005.

[29] S. Wright, *Variability within and among Natural Populations. Evolution and the Genetics of Populations*, University of Chicago Press, Chicago, USA, 1978.

[30] X. Yan, H. Zhao, X. Liu et al., "Phenotypic traits and diversity of different leaf shape accessions of the wild soybean (*Glycine soja* Sieb. et Zucc.) in China," *Canadian Journal of Plant Science*, vol. 94, no. 2, pp. 397–404, 2014.

[31] L. Sun, H. K. Zhao, F. Zhao, and Y. S. Dong, "Analysis of genetic diversity of *Glycine soja* in Northeast China," *Soybean Science*, vol. 34, pp. 355–360, 2015.

[32] D. Yanlai, Z. Tuanjie, and G. Junyi, "Genetic diversity and ecological differentiation of Chinese annual wild soybean (*Glycine soja*)," *Biodiversity Science*, vol. 16, no. 2, pp. 133–142, 2008.

[33] Y. N. Liu, L. I. Xiang-Hua, and K. J. Wang, "Analysis of the genetic variability for the mini-core collection of Chinese wild soybean (*Glycine soja*) collection in the National Gene Bank based on SSR markers," *Journal of Plant Genetic Resources*, vol. 10, pp. 211–217, 2009.

[34] H. Fan, T. J. Zhao, Y. L. Ding, G. N. Xing, and J. Y. Gai, "Genetic analysis of the characteristics and geographic differentiation of Chinese wild soybean population," *Scientia Agricultura Sinica*, vol. 45, pp. 414–425, 2012.

[35] O. Frankel, *Genetic Perspectives of Germplasm Conservation. Genetic Manipulation: Impact on Man and Society*, Cambridge University Press, Cambridge, UK, 1984.

[36] A. H. D. Brown, "Core collections: a practical approach to genetic resources management," *Genome*, vol. 31, no. 2, pp. 818–824, 1989.

[37] M. A. Nawaz, S. H. Yang, H. M. Rehman et al., "Genetic diversity and population structure of Korean wild soybean (*Glycine soja* Sieb. and Zucc.) inferred from microsatellite markers," *Biochemical Systematics and Ecology*, vol. 71, pp. 87–96, 2017.

[38] L. F. Hunneke, "Ecological implications of genetic variation in plant populations," in *Genetics and Conservation of Rare Plants*, D. A. Falk and K. E. Holsinger, Eds., pp. 31–44, Oxford University Press, New York, 1991.

[39] X. Li, K. Wang, F. Li, and M. Yan, "Research progress of wild soybean (*Glycine soja*) and suggestions for improving its effective utilization and protection," *Soybean Science*, vol. 24, pp. 305–309, 2005.

[40] Y. T. Kiang, Y. C. Chiang, and N. Kaizuma, "Genetic diversity in natural populations of wild soybean in Iwate Prefecture, Japan," *Journal of Heredity*, vol. 83, no. 5, pp. 325–329, 1992.

[41] W. An, H. K. Zhao, S. Ying et al., "Genetic diversity in annual wild soybean (*Glycine soja* Sieb. et Zucc.) and cultivated soybean (g. Max. Merr.) from different latitudes in China," *Pakistan Journal of Botany*, vol. 41, no. 5, pp. 2229–2242, 2009.

[42] R. Z. Qu, L. Hou, H. L. Lü, and H. Y. Li, "The gene flow of population genetic structure," *Hereditas*, vol. 26, no. 3, pp. 377–382, 2004.

[43] J. L. Hamrick and M. J. W. Godt, "Allozyme diversity in plant species," in *Plant Population Genetics, Breeding, and Genetic Resources*, A. H. D. Brown, M. T. Clegg, A. L. Kahler, and B. S. Weir, Eds., Sinauer Associates Inc. Sunderland, Massachusetts, USA, 1990.

[44] R. Zhao, "Exploring basic problems in conservation genetics of wild soybean (*Glycine soja* L.). Ph D thesis," Fudan University, China, 2007.

[45] S.-L. He, Y.-S. Wang, D.-Z. Li, and T.-S. Yi, "Environmental and historical determinants of patterns of genetic differentiation in wild soybean (*Glycine soja* Sieb. et Zucc)," *Scientific Reports*, vol. 6, no. 1, article 22795, 2016.

13

The Transcriptional Network Structure of a Myeloid Cell: A Computational Approach

Jesús Espinal-Enríquez,[1,2] Daniel González-Terán,[1] and Enrique Hernández-Lemus[1,2]

[1]*Computational Genomics Division, National Institute of Genomic Medicine, 14610 México City, Mexico*
[2]*Centro de Ciencias de la Complejidad, Universidad Nacional Autónoma de México, 04510 México City, Mexico*

Correspondence should be addressed to Enrique Hernández-Lemus; ehernandez@inmegen.gob.mx

Academic Editor: Graziano Pesole

Understanding the general principles underlying genetic regulation in eukaryotes is an incomplete and challenging endeavor. The lack of experimental information regarding the regulation of the whole set of transcription factors and their targets in different cell types is one of the main reasons to this incompleteness. So far, there is a small set of curated known interactions between transcription factors and their downstream genes. Here, we built a transcription factor network for human monocytic THP-1 myeloid cells based on the experimentally curated *FANTOM4* database where nodes are genes and the experimental interactions correspond to links. We present the topological parameters which define the network as well as some global structural features and introduce a relative inuence parameter to quantify the relevance of a transcription factor in the context of induction of a phenotype. Genes like ZHX2, ADNP, or SMAD6 seem to be highly regulated to avoid an avalanche transcription event. We compare these results with those of *RegulonDB*, a highly curated transcriptional network for the prokaryotic organism *E. coli*, finding similarities between general hallmarks on both transcriptional programs. We believe that an approach, such as the one shown here, could help to understand the one regulation of transcription in eukaryotic cells.

1. Introduction

Gene regulation is a key player in the development of living systems. Interactions amongst genes are critical to direct tissue-specific gene expression [1]. A dysfunctional process results in altered physiology, giving rise to malformations and diseases such as cancer. The gene regulatory networks that control gene expression are usually composed by several thousands of genes which are transcribed and translated to produce proteins that have a function in the cell. On the other hand, a reduced set of genes, called *transcription factors* (TFs), has the role of regulating the transcriptional program of the cells. These TFs enhance or repress the expression of other genes, which may be also TFs, or else *target genes.*

By looking at the number of genes that a particular TF could regulate, a hierarchical structure can be observed. In this sense, genes like MEF2C, mTOR, MYB, FOXM1, GATA3, FOXP3, BCL6, MNDA, POU2AF1, MEF2C, or SMAD3 have been reported to potentially regulate more than one thousand genes, which are transcriptional master regulators (TMRs) [2, 3]. The hierarchical character, observed in TF-driven regulatory networks, is likely due to the number of target genes that a TF may have. Another issue related to the relevance of such TFs is whether a given TF is regulated or not by another TF, in which case, a shadow effect could appear, indicating the primacy of a given TF above others [4, 5]. Deregulation of these TFs has been related to the development of cancer and other diseases [2].

Comprehensive studies regarding the extent of influence of TFs and the targets of the TFs have been made [6]. In such studies, phenomena like feedback, feed-forward loops, and other biological motifs [7] have been found. The said phenomena are indicative of a sophisticated machinery involved in the regulation of gene transcription: highly connected TFs are also regulated by others, which may (or may not) be highly connected. Hence, the importance of TFs in the whole regulatory program lies not only in the out-degree (the number of genes regulated by the TF) but also in the in-degree

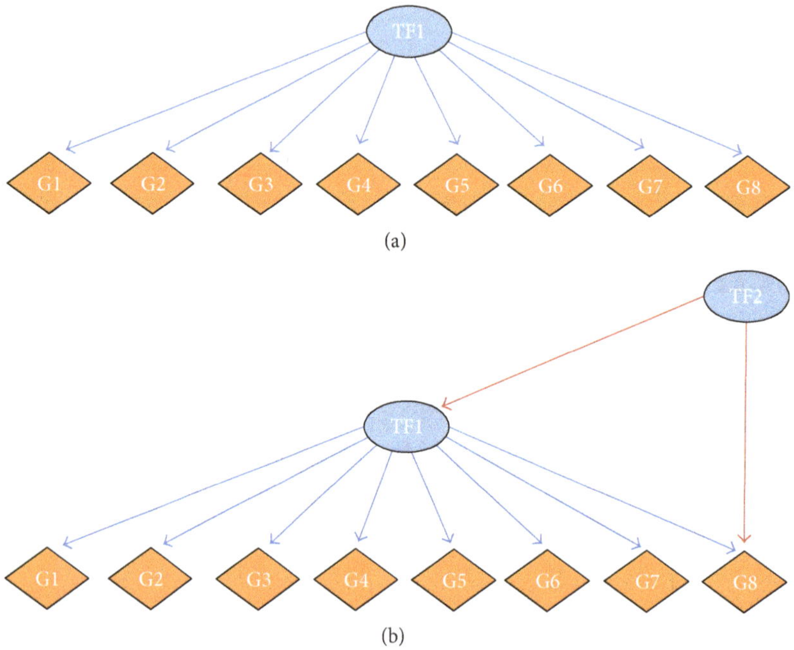

Figure 1: Representative schemes of TFs with high out-degree. (a) The TF1 is regulating its 8 target genes and it is not regulated by others. (b) The TF1 is regulating 8 targets, but at the same time, it is being regulated by TF2, which in turn is also regulating other target genes (G8).

(how many genes regulate the TF) (Figure 1). It is precisely the interplay of in- and out-degree distributions which shapes ultimately the delicate mechanisms of gene expression control, encoded in the topological structure of the transcriptional regulatory network.

The discovery of general patterns of cooperativity and coregulation for the transcriptional regulation program of eukaryotic cells will help to understand the control of genome-wide expression and how it influences the establishment of phenotypes. To this end, here, we develop a systematic strategy for the data mining curation of the whole set of interactions in *FANTOM4 Edge Express*: a comprehensive and authoritative catalog of transcriptional regulatory interactions in THP1 myeloid cells [1, 8].

We constructed a gene regulatory network for the whole set of genes and their interactions by developing a mining tool for the FANTOM4 database, adding each gene present in the said database as a node to a network, where the strength of connections between nodes corresponds to the intensity, or confidence, of the experimental interaction between any pair of genes. We obtained a network consisting in 9090 genes and 234,913 links. This transcriptional regulatory network contains a number of genes influencing many downstream genes, that is, a few genes are able to control the majority of the genes. As previously mentioned, those TFs are also regulated by other TFs. Taking this into account, we constructed a TF subnetwork, which contains 295 TF genes and 8483 links.

Acknowledging the relevance that in-degree and out-degree of TFs acquire in the context of gene regulation, we devise a *relative influence* parameter (see the corresponding subsection in Materials & Methods), related with the ratio of out-degree/in-degree of all TFs present in the network, thus indicating the relative influence of the gene (and its regulators) in terms of how many targets the said gene is regulating and the number of genes which directly regulate it.

We finally present a network visualization highlighting the importance of those genes. Genes like NRF1, SPIB, GABPA, or TFAP2A present a high RI since they do not have regulators within the database. This could be indicative of a basal and transcription-independent level of regulation, being these genes a kind of default transcriptional regulators. On the other hand, genes like ZHX2, ADNP, or SMAD6 present a very high in-degree, while their out-degree is relatively low. Considering this, we argue that these genes are also relevant, acting as locks avoiding a *transcriptional avalanche*, that is, they can allow the transcription of genes related to rapid cell division or give place to differentiation of the cell. The lack of regulation of them could lead to diseases such as cancer or malformation during development.

As mentioned before, the network also contains experimentally obtained weighted values ($S_{i,j}$) of the interactions between any pair of genes which indicates whether the effect of a gene over any other connected to it is positive (activator) or negative (repressor), that is, there are transcription factors whose "preferential" activity is either of an activating or repressing nature. This fact may follow from the very physicochemical structure of the related protein, its interactions with other proteins and with certain regions in the genome that include, but are not restricted to, the promoter regions on their target genes. Of course, these functions are phenotype dependent and may be even context dependent.

In terms of transcriptional regulation, a positive interaction represents that the source gene (TF) promotes the transcription of the target gene (effector gene). Analogously, a negative interaction will represent that the TF inhibits the transcription of the target gene. The algebraic sum of all interactions of each gene allows us to observe that some TFs are mostly activators and others are mostly repressors.

In the first case, we have for instance NFYA, NFYB, and NFYC; whereas in the latter case, we have TFAP2A, TFAP2C, and MAZ. Arguably, regulation of those genes is fundamental to initiate or terminate a transcriptional cascade, likely related to events of differentiation or cell division. Experiments along this way are necessary. We argue that this framework may help to understand the importance of TFs in terms of the ratio between in- and out-degree, which could give new insights regarding the gene regulation.

2. Materials and Methods

Reproducibility of results and methodological clarity are fundamental in all scientific endeavors, but in the case of computational biology approaches, they gain even more relevance. In the present section, we will present detailed accounts of the methods used here, and in some cases (most notably, when we introduce novel concepts), we will even write down detailed calculations. Further details and custom-made computer code for this project are available at the following link: https://github.com/CSB-IG/fantomine.

2.1. Network Construction. We mined the FANTOM4 database [8] to construct a gene regulatory network. This database is based on genome-wide dynamics of transcription start site usage in the PMA-stimulated human monocytic cell line THP-1. To this end, we made a systematic approach to search all genes present in the database. We also tracked the genes which regulate the first one as well as the genes regulated by the searched gene. If a gene has no regulators, its in-degree is zero. Analogously, if the searched gene does not have target genes, its out-degree would be zero.

An important source of information about this network is the strength of the interaction between any pair of genes. This interaction is quantified by both experiments and sequence-based transcription factor binding site (TFBS) predictions. The range of intensities $S_{i,j}$ is a Z-score whose dynamic range goes from −10 to 12. With this experimentally obtained interaction value, we constructed a weighted and directed network of the FANTOM4-based gene regulatory network.

The network was built by using a specially devised crawler, following a breadth-first search strategy to find the connections between all genes (promoters and targets). The algorithm begins with a random gene and began to *walk* in the FANTOM EDGE-DB; for each promoter and target of that gene, it adds them to a queue of genes to explore and to a dictionary of the genes explored. If the current gene is new in the dictionary, it adds it to the database and the interaction (if it is a promoter or target); otherwise, it only verifies if the interaction is new or updates the weight (it only takes the biggest value). Since some of the predicted TFBS have different values for the same target depending on the promoter region, we decided to use the highest value for all interactions (for further details regarding the interaction values, please see [8] and the FANTOM4 website: http://fantom.gsc.riken.jp/4/).

As an example of the last sentence, the SRF gene is regulated by SOX2, but this regulation have three different values in the FANTOM database, depending on the promoter region: 4.857, 0.834, and 0.845. We decided to establish a link with the highest value assuming that that interaction is the most plausible in an ideal context.

We also calculated different node centrality measures of the resulting network, such as their in-degree and out-degree distributions. The whole network depicted in Figure 2 is visualized by using Cytoscape v.3.2.2.

2.2. Relative Inuence. To have a useful measure to retrieve information regarding the regulatory balance of each gene in the context of transcriptional process, we establish a parameter of *relative influence* RI_n for each gene n in the network. This parameter reflects the fact that there are some gene regulators that control the transcription of many targets but are in turn regulated by many genes, while other regulators may possess a smaller number of targets but are also regulated by fewer genes and thus may be of similar relative influence on the general transcriptional regulatory program. The RI_n is then obtained as follows:

$$\mathrm{RI}_n = \frac{\mathrm{outDeg}_n}{\mathrm{inDeg}_n + \mathrm{outDeg}_n} - \frac{\epsilon + \mathrm{inDeg}_n}{\epsilon + \mathrm{outDeg}_n}, \quad (1)$$

where outDeg_n and inDeg_n are the out-degree and in-degree of the gene n, respectively. ϵ is a small variable (10^{-3}) to avoid division by zero. We define all negative RI_n values to be set equal to 0.

For the sake of clarity and result reproducibility, we will outline an explicit calculation as follows: For instance, the gene GABPB2 has 232 regulators (in-degree, inDeg) but 2829 targets (out-degree, outDeg) in the database. Hence, $\mathrm{RI}_{\mathrm{GABPB2}}$ is calculated as follows:

$$\mathrm{RI}_{\mathrm{GABPB2}} = \frac{2829}{232 + 2829} - \frac{0.001 + 232}{0.001 + 2829} = 0.842196754. \quad (2)$$

Since most genes have zero out-degree, that is, they have no target genes, we constructed a subnetwork which contains only TFs; those are, in general terms, the genes which define the way in which the network is regulated.

2.3. The Algebraic Sum of Interactions. Arguably, the interaction strength ($S_{i,j}$) of the links in a transcriptional network is an abstract and important parameter to quantify the activity of a TFBS recognition motif on a given TF against all its potential targets, providing information about TF interactions at a gene-by-gene basis. This fact may mask the importance of certain TFs in the transcriptional regulatory programs of the cell.

In order to categorize the importance of different TFs in this network, this value ($S_{i,j}$) can be used. A TF may be a transcriptional activator or repressor or a combination of both

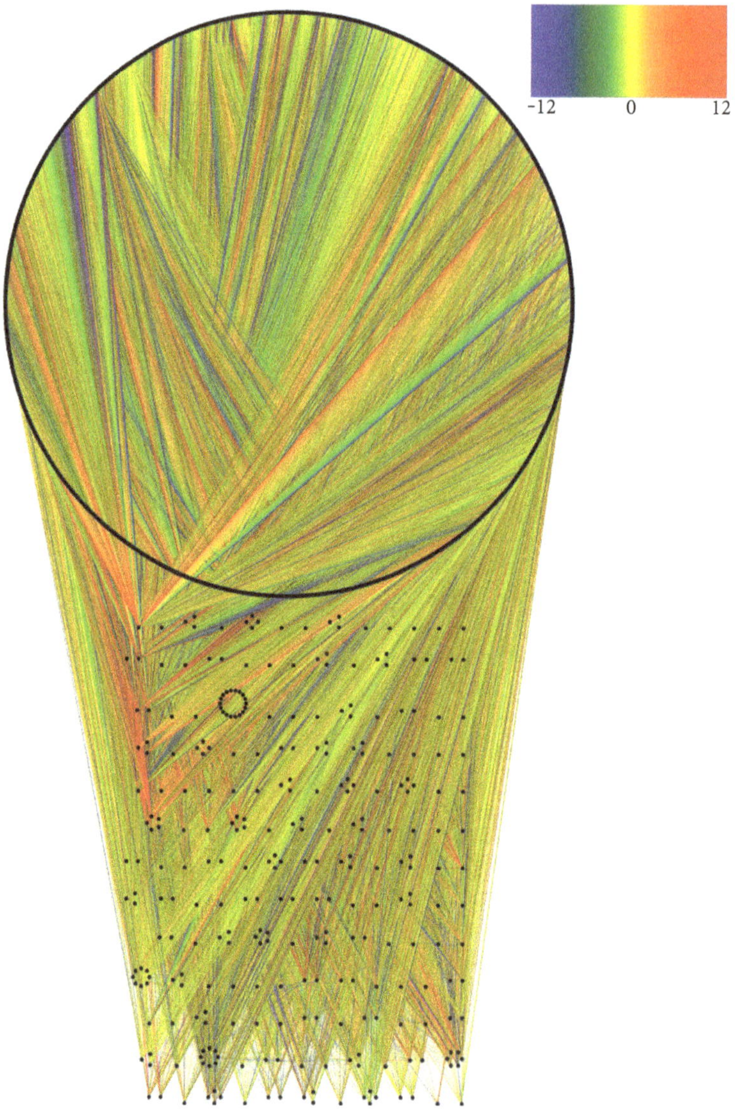

Figure 2: Transcription factor network constructed from the *FANTOM4 Edge Express* database. In this representation, nodes are represented by black dots. The color of the links depends on the value of the interaction: blue and green are repressions; yellow, orange, and red are for activations. The circle on top of the figure corresponds to target genes, that is, their out-degree = 0.

(for different targets and/or under different circumstances). To clarify this idea, we classify the TFs in terms of the total value of their interaction strength, that is, we sum all $S_{i,j}$ for each TF_i and observe if the value was positive or negative, under the assumption of an additive linear model. A positive value means the majority of their interactions are positive; thus, TF_i is an *overall activator*. On the other hand, a negative value indicates that the majority of their interactions are for repression; TF_i is acting then mostly as an *overall repressor*. As we have already stated, such terms are context and phenotype dependent, so that a TF whose main function is that of an activator in the context of myeloid cells, it well may be a repressor on a different cell type or cellular context.

Again, as an example, we will outline the explicit calculation for one instance of algebraic sum: TFAP2B has 687 negatively regulated targets and 965 positively regulated ones. The algebraic sum of all their targets is as follows: $\sum_j S_{\text{TFAP2B},j} = 965 - 687 = 278$, which converts TFAP2B in a strong *overall activator*.

3. Results

3.1. Transcription Factor Network. The TF network is shown in Figure 2. 9090 genes with 234,970 interactions are depicted. From that network, we can observe several facts: The network is depicted ranking genes according to the *relative influence*. Upper part genes have a higher RI, meanwhile the lowest RIs are on the lower part. The black circle at the upper part of the network represents those genes with zero out-degree, that is, effector genes. Those genes have RI = 0 by definition. The interactions between genes are weighted according to their respective interaction strength and colored accordingly with a continuous scale, based on the Z-scores ($S_{i,j}$) for activity expression correlation: this way, the blue and green lines indicate inhibition from the source gene

TABLE 1: The top 10 highest RI values of TFs in the FANTOM network. The last column shows the average gene expression as given by the number of transcript counts (copy number) at $t = 0$, for each gene.

Gene	In-degree	Out-degree	RI	Avg. transcript count
NRF1	0	2175	0.99999954	1628.215
SPIB	0	1778	0.999999438	316.215
TFAP2A	0	1721	0.999999419	293.865
MYOD1	0	1713	0.999999416	0.02
TFAP2B	0	1652	0.999999395	0.03
ARNT2	0	1627	0.999999385	0.01
SNAI2	0	1543	0.999999352	7.485
MYOG	0	1435	0.999999303	203.985
MYF5	0	1435	0.999999303	0.985
MYF6	0	1435	0.999999303	0.265

TABLE 2: The bottom 10 lowest RI values of TFs in the FANTOM network. The last column shows the average gene expression as given by the number of transcript counts (copy number) at $t = 0$, for each gene.

Gene	In-degree	Out-degree	RI	Avg. transcript count
RUNX1	155	556	0.517749925	9382.91
ARID5B	114	392	0.49206167	12.885
TFAP4	190	632	0.483369442	642.65
POU6F1	87	268	0.442040072	229.68
ZIC2	135	407	0.438502347	10132.46
GTF2A1	100	253	0.357562642	1219.94
POU3F1	80	198	0.331108025	4.585
SMAD6	175	408	0.289214257	2936.75
ADNP	132	292	0.256880007	651.685

to the target, whereas yellow and red lines represent activation from the source gene to the target (see color scale inset in Figure 2).

We can see, for instance, how a somewhat small set of molecules is responsible for the concerted regulation of the whole transcriptional activity of the cells. The also important fact is that these molecules carry on their regulatory function by jointly regulating themselves. In this network, genes with the highest out-degree are SP1, MAZ, ELF4, ELF1, ELF2, SPI1, ELK1, ELK4, GABPA, and GABPB2. The subnetwork containing only the targets of those genes has 140,318 interactions with 7913 out of the total of genes.

Regarding the RI, the top10 out-degree genes (which are in fact TFs) have high values; however, they do not have the highest ones, since they are also regulated by other TFs. Table 1 shows those nodes with the highest RI, as well as their experimentally observed expression values at the starting of the experiments. The expression values for the whole time series are provided as a Supplementary material available online at https://doi.org/10.1155/2017/4858173.

As revealed by the RI parameter, TFs with the lowest values have a high out-degree, but at the same time, they present a high in-degree, such is the case of ZHX2, ADNP, SMAD6, POU3F1, GTF2A1, ZIC2, POU6F1, TFAP4, ARID5B, or RUNX1. In all cases, they have (at least) twice more targets than regulators. Table 2 shows the lowest RI genes.

The *relative influence* is a simple metric which provides relevant information about the regulatory activity of a transcription factor. Interestingly, even when a single instance of RI is not enough to give insight on coregulation and gene expression patterns, the full, genome-wide distribution of RI values does it so. A closer look at the top RI molecules as presented in Tables 1 and 2 unveiled interesting patterns. For instance, amongst the TF genes that are not under regulation (at this level of description, of course, see Table 1), we can find thousands of gene targets.

If we recall that the total number of genes in this study is about 9000, then having TFs regulating around 2000 of these is indeed a powerful indicator of strong coregulation. On the other hand, it is also noticeable that the TF with the lowest RI (ZHX2) although being tightly regulated (with a total of 151 regulators) is able to participate in the regulation of 292 genes, evidencing the cooperative effect in TF regulation.

3.2. Transcription Factor Subnetwork. After eliminating all the effector genes (those genes with zero out-degree), the network is drastically reduced (295 genes and 8483 links between them). From this network, it can be observed that some genes are mainly activators meanwhile other TFs are inhibitors. This global activating/inhibiting nature of TFs will be discussed below. Regarding the structure of the subnetwork, it is interesting that several genes are highly regulated, even though they are transcription factors.

We can also see that by considering only the network formed by coregulated transcription factors, it is possible to unveil the presence of modules conformed by groups of TFs that not only regulate other target genes (absent in this network visualization) but also regulate each other. This phenomenon of regulatory modularization is an example of a phenomenon that cannot be seen from the FANTOM4 database consulted on a gene-by-gene basis but provides new biological insight on the whole genome regulatory patterns in this cell lineage.

3.3. Some TFs Are Overall Repressors Meanwhile Others Are Activators. Taking into account the overall activating or inhibiting nature of TFs (in the context of the cell type and phenotype under consideration), we search in the network those genes with extreme positive and negative values of the sum of interaction strengths ($S_{i,j}$) for each TF: They are the most important activators as well as the most important repressors of the transcriptional program. In Figure 3, we show the top 3 activators and also the top 3 repressors along with their targets. Those genes are NFYB, NFYC, and NFYA for activators and MAZ, TFAP2C, and TFAP2A for repressors. The interplay between them may be of importance in shaping the phenotype. Regarding the MAZ gene, it has also a high RI (0.984): This means that it regulates thousands of genes but it is regulated by just a few. It mainly represses the genes that it regulates.

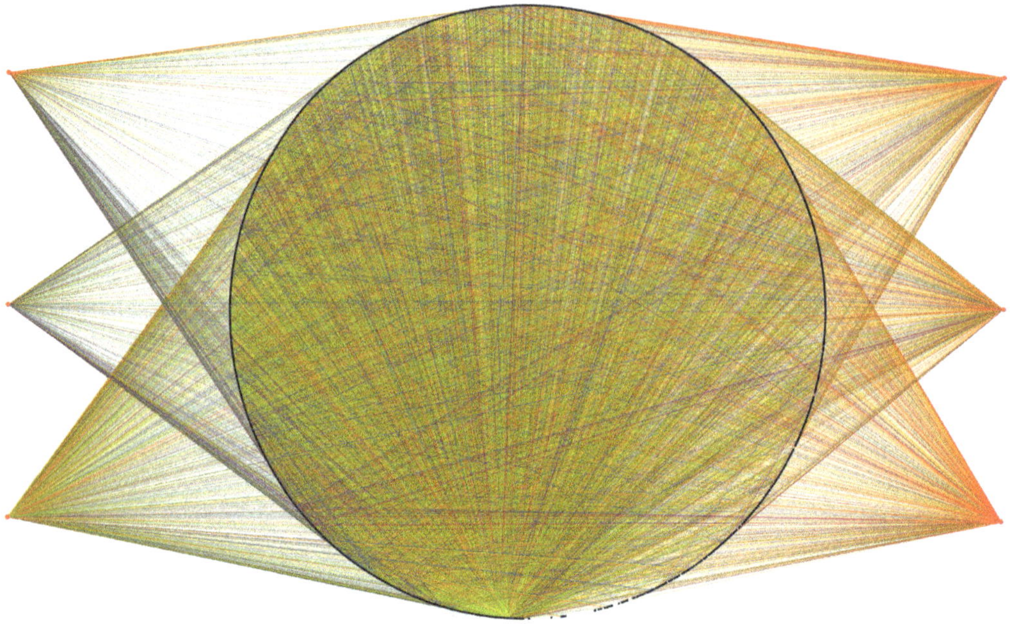

Figure 3: Top 3 overall activating TFs and top 3 overall repressing TFs. On the left side, the three genes which have more inhibiting links—TFAP2C, TFAP2A, and MAZ—are depicted. At the right side, the top 3 activating genes are shown: NFYA, NFYB, and NFYC. It can be observed that the predominance of the green and blue links at the left, meanwhile on the right side, the orange and red links, are more frequent. In the center of this network, we can see the first neighbors of those 6 genes. Links are colored as in Figures 2 and 5.

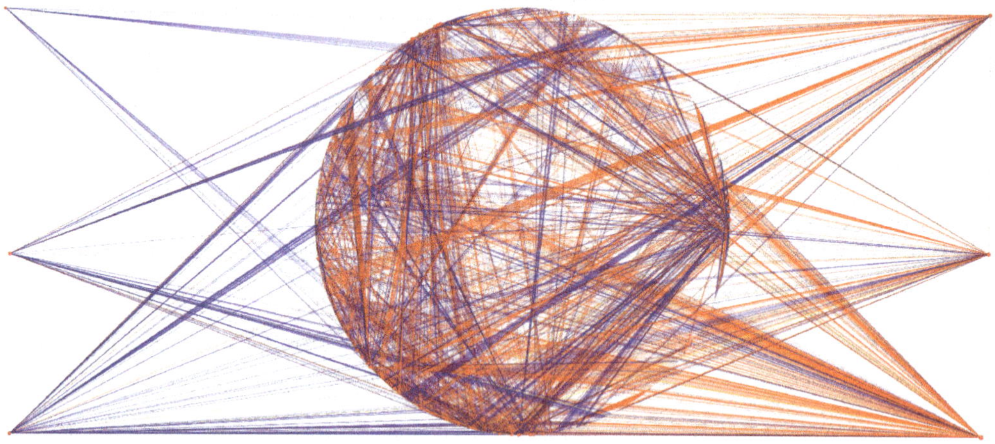

Figure 4: Top 3 overall activating TFs and top 3 overall repressing TFs in the *E. coli* network. On the left side, the three genes which have more inhibiting links—Crp, Fnr, and Fis—are depicted. At the right side, the top 3 activating genes are shown: Narl, Phob, and Lrp. It can be observed (analogous to the human transcriptional network of Figure 5) that the predominance of the blue links at the left, meanwhile on the right side the red links, are more frequent. In the center of this network, the first neighbors of those 6 genes are shown. It is also possible to observe some small disconnected components at the upper part of the figure.

3.4. The Transcriptional Network Structure of a Prokaryotic Cell: A Comparison. The topological and functional features, revealed in the transcriptional network of the myeloid cell with the approach performed here, were compared with a prokaryotic model, in order to provide a deeper understanding of the transcriptional regulatory program at the genome-wide level. The prokaryotic network architecture was obtained with the same methodology that we used for the eukaryotic model, but we obtained information about the genetic interactions from RegulonDB [9], a rigorously curated database of a genome-wide transcriptional regulation in the bacteria *Escherichia coli*. This database contains information regarding the type of interaction between genes, positive, negative, or dual, as well as the direction of the said interactions. We decided to analyze the whole transcriptional network, observe the number of transcription factors,

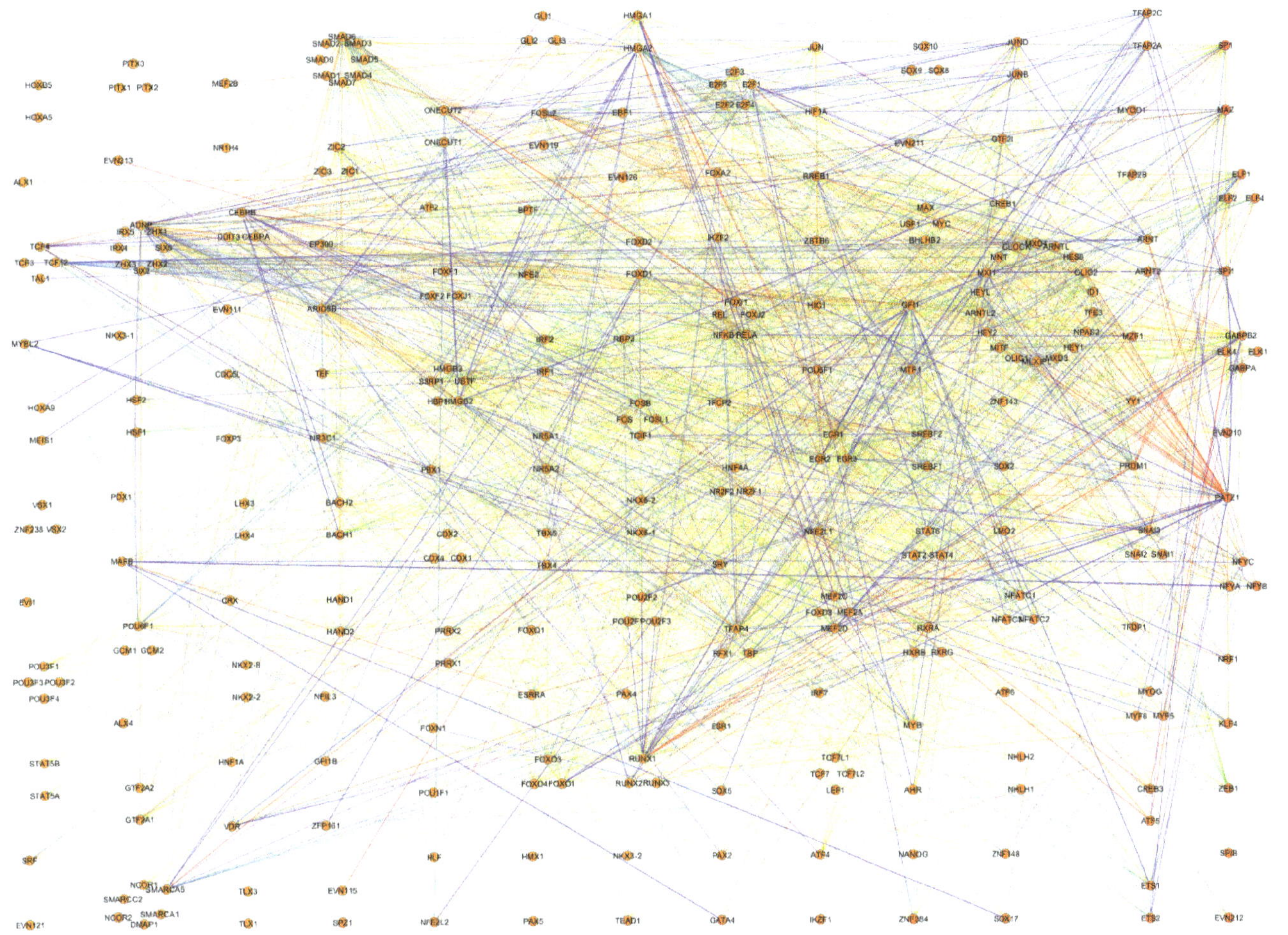

FIGURE 5: Transcription factor subnetwork. In this graph, the 298 TFs of the FANTOM4 network are depicted. Genes are sorted according to the gene out-degree (top left-to-bottom right); set of nodes describing a circle has the same out-degree. The color code of links is the same as in Figure 2.

calculate the RI for each gene, and obtain the overall activators and repressors.

This directed network contains 1988 genes and 4414 interactions. 202 of those genes correspond to transcription factors. In Figure 4, we show the transcriptional network of *E. coli*, depicting the links according to its type: blue for inhibition and red for activation. The top 3 overall activators and repressors are separated to illustrate that the general effect that these transcription factors exert on their targets is analogous to Figure 5. Crp, Fnr, and Fis genes are the top 3 repressors, meanwhile Narl, Phob, and Lrp correspond to the top 3 overall activators. Table 3 shows the top 5 and bottom 5 RI genes of the transcriptional network.

We also constructed the TF-only network for *E. coli*. This network is depicted in Figure 6. As it can be observed, coregulation of those TFs is evident.

TABLE 3: The top 5 highest and lowest values of TFs in the RegulonDB *E. coli* network.

Gene	In-degree	Out-degree	RI
Ihf	0	249	0.999999598
H-ns	0	186	0.999999462
NsrR	0	83	0.999998795
Flhdc	0	80	0.99999875
Narp	0	65	0.999998462
Uvry	1	2	0.166641668
Gadx	15	29	0.141847865
Dpia	6	11	0.101600146
Glcc	4	7	0.064928943
Gutm	4	7	0.064928943

4. Discussion

In this work, we developed a systematic search of the FANTOM4 database, which took into account several experimentally proven interactions between genes and transcription factors. We constructed a network where the genes are nodes and their interactions correspond to links. The network is directed because it takes into account transcription factor binding sites to establish a relationship in which the TF is regulating to another gene (which could actually be another

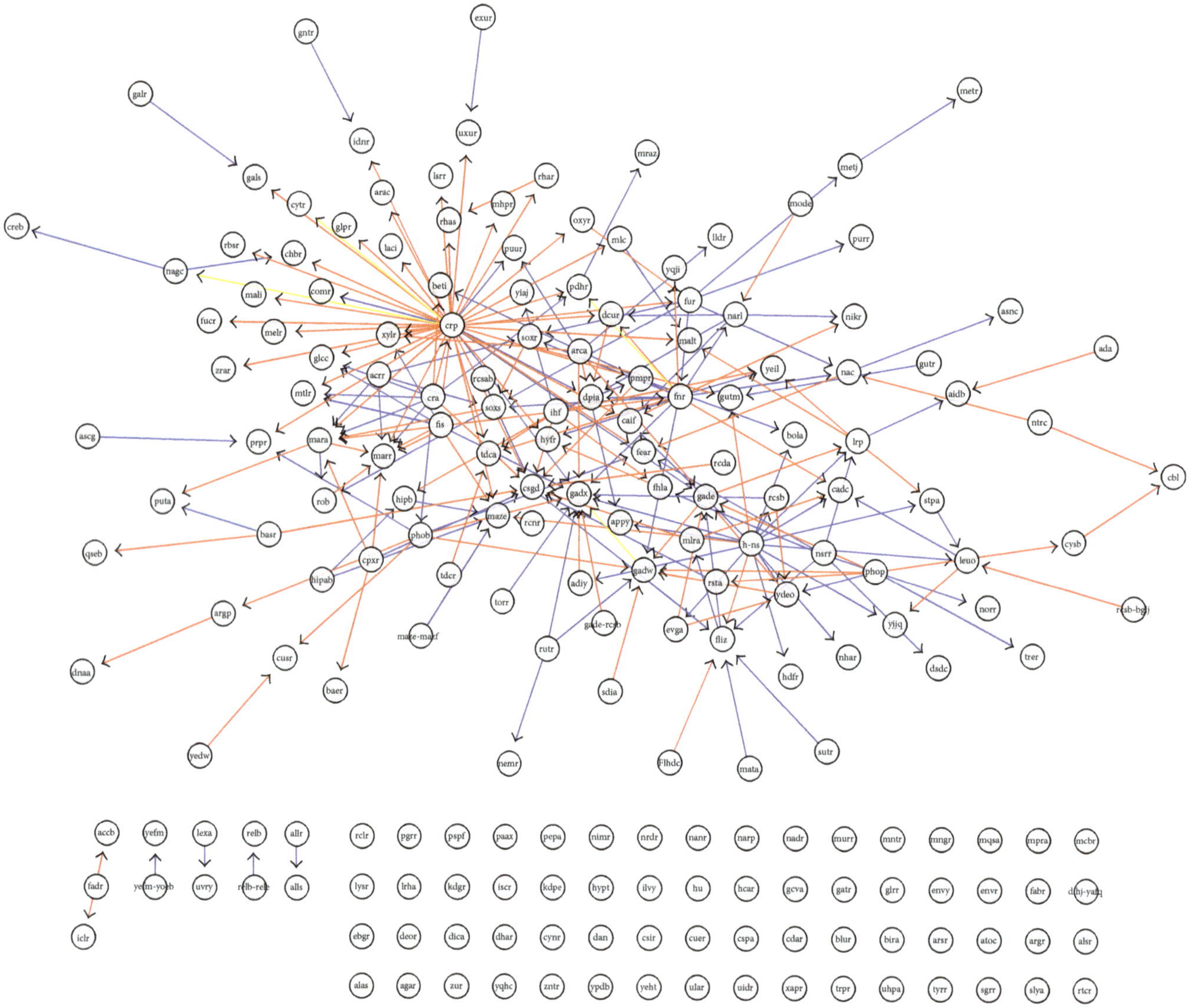

FIGURE 6: Transcription factor subnetwork. In this graph, the 203 TFs of the *E. coli* RegulonDB network are depicted. The color code of links is the same as that in Figure 4. As it was observed in the whole network, in this TF subnetwork, some disconnected components appear.

transcription factor). We found several nodes which do not regulate other genes, but they are highly regulated. On the other hand, there is a small group of TFs which are regulating several other genes, but they are not regulated by any gene. This could allow us to hypothesize that those genes are master regulators because they could be inducers of particular phenotypes, as is the case for NRF1, SPIB, or TFAP2A.

4.1. The Highest RI Genes May Have Crucial Roles in Cell Specificity. Nuclear respiratory factor 1 (NRF1) [10, 11] activates the expression of crucial metabolic genes related to responses to oxidative stress, endoplasmic reticulum stress, xenobiotic stress, and inflammation [12–14]. NRF1 has also been found misregulated in different carcinomas [15–18]. Despite NRF1 being a key player in the induction of several functions in the cell, there are no reports of transcriptional regulators of NRF1. It is regulated at the translational [19] and posttranslational [19–21] levels. This information reinforces our findings regarding the transcriptional independence of NRF1 as well as the relevance that it acquires in the context of the regulatory network.

The second highest RI belongs to the SPiB gene. The spleen focus-forming virus integration site [22] gene is part of the ETS transcription factor family. It is involved in differentiation [23], immune response [22, 24], apoptosis [25], and activation of early viral expression [26], amongst others. Finally, the third highest RI gene is TFAP2A. This transcription factor is known as a tumor suppressor gene. Decreased expression of this gene has been related to many neoplasms [27–29], as well as other diseases, such as the brachio-oculo-facial syndrome [30, 31].

By observing the relevance of these three genes in the maintenance of correct cell behavior, it is interesting to observe that there are no regulators in our network for those genes. It could be due to the fact that the experimentally curated network is not complete or the available information regarding the regulators is still under construction. However, it is remarkable that the high number of targets contrasting with the number of regulators of them is in instances such as this, where hierarchy acquires relevance. It is worth mentioning that this is a transcriptional network. We are not looking at posttranslational modifications, where these genes could be regulated. However, the transcriptional relevance of them is evidenced by this approach.

4.2. Misregulation of the Lowest RI Genes Is Associated with Several Malformations. On the other hand, the three lowest RI genes, ZHX2, ADNP, and SMAD6 are highly regulated; regulation of them is probably more critical in order to avoid transcriptional avalanches. Let us recall that these genes, even though they are highly regulated (higher in-degree counts), are still active transcription factors (see Table 2). In the same sense, mutations of these genes could be involved in the development of several and important diseases. For instance, ZHX2 is considered as a transcriptional repressor [32], which binds the promoter regions, thus regulating transcription of their target genes. This TF also suppresses glypican 3 (GPC3) transcription [33]. Furthermore, ZHX2 has been found as a tumor suppressor [34]. This gene is transcriptionally suppressed by MSX1 and XBP1 [35]. This downregulation is crucial for progression of Hodgkin lymphoma. Concomitantly, there are at least 40 transcription factor binding sites downstream ZHX2 gene [35] which regulate its expression.

The second lowest RI belongs to the ADNP gene, which is essential for brain formation and correct neural development [36]. Mutations in this gene have been related to neuronal disorders such as autism [37, 38], Alzheimer's disease [39], or schizophrenia [40]. Moreover, ADNP interacts with HP1 regulating chromatin remodeling during embryogenesis [41]. Total absence of ADNP is lethal, thus indicating its crucial role in the regulation of transcriptional programs.

Finally, SMAD6 is a signal transducer that modulates multiple signaling pathways, such as the BMP and TGF-beta/activin signaling [42], erythropoiesis [43], or cell cycle [44]. Incorrect regulation of this gene is related to lung adenocarcinoma [42, 45], oral squamous cell carcinoma [46], ovarian cancer [47], or cardiovascular malformation [48].

The RI, introduced here to study the network, remarks the relevance of TFs with the highest values in terms of its master role as transcription factors (since they have no upstream transcriptional regulators); furthermore, this value unveils the importance of the lowest scored genes: those genes need to be highly regulated in order to maintain its correct transcriptional behavior. As it has been observed experimentally, misregulation of the lowest RI genes may result in lethal phenotypes or cancer.

The approach developed here may help to understand genomic regulation. The RI parameter introduced provides insight regarding the influence and importance of the genes in the context of maintenance of a particular phenotype.

4.3. Overall Activators and Repressors May Define Transcriptional Programs. Finally, the activation or inhibitory nature of TFs is an important field of investigation. With an approach such as the one presented here, research can be guided to unveil the overall nature of certain TFs, by observing whether its overall effect is activation or inhibition. A TF, which positively regulates hundreds of downstream targets, could induce a particular phenotype by activating the said targets. On the contrary, if the majority of the targets is inhibited by the TF, this gene could prevent or stop a particular transcriptional program.

The following TFs are the top 3 overall activators. Interestingly, this set corresponds precisely with the so-called NF-Y complex: NFYA, NFYB, and NFYC, whose role is to bind a sequence in DNA to start the transcription process. They are involved in several basal activities, such as expression of human proteasome genes [49], transcriptional cascade via CDCA8 gene [50], and remodeling of chromatin [51]. Furthermore, the NFY complex has been associated with the coexpression of other TFs to start transcriptional cascades [52]. However, the separated subunit NFYB also can be an inhibitor of DNA topoisomerase II-α [53], indicating that the role of this complex is not restricted to be an activator, revealing thus another aspect of the complexity of the transcriptional program in eukaryotic cells.

As mentioned above, the lowest RI gene in the network is ZHX2. This TF interacts with the A subunit of nuclear factor-Y (NFYA). For instance, ZHX2 represses activation of MDR1 transcription mediated by NFYA [54]. Interestingly enough is the fact that during liver carcinoma, the normal transcriptional program of ZHX2 is highly altered. ZHX2 represses NFYA during liver carcinoma [54]. Taking into account the fact that NFYA is the most important overall activator in the network, we can argue that under a repression of NFYA via ZHX2, the consequence will be a general inhibition of the transcriptional program, which may result for instance, in the progression of liver carcinoma. This example highlights how crucial is the correct control of interconnections in this transcriptional network.

The following are the top 3 overall inhibitor genes and might be involved in the control of transcription by avoiding anomalous events: MAZ, TFAP2A, and TFAP2B. The Myc-associated zinc-finger protein (MAZ) was identified as participating in breast cancer cells by interacting with SAF-1 and inducing transcription of Ras [55]. MAZ also has a role in prostate cancer [56] by interacting with the androgen receptor. On the other hand, TFAP2A, B genes, known tumor suppressor genes, participate in the reduction of glioma progression, by downregulation of Bcl-xl, Bcl-2, c-IAP2, and survivin [27]. These genes have been encountered decreased in several types of cancer, glioma [57], prostate cancer [56], breast cancer [58], or testicular cancer [59]. This is a clear example that the absence of its transcriptional inhibition generates dramatic changes in phenotype.

4.4. Similarities with the Prokaryotic Cell Transcriptional Program. Regarding the RI in the *E. coli* network, the higher one, Ihf (integration host factor), plays a crucial role in the survival of the cell, induction of acid resistance, and expression of several other factors [60–62]. H-ns acts on DNA binding of RNA polymerase [63]. The NSrR gene is a major transcriptional repressor in response to iron and also a negative regulator of motility [64–66]. On the other hand, the lowest RI gene is gutM, a crucial transcription factor involved in the phosphotransferase system [67]. Mutations on those genes have a direct impact on the resulting phenotype.

To highlight the importance of the implications that the concepts of RI and the overall activator/repressor may have in the regulatory program, we provide a functional comparison of these measures in the prokaryotic transcriptional network; for this purpose, we can observe (for instance) the Fis gene in *E. coli*. This gene is the second most important overall repressor. Fis acts repressing the Crp gene, which is the most connected gene in the *E. coli* genome. Fis in turn is regulated positively by Ihf, the gene with the highest RI of *E. coli*. With this example, we highlight that the transcriptional regulation of the most influential genes could determine the phenotype of a cell via the transcriptional cascades generated by activation or repression of those influential genes.

4.5. Final Considerations. With this approach, we present a hierarchical transcription network built from a highly curated database, containing the values of the interactions between TFs and their targets. We observed that the large majority of genes is controlled by just a few TFs. Those few TFs are strongly coregulated between them, which is translated into a fine tuning in the transcription process. A way to quantify this is by the relative influence parameter (RI), in which the lesser regulated genes and highly regulated ones are relevant for global transcriptional control.

Finally, the extent to which genes are regulated is also important, whether regulation is positive or negative. The negative interaction means that the TF is a repressor of the target, meanwhile a positive interaction represents an activation given by the TF. A global value of each TF in terms of their regulatory values is presented. We observe that NFY subunits are overall activators, meanwhile MAZ and TFAP2A and B are overall repressors. Those genes must be important in the context of transcription. Arguably, regulation of those genes presented in this discussion is fundamental to initiate or terminate a transcriptional cascade, likely related to events of differentiation or cell division. Experiments along this way are necessary. We argue that this framework may help to understand the importance of TFs in terms of the ratio between in- and out-degree as well as their overall effect on targets, which could in turn give new insights regarding the gene regulation.

Conflicts of Interest

The authors declare that they have no conflicts of interest.

Acknowledgments

The research leading to these results has received funding from Grant no. 179431/2012 from the Consejo Nacional de Ciencia y Tecnología (CONA-CyT), as well as a federal funding from the National Institute of Genomic Medicine (INMEGEN). One of the authors (Enrique Hernández-Lemus) also acknowledges support from the 2016 Marcos Moshinsky Research Chair in the Physical Sciences.

References

[1] T. Ravasi, H. Suzuki, C. V. Cannistraci et al., "An atlas of combinatorial transcriptional regulation in mouse and man," *Cell*, vol. 140, no. 5, pp. 744–752, 2010.

[2] K. Baca-López, M. Mayorga, A. Hidalgo-Miranda, N. Gutiérrez-Nájera, and E. Hernández-Lemus, "The role of master regulators in the metabolic/transcriptional coupling in breast carcinomas," *PLoS One*, vol. 7, no. 8, article e42678, 2012.

[3] J. H. Woo, Y. Shimoni, W. S. Yang et al., "Elucidating compound mechanism of action by network perturbation analysis," *Cell*, vol. 162, no. 2, pp. 441–451, 2015.

[4] H. Tovar, R. García-Herrera, J. Espinal-Enríquez, and E. Hernández-Lemus, "Transcriptional master regulator analysis in breast cancer genetic networks," *Computational Biology and Chemistry*, vol. 59, Part B, pp. 67–77, 2015.

[5] A. A. Margolin, I. Nemenman, K. Basso et al., "ARACNE: an algorithm for the reconstruction of gene regulatory networks in a mammalian cellular context," *BMC Bioinformatics*, vol. 7, Supplement 1, p. S7, 2006.

[6] J. M. Vaquerizas, S. K. Kummerfeld, S. A. Teichmann, and N. M. Luscombe, "A census of human transcription factors: function, expression and evolution," *Nature Reviews Genetics*, vol. 10, no. 4, pp. 252–263, 2009.

[7] U. Alon, "Network motifs: theory and experimental approaches," *Nature Reviews Genetics*, vol. 8, no. 6, pp. 450–461, 2007.

[8] FANTOM Consortium, H. Suzuki, A. R. Forrest et al., "The transcriptional network that controls growth arrest and differentiation in a human myeloid leukemia cell line," *Nature Genetics*, vol. 41, no. 5, pp. 553–562, 2009.

[9] S. Gama-Castro, H. Salgado, A. Santos-Zavaleta et al., "RegulonDB version 9.0: high-level integration of gene regulation, coexpression, motif clustering and beyond," *Nucleic Acids Research*, vol. 44, no. D1, pp. D133–D143, 2015.

[10] J. Y. Chan, X. L. Han, and Y. W. Kan, "Cloning of Nrf1, an NF-E2-related transcription factor, by genetic selection in yeast," *Proceedings of the National Academy of Sciences*, vol. 90, no. 23, pp. 11371–11375, 1993.

[11] N. C. Andrews, H. Erdjument-Bromage, M. B. Davidson, P. Tempst, and S. H. Orkin, "Erythroid transcription factor NF-E2 is a haematopoietic-specific basic-leucine zipper protein," *Nature*, vol. 362, no. 6422, pp. 722–728, 1993.

[12] M. Bugno, M. Daniel, N. L. Chepelev, and W. G. Willmore, "Changing gears in Nrf1 research, from mechanisms of regulation to its role in disease and prevention," *Biochimica et Biophysica Acta (BBA)-Gene Regulatory Mechanisms*, vol. 1849, no. 10, pp. 1260–1276, 2015.

[13] M. Mohrin, J. Shin, Y. Liu et al., "A mitochondrial UPR-mediated metabolic checkpoint regulates hematopoietic stem cell aging," *Science*, vol. 347, no. 6228, pp. 1374–1377, 2015.

[14] D. V. Ho and J. Y. Chan, "Induction of herpud1 expression by ER stress is regulated by Nrf1," *FEBS Letters*, vol. 589, no. 5, pp. 615–620, 2015.

[15] V. O. Okoh, N. A. Garba, R. B. Penney et al., "Redox signalling to nuclear regulatory proteins by reactive oxygen species

contributes to oestrogen-induced growth of breast cancer cells," *British Journal of Cancer*, vol. 112, no. 10, pp. 1687–1702, 2015.

[16] M. Biswas, D. Phan, M. Watanabe, and J. Y. Chan, "The Fbw7 tumor suppressor regulates nuclear factor E2-related factor 1 transcription factor turnover through proteasome-mediated proteolysis," *Journal of Biological Chemistry*, vol. 286, no. 45, pp. 39282–39289, 2011.

[17] J. Zhang, X. Zheng, and Q. Zhang, "Egln2 positively regulates mitochondrial function in breast cancer," *Molecular & Cellular Oncology*, vol. 3, no. 2, article e1120845, 2015.

[18] L. Zhao, M. Tang, Z. Hu et al., "mir-504 mediated down-regulation of nuclear respiratory factor 1 leads to radio-resistance in nasopharyngeal carcinoma," *Oncotarget*, vol. 6, no. 18, pp. 15995–16018, 2015.

[19] Y. Zhang and J. D. Hayes, "The membrane-topogenic vectorial behaviour of Nrf1 controls its post-translational modification and transactivation activity," *Scientific Reports*, vol. 3, no. 1, article 2006, 2013.

[20] Y. Zhang, J. M. Lucocq, and J. D. Hayes, "The Nrf1 CNC/bZIP protein is a nuclear envelope-bound transcription factor that is activated by t-butyl hydroquinone but not by endoplasmic reticulum stressors," *Biochemical Journal*, vol. 418, no. 2, pp. 293–310, 2009.

[21] Y. Zhang, Y. Ren, S. Li, and J. D. Hayes, "Transcription factor Nrf1 is topologically repartitioned across membranes to enable target gene transactivation through its acidic glucose-responsive domains," *PLoS One*, vol. 9, no. 4, article e93458, 2014.

[22] S. Gallant and G. Gilkeson, "ETS transcription factors and regulation of immunity," *Archivum Immunologiae et Therapiae Experimentalis*, vol. 54, no. 3, pp. 149–163, 2006.

[23] Y. Zhou, X. Liu, L. Xu et al., "Transcriptional repression of plasma cell differentiation is orchestrated by aberrant overexpression of the ETS factor SPIB in Waldenström macroglobulinaemia," *British Journal of Haematology*, vol. 166, no. 5, pp. 677–689, 2014.

[24] Y. Yang, A. L. Shaffer 3rd, N. C. Emre et al., "Exploiting synthetic lethality for the therapy of ABC diffuse large B cell lymphoma," *Cancer Cell*, vol. 21, no. 6, pp. 723–737, 2012.

[25] J. J. Karrich, M. Balzarolo, H. Schmidlin et al., "The transcription factor Spi-B regulates human plasmacytoid dendritic cell survival through direct induction of the antiapoptotic gene BCL2-A1," *Blood*, vol. 119, no. 22, pp. 5191–5200, 2012.

[26] L. J. Marshall, L. D. Moore, M. M. Mirsky, and E. O. Major, "JC virus promoter/enhancers contain TATA box-associated Spi-B-binding sites that support early viral gene expression in primary astrocytes," *Journal of General Virology*, vol. 93, Part 3, pp. 651–661, 2012.

[27] W. Su, J. Xia, X. Chen et al., "Ectopic expression of AP-2α transcription factor suppresses glioma progression," *International Journal of Clinical and Experimental Pathology*, vol. 7, no. 12, pp. 8666–8674, 2014.

[28] A. R. Hallberg, S. U. Vorrink, D. R. Hudachek et al., "Aberrant CpG methylation of the TFAP2A gene constitutes a mechanism for loss of TFAP2A expression in human metastatic melanoma," *Epigenetics*, vol. 9, no. 12, pp. 1641–1647, 2014.

[29] D. Shi, F. Xie, Y. Zhang et al., "TFAP2A regulates nasopharyngeal carcinoma growth and survival by targeting HIF-1α signaling pathway," *Cancer Prevention Research*, vol. 7, no. 2, pp. 266–277, 2014.

[30] H. Li, R. Sheridan, and T. Williams, "Analysis of TFAP2A mutations in branchio-oculo-facial syndrome indicates functional complexity within the AP-2α DNA-binding domain," *Human Molecular Genetics*, vol. 22, no. 16, pp. 3195–3206, 2013.

[31] T. I. Meshcheryakova, R. A. Zinchenko, T. A. Vasilyeva et al., "A clinical and molecular analysis of branchio-oculo-facial syndrome patients in Russia revealed new mutations in TFAP2A," *Annals of Human Genetics*, vol. 79, no. 2, pp. 148–152, 2015.

[32] H. Kawata, K. Yamada, Z. Shou et al., "Zinc-fingers and homeoboxes (ZHX) 2, a novel member of the ZHX family, functions as a transcriptional repressor," *Biochemical Journal*, vol. 373, Part 3, pp. 747–757, 2003.

[33] F. Luan, P. Liu, H. Ma et al., "Reduced nucleic ZHX2 involves in oncogenic activation of glypican 3 in human hepatocellular carcinoma," *The International Journal of Biochemistry & Cell Biology*, vol. 55, pp. 129–135, 2014.

[34] S. Nagel, B. Schneider, A. Rosenwald et al., "t(4;8)(q27;q24) in Hodgkin lymphoma cells targets phosphodiesterase PDE5A and homeobox gene ZHX2," *Genes, Chromosomes and Cancer*, vol. 50, no. 12, pp. 996–1009, 2011.

[35] S. Nagel, B. Schneider, C. Meyer, M. Kaufmann, H. G. Drexler, and R. A. Macleod, "Transcriptional deregulation of homeobox gene ZHX2 in Hodgkin lymphoma," *Leukemia research*, vol. 36, no. 5, pp. 646–655, 2012.

[36] I. Gozes, "Activity-dependent neuroprotective protein: from gene to drug candidate," *Pharmacology & Therapeutics*, vol. 114, no. 2, pp. 146–154, 2007.

[37] G. Vandeweyer, C. Helsmoortel, A. Van Dijck et al., "The transcriptional regulator ADNP links the BAF (SWI/SNF) complexes with autism," *American Journal of Medical Genetics Part C: Seminars in Medical Genetics*, vol. 166, no. 3, pp. 315–326, 2014.

[38] C. Helsmoortel, A. T. Vulto-van Silfhout, B. P. Coe et al., "A SWI/SNF-related autism syndrome caused by de novo mutations in ADNP," *Nature Genetics*, vol. 46, no. 4, pp. 380–384, 2014.

[39] M. H. Yang, Y. H. Yang, C. Y. Lu et al., "Activity-dependent neuroprotector homeobox protein: a candidate protein identified in serum as diagnostic biomarker for Alzheimer's disease," *Journal of Proteomics*, vol. 75, no. 12, pp. 3617–3629, 2012.

[40] A. Merenlender-Wagner, A. Malishkevich, Z. Shemer et al., "Autophagy has a key role in the pathophysiology of schizophrenia," *Molecular Psychiatry*, vol. 20, no. 1, pp. 126–132, 2015.

[41] S. Mandel, G. Rechavi, and I. Gozes, "Activity-dependent neuroprotective protein (ADNP) differentially interacts with chromatin to regulate genes essential for embryogenesis," *Developmental Biology*, vol. 303, no. 2, pp. 814–824, 2007.

[42] H. S. Jeon, T. Dracheva, S. H. Yang et al., "SMAD6 contributes to patient survival in non-small cell lung cancer and its knock-down reestablishes TGF-β homeostasis in lung cancer cells," *Cancer Research*, vol. 68, no. 23, pp. 9686–9692, 2008.

[43] Y. J. Kang, J. W. Shin, J. H. Yoon et al., "Inhibition of erythropoiesis by Smad6 in human cord blood hematopoietic stem cells," *Biochemical and Biophysical Research Communications*, vol. 423, no. 4, pp. 750–756, 2012.

[44] K. Pardali, M. Kowanetz, C. H. Heldin, and A. Moustakas, "Smad pathway-specific transcriptional regulation of the cell

cycle inhibitor p21WAF1/Cip1," *Journal of Cellular Physiology*, vol. 204, no. 1, pp. 260–272, 2005.

[45] E. Frullanti, F. Colombo, F. S. Falvella et al., "Association of lung adenocarcinoma clinical stage with gene expression pattern in noninvolved lung tissue," *International Journal of Cancer*, vol. 131, no. 5, pp. E643–E648, 2012.

[46] F. R. Mangone, F. Walder, S. Maistro et al., "Smad2 and Smad6 as predictors of overall survival in oral squamous cell carcinoma patients," *Molecular Cancer*, vol. 9, no. 1, p. 1, 2010.

[47] J. Yin, K. Lu, J. Lin et al., "Genetic variants in TGF-β pathway are associated with ovarian cancer risk," *PLoS One*, vol. 6, no. 9, article e25559, 2011.

[48] H. L. Tan, E. Glen, A. Töpf et al., "Nonsynonymous variants in the SMAD6 gene predispose to congenital cardiovascular malformation," *Human Mutation*, vol. 33, no. 4, pp. 720–727, 2012.

[49] H. Xu, J. Fu, S. W. Ha et al., "The CCAAT box-binding transcription factor NF-Y regulates basal expression of human proteasome genes," *Biochimica et Biophysica Acta (BBA)-Molecular Cell Research*, vol. 1823, no. 4, pp. 818–825, 2012.

[50] C. Dai, C. X. Miao, X. M. Xu et al., "Transcriptional activation of human CDCA8 gene regulated by transcription factor NF-Y in embryonic stem cells and cancer cells," *Journal of Biological Chemistry*, vol. 290, no. 37, pp. 22423–22434, 2015.

[51] C. Romier, F. Cocchiarella, R. Mantovani, and D. Moras, "The NF-YB/NF-YC structure gives insight into DNA binding and transcription regulation by CCAAT factor NF-Y," *Journal of Biological Chemistry*, vol. 278, no. 2, pp. 1336–1345, 2003.

[52] J. D. Fleming, G. Pavesi, P. Benatti, C. Imbriano, R. Mantovani, and K. Struhl, "NF-Y coassociates with FOS at promoters, enhancers, repetitive elements, and inactive chromatin regions, and is stereo-positioned with growth-controlling transcription factors," *Genome Research*, vol. 23, no. 8, pp. 1195–1209, 2013.

[53] C. F. Chen, X. He, A. D. Arslan et al., "Novel regulation of nuclear factor-YB by miR-485-3p affects the expression of DNA topoisomerase IIα and drug responsiveness," *Molecular Pharmacology*, vol. 79, no. 4, pp. 735–741, 2011.

[54] H. Ma, X. Yue, L. Gao et al., "ZHX2 enhances the cytotoxicity of chemotherapeutic drugs in liver tumor cells by repressing MDR1 via interfering with NF-YA," *Oncotarget*, vol. 6, no. 2, p. 1049, 2015.

[55] A. Ray and B. K. Ray, "Induction of Ras by SAF-1/MAZ through a feed-forward loop promotes angiogenesis in breast cancer," *Cancer Medicine*, vol. 4, no. 2, pp. 224–234, 2015.

[56] L. Jiao, Y. Li, D. Shen et al., "The prostate cancer-up-regulated myc-associated zinc-finger protein (MAZ) modulates proliferation and metastasis through reciprocal regulation of androgen receptor," *Medical Oncology*, vol. 30, no. 2, pp. 1–8, 2013.

[57] R. Britto, S. Umesh, A. S. Hegde et al., "Shift in AP-2α localization characterizes astrocytoma progression," *Cancer Biology & Therapy*, vol. 6, no. 3, pp. 413–418, 2007.

[58] B. C. Turner, J. Zhang, A. A. Gumbs et al., "Expression of AP-2 transcription factors in human breast cancer correlates with the regulation of multiple growth factor signalling pathways," *Cancer Research*, vol. 58, no. 23, pp. 5466–5472, 1998.

[59] C. E. Hoei-Hansen, J. E. Nielsen, K. Almstrup et al., "Transcription factor AP-2 is a developmentally regulated marker of testicular carcinoma in situ and germ cell tumors," *Clinical Cancer Research*, vol. 10, no. 24, pp. 8521–8530, 2004.

[60] T. Nystorm, "Glucose starvation stimulon of *Escherichia coli*: role of integration host factor in starvation survival and growth phase-dependent protein synthesis," *Journal of Bacteriology*, vol. 177, no. 19, pp. 5707–5710, 1995.

[61] H. Bi and C. Zhang, "Integration host factor is required for the induction of acid resistance in *Escherichia coli*," *Current Microbiology*, vol. 69, no. 2, pp. 218–224, 2014.

[62] D. Charlier, M. Roovers, D. Gigot, N. Huysveld, A. Piérard, and N. Glansdorff, "Integration host factor (IHF) modulates the expression of the pyrimidine-specific promoter of the carAB operons of Escherichia coli K_{12} and *Salmonella typhimurium* LT_2," *Molecular and General Genetics MGG*, vol. 237, no. 1-2, pp. 273–286, 1993.

[63] S. S. Singh and D. C. Grainger, "H-NS can facilitate specific DNA-binding by RNA polymerase in AT-rich gene regulatory regions," *PLoS Genetics*, vol. 9, no. 6, article e1003589, 2013.

[64] N. P. Tucker, M. G. Hicks, T. A. Clarke et al., "The transcriptional repressor protein NsrR senses nitric oxide directly via a [2Fe-2S] cluster," *PLoS One*, vol. 3, no. 11, article e3623, 2008.

[65] D. F. Browning, D. J. Lee, S. Spiro, and S. J. Busby, "Down-regulation of the *Escherichia coli* K-12 nrf promoter by binding of the NsrR nitric oxide-sensing transcription repressor to an upstream site," *Journal of Bacteriology*, vol. 192, no. 14, pp. 3824–3828, 2010.

[66] J. D. Partridge, D. M. Bodenmiller, M. S. Humphrys, and S. Spiro, "NsrR targets in the *Escherichia coli* genome: new insights into DNA sequence requirements for binding and a role for NsrR in the regulation of motility," *Molecular Microbiology*, vol. 73, no. 4, pp. 680–694, 2009.

[67] M. Yamada and M. H. Saier Jr., "Physical and genetic characterization of the glucitol operon in *Escherichia coli*," *Journal of Bacteriology*, vol. 169, no. 7, pp. 2990–2994, 1987.

14

Identification and Characterization of the Diverse Stress-Responsive *R2R3*-RMYB Transcription Factor from *Hibiscus sabdariffa* L.

Bahaeldeen Babikar Mohamed,[1,2] Beenish Aftab,[1] Muhammad Bilal Sarwar,[1] Bushra Rashid,[1] Zarnab Ahmad,[1] Sameera Hassan,[1] and Tayyab Husnain[1]

[1]*Center of Excellence in Molecular Biology, University of the Punjab, 87-West Canal Bank Road, Thokar Niaz Baig, Lahore 53700, Pakistan*
[2]*National Center for Research (NCR), Ministry of Sciences and Telecommunications, Khartoum, Sudan*

Correspondence should be addressed to Bushra Rashid; bushra.cemb@pu.edu.pk

Academic Editor: Ferenc Olasz

Various regulatory proteins play a fundamental role to manage the healthy plant growth under stress conditions. Differential display reverse transcriptase PCR and random amplification of cDNA ends (RACE) was used to explore the osmotic stress-responsive transcripts. We identified and characterized the salt stress-responsive R2R3 type RMYB transcription factor from *Hibiscus sabdariffa* which has an open reading frame of 690 bp, encoding 229 long chain amino acids. In silico analysis confirmed the conserved R2 and R3 domain as well as an NLS-1 localization site. The deduced amino acids of RMYB shared 83, 81, 80, 79, 72, 71, and 66% homology with *Arabidopsis thaliana*, *Glycine max*, *Oryza sativa*, *Zea maize*, *Malus domestica*, *Populus tremula* × *Populus alba*, and *Medicago sativa* specific MYB family, respectively. We observed the gene upregulation in stem, leaf, and root tissue in response to abiotic stress. Furthermore, RMYB gene was cloned into plant expression vector under CaMV35S promoter and transformed to *Gossypium hirsutum*: a local cotton cultivar. Overexpression of RMYB was observed in transgenic plants under abiotic stresses which further suggests its regulatory role in response to stressful conditions. The RMYB transcription factor-overexpressing in transgenic cotton plants may be used as potential agent for the development of stress tolerant crop cultivars.

1. Introduction

Drought and salinity are the most devastating environmental stresses, which cause major reductions in crop productivity and quality. Worldwide, about 20% of the total cultivated land and 33% of irrigated agricultural land are affected by salinity with 10% annual increase due to saline water supply and poor agricultural practices [1]. These statistics make salinity a mighty abiotic stress factor that limits the plant growth particularly root to shoot ratio due to limited photosynthetic activity. Understanding the salinization impact on root tissue is critical, as it directly bears the excessive ion exchange imbalance in surrounding areas and it also performs other numerous vital functions like holding of the plant in the soil, water, and nutrient uptake and ionic homeostasis [2]. Regulation of gene expression under salinity stress includes a wide array of mechanisms that are used by plants to up or downregulate the production of specific gene products (mRNA to protein). These genes involve three major categories: (i) those that are involved in signaling cascade and in transcriptional control (transcription factor), such as MyC, MAP, and *SOS* kinases phospholipids and transcription factors such as *HSF and CBF/DREB*, MYB, bZIp, bHLH, WRKY, and NAC transcription factors; (ii) those that function directly in the protection of membranes and proteins, such as heat shock protein (*HSPs*) and chaperones, and late embryogenesis abundant (*LEA*) proteins; and (iii) those that are involved in water and ion transport such as aquaporins and ion transporters [3]. The most important regulators of genes and gene clusters are the transcription factors. Their structure of DNA-binding domain is considered for classification into families and subfamilies. Many families like bZIP, WRKY, AP2, NAC, C_2H_2 zinc finger, and MYB TFs have been deeply studied to play a regulatory role in a stressful environment.

MYB protein constitutes one of the largest TF groups that are functionally diverse and found in all eukaryotes. Currently, some of the MYB proteins have been identified in *Arabidopsis*, rice (*Oryza sativa*) [4], maize (*Zea mays*), and cotton (*Gossypium hirsutum*) genome [5]. Most of the plant MYB TFs belong to the R2R3-type and play an active role in (1) primary and secondary metabolites, (2) cell fate and identity, (3) developmental processes, and (4) response to biotic and abiotic stresses, while others like R1/2-MYB proteins are mainly involved in regulation of circadian rhythms. R1R2R3-MYB proteins participate in the formation of B-type cyclin, cell-cycle control, and stress tolerance [4].

Over the past decade, the role of different members of MYB has been deeply studied in response to diverse abiotic stress conditions. For instance, AtMYB2 functions in the ABA-mediated drought stress response in *Arabidopsis thaliana* [6]. AtMYB102 is found to be a key regulatory component in response to wounding, osmotic and salinity stress, and ABA application. In addition, AtMYB96 modulates ABA signaling in response to abiotic stresses in *Arabidopsis thaliana* [7]. Overexpression of *AtMYB44* has been observed to enhance the abiotic stress tolerance by maintaining the stomatal closure mechanism in transgenic *Arabidopsis thaliana* [8].

Hibiscus sabdariffa L. (English name: Red sorrel, Roselle; Arabic name: Karkade) belonging to the Malvaceae family is an annual plant with basic chromosomal number $x = 18$ ($4n = 72$). It is mostly grown in tropical and subtropical regions of Africa. Its fleshy calyx (sepals) is a good source of natural antioxidants (anthocyanin and protocatechuic acid) that protects the membrane from damage by free radicals especially lipid peroxidation [9]. DDRT-PCR is a reliable technique which is primarily based on reverse transcriptase polymerase chain reaction [10]. It has been used to isolate differentially expressed genes from different plant species under abiotic stresses [11, 12]. In this study, we have reported a novel nuclear-localized R2R3-type RMYB gene isolated from *Hibiscus sabdariffa* L. under salt stress. Up till now, MYB TFs have been identified and isolated from some land plants, but this study consists of the first report regarding the characterization of MYB TFs from the *Hibiscus sabdariffa* L. roots in response to salt stress. Overexpressing gene in transgenic plants suggests its putative role in the regulation of stress tolerance mechanism.

2. Materials and Methods

2.1. Plant Material, Stress Treatment, and RNA Isolation. Seeds of *Hibiscus sabdariffa* L. cultivar (Bulk Rahad) were grown in plastic pots (25×30 cm) with composite soil mixture (peat, soil, and sand) in equal proportion (1 : 1 : 1) in the greenhouse at 25 ± 2°C with relative humidity near 50% [9]. One month after germination, 100 mM NaCl was applied to the experimental plants. A control group of plants watered with normal tap water was also maintained as controls. After stress application, roots of treated and control plants were sampled and immediately frozen in liquid nitrogen and total RNA isolation was carried out by using phenol-based method [13]. Agarose gel electrophoresis (0.8%) was done to check the integrity of total RNA, and quantification was done by using the Nanodrop, ND-1000 spectrophotometer (NanoDrop Technologies Inc.). *DNase*1 (Sigma number AMPD1-1KT) treatment was done to remove the DNA contamination.

2.2. RMYB Gene Identification. Differential display of mRNA was performed as previously described [10], with minor modifications. Total RNA (1 μg) was reverse transcribed with an anchored oligo-dT primer by using RevertAid™ H Minus First-Strand cDNA Synthesis Kit (Fermentas, Burlington, ON) according to the manufacturer's protocol. DDRT-PCR was carried out with 15 arbitrary primers in combination with 11 anchoring primers randomly (Supplementary Table 1 available online at https://doi.org/10.1155/2017/2763259) in 25 μl volume containing 2.5 units (0.5) μl of Taq Polymerase (Invitrogen), 1 μl of each primer (10 picomole), 0.5 mM dNTPs, 2.5 μl 10 × PCR buffer, 500 ng (1 μl) cDNA, and 1 μl (50 mM) $MgCl_2$. The cycling conditions were: initial denaturation at 95°C for 2 min followed by 35 cycles of denaturation at 95°C for 45 s, annealing at 42°C for 1.30 min, extension at 72°C for 45 s, and a final elongation step at 72°C for 10 min. The amplified products were run on 16% denatured polyacrylamide gel as mentioned by Maqbool et al. [3]. Reproduced amplicon was cloned into pJET vector and sequenced by using the ABI Prism Dye Terminator kit and ABI model 3100 automated DNA Sequencer (Applied Biosystems, Foster City, CA) while homology study was done with online BLAST tool (http://blast.ncbi.nlm.nih.gov/Blast.cgi). The potential amplicon was further amplified from ends by using the GeneRacer kit (Invitrogen Life Technologies, Carlsbad, CA). Total RNA (3 μg) was used to generate RACE ready first-strand cDNA and forward gene-specific primer (5′- GCAGGATCGAG ACCTTGGCGTCTCA-3′), reverse primer (5′- TCAGAC ATCCGCGGACGGTGCTTTA-3′), and gene racer primer (5′-CGACTGGAGCACGAG-GACACTGA-3′) were used to generate the 5′ end of the gene. The PCR profile was set as initial denaturation and cycle one at 94°C for 2 min, annealing at 67°C for 30 s, and extension at 68°C for 1 min. After every five subsequent cycles, the annealing temperature was decreased by 1°C until 62°C temperature was achieved and additional 30 cycles at an annealing temperature of 62°C were carried out at the end. PCR products were resolved in 1% (*w*/*v*) agarose geland DNA fragment of the expected size was purified, cloned, and sequenced. Amplification of RACE generated full-length gene from genomic DNA was done by using the primers *RMYB-F* (5′TTACTCCCACC

TTCTGCAATTT′3) and *RMYB-R* (5′GATTGAGTAGGC ACCGAAGTTT′3) then cloned into pJET1.2 vector and sequenced.

Homology studies were carried out by using online BLAST (Basic Local Alignment Search Tool) and Pairwise alignment algorithm (http://www.ebi.ac.uk/emboss/align) programs. Exon/intron boundaries, untranslated regions (UTRs), and poly-A tail regions were located by softberry server (http://www.softberry.com/berry.phtml). The conceptual translation of nucleotide sequence was done by using the Open Reading Frame Finder (ORF) program (http://www.ncbi.nlm.nih.gov/gorf/gorf.html). Molecular weight and subcellular localization were carried out at ExPASy server (http://expasy.org). PSIPRED software was used to predict the RMYB gene secondary structure. Transmembrane helix of the RMYB was calculated by MEMSAT3 and MEMSATVSVM software while the PredictProtein server was used to understand the solvent accessibility of the protein. Multiple sequence alignment was performed by using the CLUSTALW with default parameters through EMBnet (http://www.ch.embnet.org/software/ClustalW.html). Phylogenetic relationship was constructed by using the CLC Genomics Workbench software with R2R3-MYB publicly available amino acid sequence of MYB-like 102 *Arabidopsis thaliana* (NP_567626), transcription factor *MYB*39 of *Oryza sativa* Japonica group (XP_015645522), *Glycine max MYB* transcription factor *MYB*77 (ABH02890), *Medicago sativa MYB* transcription factor (XP_003608268), *MYB* DNA-binding domain superfamily protein from *Zea mays* (DAA63102), *Malus domestica* (NP_001281023), and *MYB* transcription factor from *Populus tremula* × *Populus alba* (AAN05422). Structural modeling was carried out by homology modeling servers; Phyre2 (http://www.sbg.bio.ic.ac.uk/phyre2/html/page.cgi?id=index), Jigsaw (http://bmm.cancerresearchuk.org/~3djigsaw/), SWISS-MODEL (http://swissmodel.expasy.org/interactive), I-Tasser (http://zhanglab.ccmb.med.umich.edu/I-TASSER/), and RaptorX (http://raptorx.uchicago.edu/). Those models were further inspected by the Phi/Psi Ramachandran plot and 3D ID score was calculated at PROCHECK server (http://services.mbi.ucla.edu/SAVES/). Active binding amino acid residues within SANT domains were predicted by the online metaDBSite database [14]. PyMOL-Educational software (http://www.pymol.org/) was used to visualize the predicted parameters in 3D structure. Motif Scan (http://myhits.isb-sib.ch/cgi-bin/motif_scan) web server was used to study the primary MYB protein for functionally active sites such as N-glycosylation sites, protein Kinases sites, and sites for myristoylation and phosphorylation.

2.3. Expression Analysis of RMYB Transcript by Real-Time PCR. The response of RMYB transcript at tissue-specific level (root, leaf, and stem) was studied under drought, salt, and cold stress in the greenhouse. Gene-specific primers RMYB-F (5-CACATTCCAAGGCTGGATCT-3), RMYB-R (5-CAACCGGACCATCATCTTCT-3) were used for this purpose. Salt stress was applied as described in the previous section; cold stress was applied by keeping the plants at 4°C, while drought stress was applied by cessation of water while maintaining control plants without any stress treatment to normalize the relative fold data. Total RNA was extracted from leaf, stem, and root tissue of both control and stressed plants as described before. IQ5 cycler (Bio-Rad, Hercules, CA) was used with IQ™ SYBR_Green Super-mix dye (Bio-Rad), while qPCR cycling conditions were set an initial denaturation at 95°C for 5 min followed by 40 cycles of denaturation at 94°C for 30 s, annealing at 60°C for 30 s with extension at 72°C for 40 s and a final elongation at 72°C for 10 min. Melting curve analysis was performed by continuous monitoring of the fluorescence between 60°C and 95°C with 0.5°C increment after every 30 s. Statistical analysis was performed by using the Relative Expression Search Tool (REST) (http://rest-2009.gene-quantification.info/). *β-Actin* gene expression was used for data normalization. Analysis of variance (ANOVA) was performed by using STATISTICA 8.1 data analysis software to observe the significant difference of RMYB expression in different plant tissues.

2.4. Plant Expression Vector Construction and Genetic Transformation. Full-length sequence primers with *Xho*I restriction sites (*Xho*I-F 5′-GGCTCGAGATGTCGAAACC TCCTCCTCCT-3′, *Xho*I-R 5′-GGCTCGAGTTACCTGGG ACCGGAAAGC-3′) at 5′ ends were used to amplify the open reading frame of RMYB transcript which introduced the *Xho*I restriction site within up and downstream of the start and stop codon. The amplified product and pCAMBIA-1301 vector were double digested with *Xho*I individually for the development of overhang region and further ligated by using T4 DNA Ligase (Invitrogen) according to the manufacturer's protocol. Cloning and orientation of RMYB transcript were confirmed via restriction digestion and orientation primers. Mature embryos of the local cotton variety CIM-496 were used for the transformation of RMYB gene through *Agrobacterium*-mediated transformation method [15] with some modifications. Nontransformed plants were maintained as control. GUS-positive plants were acclimatized and shifted to a greenhouse for further analysis.

2.5. Stress Tolerance and Expression Analysis of Transgenic Plants. Transgenic plants were screened out through PCR by using a forward primer for vector and reverse gene-specific primer. Drought, salt, and cold stress were applied to the transgenic as well as nontransgenic plants after 3 weeks of shifting of plants to the greenhouse. 100 mM NaCl stress was applied for 5 days, while plants were dehydrated up to 5 days to apply the drought stress. For cold stress, the transgenic plants were kept in the cold room at 4°C for 5 h (each stress was applied to a separate group of plants). *β-Actin* gene-specific primers, *β-actin-F* (5′ TGGGGCTACTCTC AAAGGGTTG′3) and *β-actin -R* (5′ TGAGAAATTGCTG AAGCCGAAA ′3), were used as internal control to normalize the data. Gene expression was calculated as a fold-increase relative to the same transcript under control conditions with the comparative Ct method by using the Relative Expression Search Tool (REST) based on the

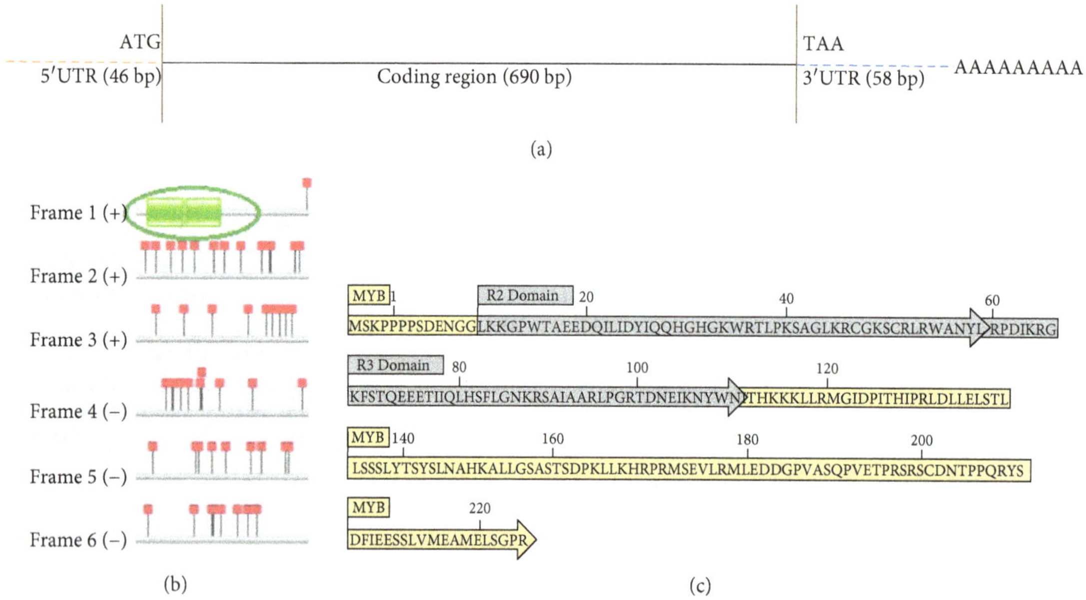

Figure 1: (a) Schematic representation of the full-length RMYB gene layout after RACE analysis. Solid black lines indicate coding region (690 bp), orange color dotted lines indicate 5′ UTR region, and blue dotted lines indicate 3′ UTR region while the sequence of AAAAAA indicates the poly-A tail region of the translated product of the gene. (b) The green box represents the two active SANT/MYB domains in the cDNA sequence as predicted by the Pfam protein family database. (c) The complete schematic presentation of the translated product of RMYB gene. The gray colored amino acids represent the R2R3 MYB DNA-binding domains.

$2^{-\Delta ct}$. Standard deviation was calculated to show the variation in the replicates.

$$\text{Relative expression ratio} = \frac{(E_{\text{target}})^{\text{Cttarget (control−sample)}}}{(E_{\text{ref}})^{\text{Ctref (control−sample)}}}. \quad (1)$$

3. Results

3.1. Identification of RMYB Transcript. Different combinations of anchored and arbitrary primers generated a number of transcripts after DDRT-PCR. Out of those transcripts, only 5 potential amplicons ranging from 200 to 600 bp showed differential existence consistently. Therefore, those transcripts were isolated, cloned, sequenced, analyzed and, submitted as potential salt-responsive ESTs in NCBI GenBank database with accession numbers [JK757799.1, JK757800.1, JK757801.1, JK757802.1, and JZ152799.1]. BLAST showed significant homology with different stress-responsive genes in multiple plant species such as *Vitis vinifera* cultivar Danuta VINE-1 repeats element, gag-pol polyprotein gene, *Glycine max* F-box protein SKIP14-like, *Gossypium hirsutum* mRNA of *MYB* protein, putative retrotransposon copia, and transposon *GORGE3-like*. Transcript accession number [GenBank: JZ152799] did not show significant homology with any of the functional proteins. The transcript number JK757801.1 encoding myeloblastosis protein was amplified from 5′ and 3′ by RACE, amplifying the extra 435 bp nucleotides by adding up to the previous 225 bp long EST. The newly generated full-length amplicon was named as RMYB which encodes 794 bp long amplicon from cDNA template (GenBank: KC145282), while 690 bp long amplicon was obtained from genomic DNA source (Figure 1(a)).

3.2. Sequence Analysis. Comparative analysis of the genomic DNA and cDNA sequences of RMYB showed that the transcribed product was intron free. We found 46 bp in the 3′ UTR region and 58 bp in 5′ UTR region downstream the termination codon (TAA). Total 9 bp encompassed the poly-A tail region spanning from 785 to 794 bp on the 3′ end of the cDNA template. Analysis of active MYB/SANT domains in the cDNA sequence through Pfam protein database is shown in Figure 1(b). Translated protein of RMYB gene is composed of 229 amino acids with 5.13 pI and 56.9 kDa. More than 75% sequence similarity was observed with different plant species including gymnosperms and angiosperms with active *MYB* DNA-binding domains (Figure 1(c)). PSIPRED secondary structure prediction showed the presence of 41% helix and 59% loop in the protein sequence of the RMYB gene (Supplementary file 1-S1). PredictProtein server predicted the 72.49% as the exposed area and accessible to the solvents, 7.2% intermediately reactive, while the remaining 20.09% as nonreactive and was buried and inaccessible to the solvents (Supplementary file 1-S2). The regions between 44 to 59 amino acids spanning ATP/GTP-binding site motif at the 47–54aa position as well as C-terminal extracellular region and 61 to 82 amino acids had a C-terminal extracellular region and cytoplasmic N-terminal region, respectively (Supplementary file 1-S3). Moreover, four CK2_PHOSPHO_SITE caseine kinase II at 20–23, 73–76 position, 102–105, and 162–165aa was observed through PROSITE motif analysis. There were three motifs specific for N-myristoylation site at 47–52, 88–93, and 159–164aa position, two sites for PHOSPHO_SITE protein kinase C phosphorylation at 54–56, 195–197aa position, and one for tyrosine kinase phosphorylation at 101 to 109 aa position. Multiple sequence alignment carried out by CLUSTALW algorithm showed 66–83% similarities for amino acid residues with other plant species such as *Arabidopsis thaliana* (83%), *Glycine max* (81%), *Oryza sativa* (*Japonica* group) (80%), *Zea maize* (79%), *Malus domestica* (72%), *Populus tremula* × *Populus alba* (71%), and *Medicago sativa* (66%) (Figure 2(a)). The phylogenetic study revealed closely related *MYB* genes in *Glycine max* and *Zea mays*, while distantly related to *Medicago truncatula* and *Malus domestica*. These characteristics of RMYB specified its relationship with R2R3-*MYB* gene group family (Figure 2(b)).

3.3. Structure Prediction and Active Binding Residue Identification. The 3D structure models of RMYB gene were generated by using different online servers are shown in Supplementary file 1 (S3). Standalone generation of protein structure based on homology modeling was left out as we observed only 50% homology with *Mus musculus* (house mouse) chain A. Crystallized structure of C-*MYB* R2 in the protein data bank (PDB) provided the recommended score to start standalone modeling. The structure was generated by *RaptorX* server and evaluated by using various structural verification software suitable for further studies based on Ramachandran plot and verification of 3D program (Figure 3(a)). The educational version of Pymol visualized the 3D generated structure and PROCHECK analysis further confirmed the stability of the structure (Supplementary file 3). Ramachandran plot analysis revealed more than 93% residues in the favored region, 44% in the allowed region, while only 2.6% in the outlier region of the plot (Figure 3(b)). None of the amino acid residues have Phi/Psi angles in the disallowed region which indicates the high stability of the generated protein structure. Verified 3D verification program also showed the reliable ID score for *RaptorX* generated model. MetaDBSite server analyzed the binding affinity of amino acid residues in the DNA-binding domain. Total 39 residues were predicted to have high binding affinity for the nucleic acid (Figure 3(c)). Among those, 74.3% of the residues was in R2R3 *MYB* binding domain, while the remaining were in the variable C-terminal region (Figures 4(a) and 4(b)).

3.4. Expression of RMYB Gene in Hibiscus sabdariffa L. in Response to Abiotic Stresses. An RMYB fragment of 120 bp was amplified by real-time PCR from cDNA of *Hibiscus sabdariffa* L. Gene expression was 2.0, 1.5, and 1.0-fold in the stem, leaf, and root, respectively, under salt stress (Figure 5(a)), while the expression was 8.0, 5.0, and 3.0 times higher in response to drought, salt, and cold stressed roots respectively (Figure 5(b)). Gene expression was compared with control or nonstressed plants and data normalization was carried out by using *β-actin* gene as a housekeeping control.

3.5. Gene Cloning and Genetic Transformation. RMYB gene was cloned in the plant expression vector pCAMBIA 1301 (11.837 kb), and full-length primer amplified the complete coding sequence of 690 bp with an insertion of the *Xho*I restriction site sequence before and after the start and stop codon (Figure 6(a)). The ligation was confirmed by restriction digestion (Figure 6(b)). The orientation of insert in the vector was confirmed by orientation primers which yielded the known size of the fragment that was further confirmed by sequencing data, showing 100% homology with the RMYB gene sequence. Complete schematic representation of the cassette of the vector with fragment is shown in Figure 6(c). Complete transformation procedure can be accessed from Supplementary file 1 (S4). Histochemical GUS assay detected the transit GUS gene expression as a blue color in transgenic leaf tissues, while there was no color in the nontransgenic plants' leaves (Figure 7(a)). PCR analysis showed that the genomic DNA of transgenic plants amplified the gene of interest (Figure 7(b)). In this experiment, the overall transformation efficiency was about 0.7%. Figure 7(c) shows the transgenic plants shifted to the soil and the effect of different abiotic stresses on these plants.

3.6. Overexpression of RMYB Gene in Transgenic Plants. Variable response of gene expression was observed in transgenic cotton plants under different abiotic stresses. About 55-fold expression of the RMYB gene was detected under osmotic stress and 41-fold was overexpressed under salt stress. Gene of interest was 15-fold more expressed under the cold stress (Figure 7(d)). This overexpression in transgenic plants was compared to the nontransgenic plants.

4. Discussion

TFs play an important role in the regulation of a plant's genetic response, controlling the cascade of genes involved in abiotic stresses. Various studies have been carried out for their identification and characterization. MYB family of transcription factors is a large set of proteins that is mostly involved in an array of functions regarding growth and development, primary and secondary metabolism, cell fate regulation, and response to biotic and abiotic stresses [16]. Differential display profiling is an efficient technique to explore differentially expressed genes under altered growth conditions. The sensitivity of this technique results in good detection of genes present in low abundance (induced or repressed), and only a small amount of starting material is required [10, 17]. Several biotic and abiotic stress-responsive transcripts have been identified from different plant species by this methodology such as tomato, rice, barley, *Arabidopsis*, sunflower, and cotton. However, such type of investigation has yet not been reported in *Hibiscus sabdariffa* L. The RMYB transcript that we have identified and isolated showed more than 70% homology with MYB TF's gene family. Full-length RMYB sequence encoding 229 amino acid long chain with the single open reading frame was retrieved by RACE. In silico analysis revealed the presence of the conserved MYB DNA-binding domains spanning from 16 to 114 amino acid residues at the N-terminus. Recurrence of conserved sequence in the domain is a characteristic feature of MYB TFs (1R, R2R3, 3R, and 4R) on which its nomenclature is based. The RMYB gene revealed its homology to the R2R3 type *MYB* class which is the largest group of plant TFs [18]. R2R3 is thought to be the descendants of R1R2R3-MYB gene ancestor formed by the loss of R1 repeat

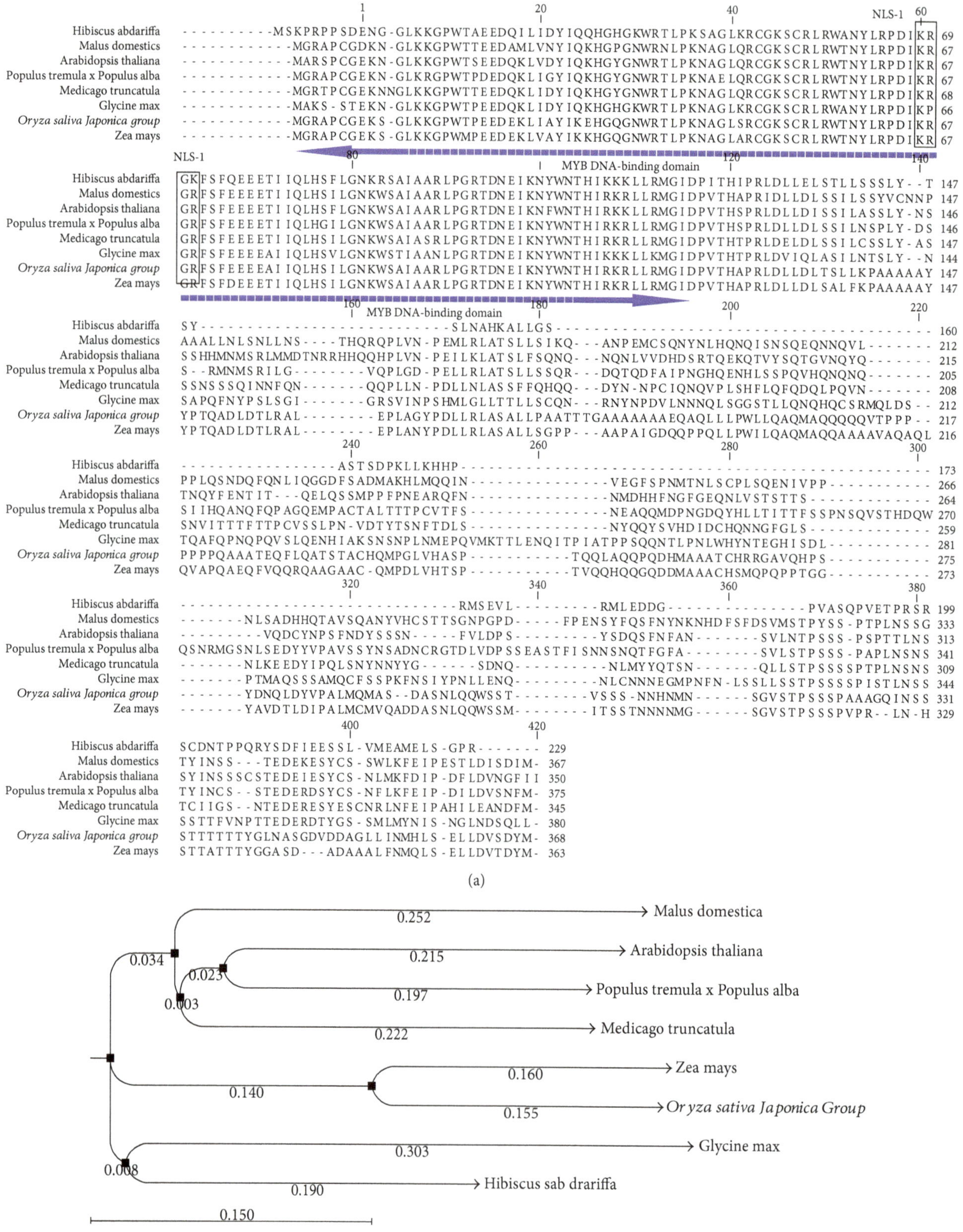

FIGURE 2: (a) Multiple sequence alignment of RMYB gene protein with other MYB gene sequences. Conserved residues are in blue color. Active functional domain span is represented by an arrow. The box covered portion represents the NLS-1. (b) Distance-based and neighbor-joining tree correlated complete RMYB amino acid sequence to full-length MYB amino acid sequences from different plant species. Bootstrap number on each node represents the significance level while the arrow symbol points out the studied species.

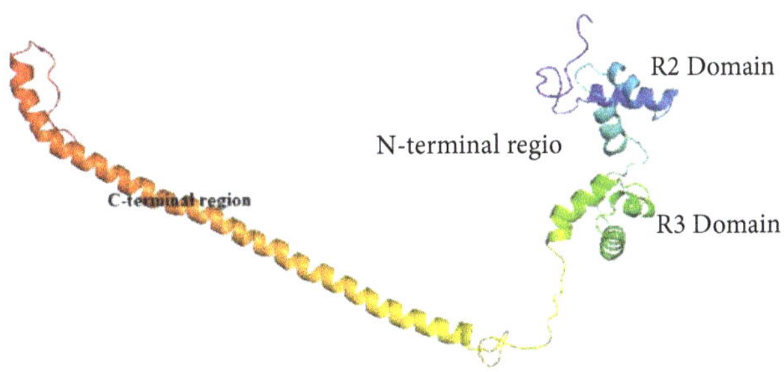

(a)

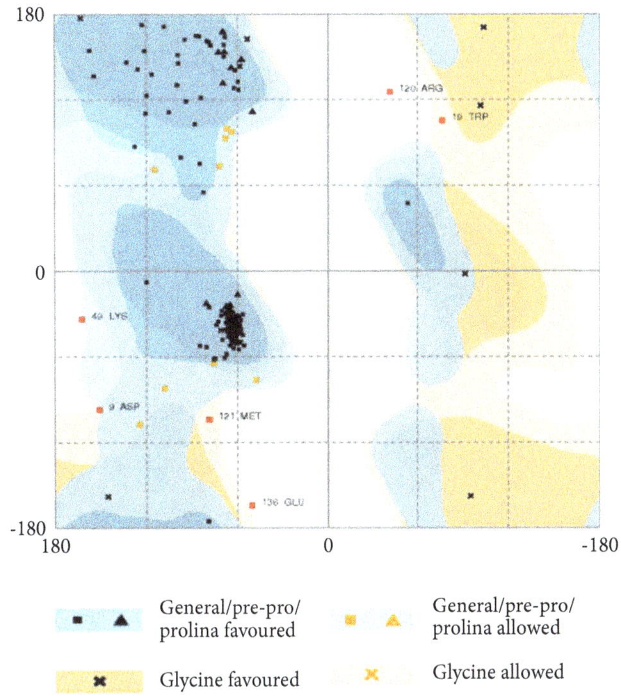

Number of residues in favoured region (~98.0% expected) : 211 (93.0%)
Number of residues in allowed region (~2.0% expected) : 10 (4.4%)
Number of residues in outlier region : 6 (2.6%)

(b)

Position	Residue	Meta DBS	Position	Residue	Meta DBS	Position	Residue	Meta DBS	Position	Residue	Meta DBS	Position	Residue	Meta DBS	Position	Residue	Meta DBS	Position	Residue	Meta DBS	Position	Residue	Meta DBS
1	M	-	31	I	-	61	N	+	91	R	+	121	M	-	151	L	-	180	R	-	211	D	-
2	S	+	32	Q	-	62	Y	+	92	S	+	122	G	-	152	N	-	181	M	-	212	F	-
3	K	+	33	Q	-	63	L	-	93	A	-	123	I	-	153	A	-	182	L	-	213	I	-
4	P	-	34	H	-	64	R	+	94	I	-	124	D	-	154	H	-	183	E	-	214	E	-
5	P	+	35	G	-	65	P	-	95	A	-	125	P	-	155	K	-	184	D	-	215	E	-
6	P	-	36	H	-	66	D	-	66	A	-	126	I	-	156	A	-	185	D	-	216	S	-
7	P	-	37	G	-	67	I	-	97	R	+	127	T	-	157	L	-	186	G	-	217	S	-
8	S	-	38	K	+	68	K	+	98	L	-	128	H	-	158	L	-	187	P	-	218	L	-
9	D	-	39	W	+	69	R	+	99	P	-	129	I	-	159	G	-	188	V	-	219	V	-
10	E	-	40	R	+	70	G	-	100	G	-	130	P	-	160	S	-	189	A	-	220	M	-
11	N	-	41	T	+	71	K	-	101	R	+	131	R	-	161	A	-	190	S	-	221	E	-
12	G	-	42	L	-	72	F	-	102	T	+	132	L	-	162	S	-	191	Q	-	222	A	-
13	G	-	43	P	-	73	S	-	103	D	-	133	D	-	163	T	-	192	P	-	223	M	-
14	L	-	44	K	+	74	T	-	104	N	+	134	L	-	164	S	-	193	V	-	224	E	-
15	K	+	45	S	+	75	Q	-	105	E	-	135	L	-	165	D	-	194	E	-	225	L	-
16	K	+	46	A	-	76	E	-	106	I	-	136	E	-	166	P	-	195	T	-	226	S	-
17	G	-	47	G	-	77	E	-	107	K	+	137	L	-	167	K	-	196	P	-	227	G	-
18	P	-	48	L	-	78	E	-	108	N	+	138	S	-	168	L	-	197	R	-	228	P	-
19	W	-	49	K	+	79	T	-	109	Y	+	139	T	-	169	L	-	198	S	-	229	R	+
20	T	-	50	R	+	80	I	-	110	W	-	140	L	-	170	K	-	199	R	-			
21	A	-	51	C	-	81	I	-	111	N	-	141	L	-	171	H	-	200	S	-			
22	E	-	52	G	-	82	Q	-	112	T	+	142	S	-	172	R	-	201	C	-			
23	E	-	53	K	+	83	L	-	113	H	+	143	S	-	173	P	-	202	D	-			
24	D	-	54	S	+	84	H	-	114	I	-	144	S	-	174	R	-	203	N	-			
25	Q	-	55	C	-	85	S	-	115	K	+	145	L	-	175	M	-	204	T	-			
26	I	-	56	R	+	86	F	-	116	K	+	146	Y	-	176	S	-	205	P	-			
27	L	-	57	L	-	87	L	-	117	K	+	147	T	-	177	E	-	206	P	-			
28	I	-	58	R	+	88	G	-	118	L	-	148	S	+	178	V	-	207	Q	-			
29	D	-	59	W	+	89	N	-	119	L	-	149	Y	-	179	L	-	208	R	-			
30	Y	-	60	A	-	90	K	-	120	R	-	150	S	-	180	R	-	209	Y	-			
																		210	S	-			

(c)

Figure 3: (a) 3D predicted structure for RMYB gene by RaptorX server, with the variable C-terminal region, conserved N-terminal region, and conserved R2 and R3 domains. (b) Ramachandran plot of RMYB gene protein revealing stability parameters. (c) metaDBSite predicted protein binding residue with the nucleic acid. Binding residues are labeled with "+" and nonbinding residues are labeled with "–" and highlighted portion represents the R2R3 domain of the RMYB gene.

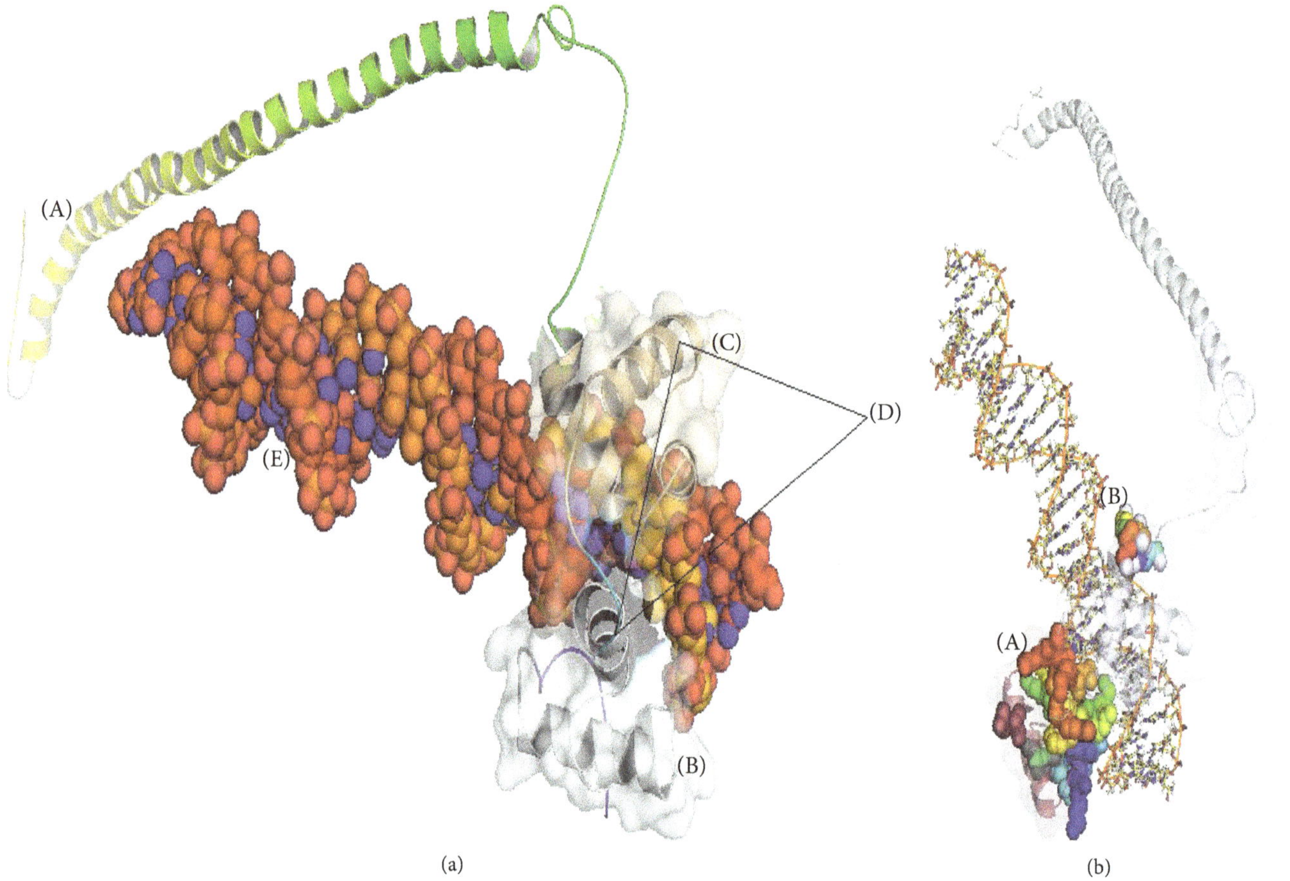

Figure 4: (a) 3D representation of the predicted binding residues of the conserved domain with the DNA (A) variable region of RMYB, (B) R2 domain of RMYB, (C) R3 domain of RMYB, and (D) conserved region among the MYB gene families. (b) Amino acid residues directly involved in binding with DNA in R2 domain (A) predicted amino acid residues for binding with DNA in R3 domain (B).

and subsequent expansion of the gene family. MYB domain is thought to be highly conserved while other regions of R2R3-MYB proteins are mostly variable [6, 19]. The nonconserved diverse sequence at the C-terminal region makes these factors gain wider and distinct features, such as control of development, secondary metabolism, hormonal regulation, and response to biotic and abiotic stresses including drought, salt, wounding, cold, and freezing. We also observed HTH similar repeats in the other MYBs [20]. RMYB gene possesses the NLS-1 signal within the DNA binding domain for transport of proteins into the nucleus like other MYB proteins such as Md*MYB*, Fv*MYB*44 and *GmMYB44* (NLS1-KGRK), *AtMYB59*, and *AtMYB48* (NLS2-RKKAQEKKR) [21–23]. The deduced translated product showed a significant homology with the other crop plants like *Arabidopsis thaliana*, *Zea mays*, *Oryza sativa*, *Glycine max*, *Medicago sativa*. [4]. The coded transcript within the DNA-binding domain of RMYB possesses a highly conserved and evenly distributed residue "w", which is important for making hydrophobic core and structural stability of the MYB repeats [24, 25].

Phylogenetic analysis of the MYB proteins has been extensively carried out in *Arabidopsis thaliana*, *Oryza sativa*, *Zea mays* and *Populus*, and their evolutionary relationship has been studied systematically [24]. The phylogenetic relationship has grouped the *Hibiscus sabdariffa* L. with *Zea mays* and *Glycine max* specific MYB's families and showed divergence from *Arabidopsis thaliana*, because, a small subset of R2R3 MYB proteins, regulating salt and dehydration genes (i.e., *AtMYB2*) does not contain the first cys residue which makes it impossible to form an S-S bond. Therefore, an alternative mechanism involving nitrosylation of cysteine-S might control its DNA binding activity. These transcription factors also undergo ubiquitination at lysine residues near or at the transcription activation domain, thereby, creating a positive influence on the transcriptional activity. Phosphorylation often has an enhancement effect on the transcriptional activity of MYBs, as observed in *Arabidopsis* and *Nicotiana tabacum* [25, 26]. This divergent relationship reveals that the MYB family underwent a rapid expansion due to gene duplication, sequence deviation, and gene conversion throughout the evolution. Such divergent clades might have been lost with the evolution of *Arabidopsis* lineage or gained along with the development of lineages of rosid and asterid. The subgroups which are closer show that the genes have originated from the same duplication event.

Functional conservation is achieved through the clustering of genes and conserved motifs outside the DNA-binding domain aiding in the identification of different functionally conserved motifs. Redundancy observed at amino

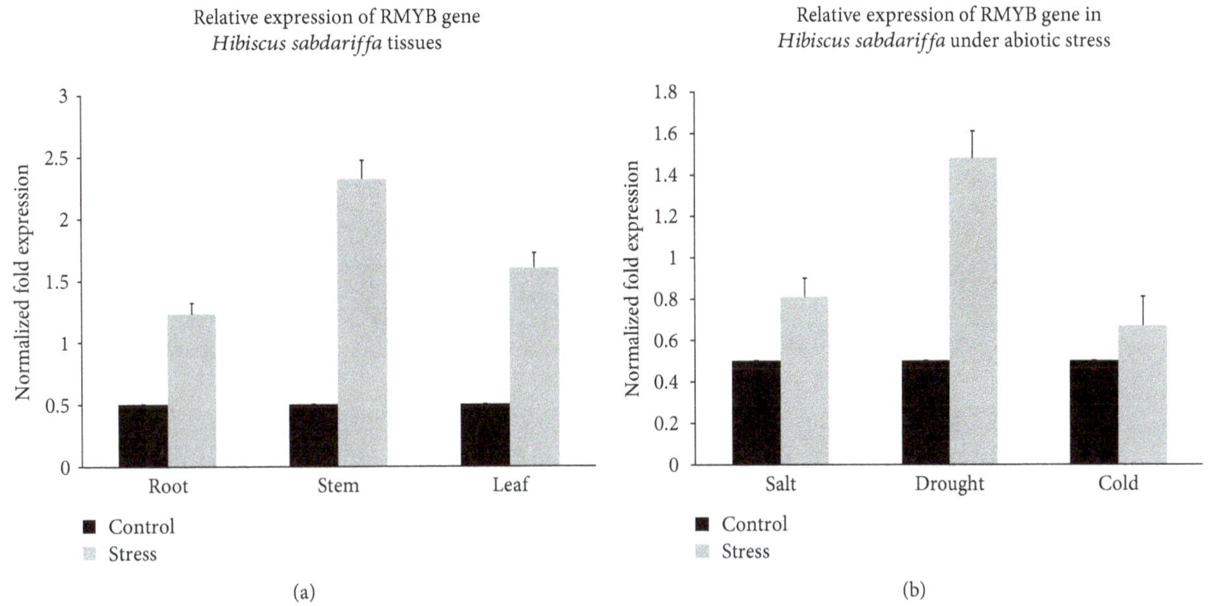

FIGURE 5: RT-PCR expression analysis of the RMYB gene in *Hibiscus sabdariffa* L. plant tissues under different abiotic stresses. (a) Effect of salt stress on the root, stem, and leaf. (b) Effect of salt, drought, and cold stress on leaves.

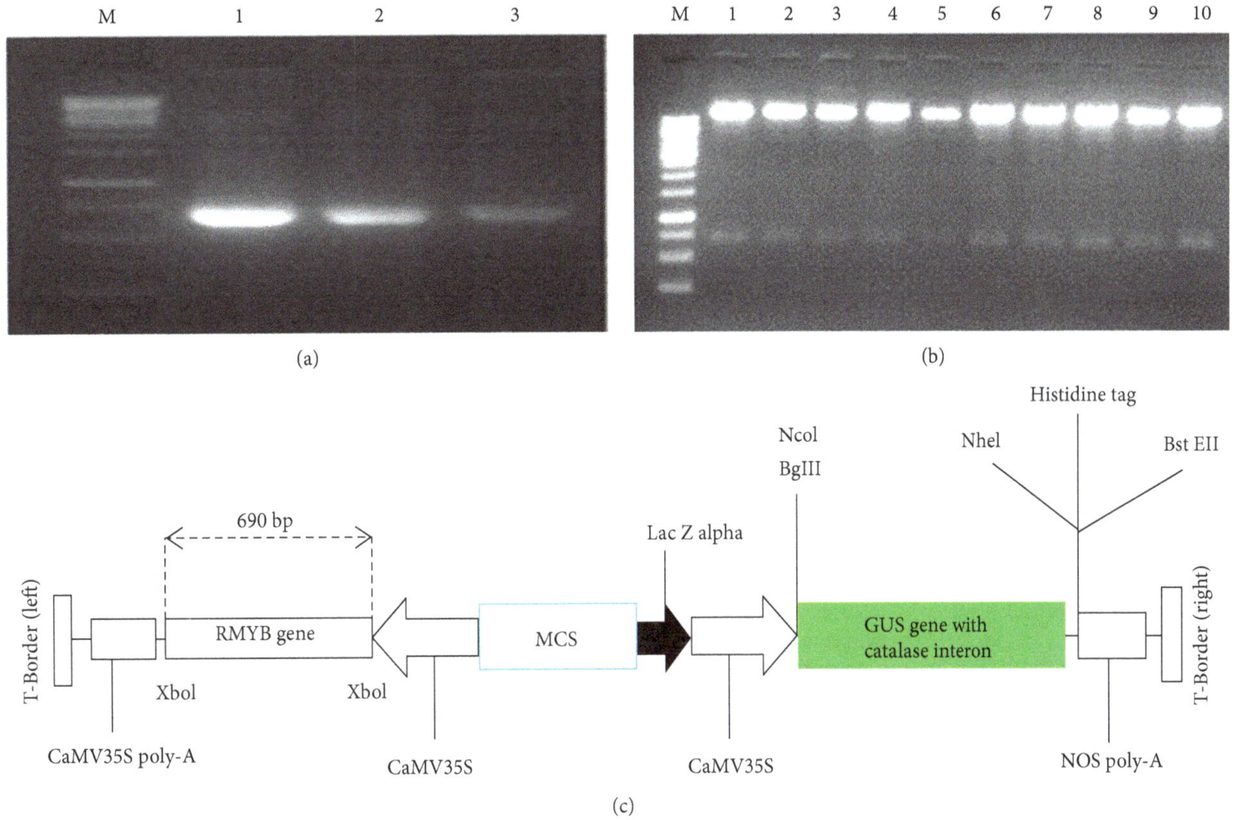

FIGURE 6: (a) Amplification of the RMYB full-length gene from salt-stressed root of *Hibiscus sabdariffa* L. with *Xho*I restriction sites. Lane M-1 kb marker, lanes 1–3 amplified RMYB product. (b) Cloning confirmation by restriction digestion (c) Schematic representation of RMYB gene in pCAMBIA expression vector in CaMV 35S promoter driven by GUS.

acid sequence level among the MYB genes, suggests that, differential/spatial expression of genes causes them to have similar molecular functions, yet, they display variable biological characteristics when mutated. So, developmental biology does not cause the redundancy, instead, during evolution, the genes with closely related functions exist and are maintained

FIGURE 7: (a) Immunohistochemically GUS assay of the transgenic cotton leaf. (b) PCR amplification of a 690 bp fragment of RMYB gene from transgenic plants, M:1 kb DNA Ladder, lane 1–7 transgenic plants. (c) Transgenic plants in pots under drought, cold, and salt stress. (d) Expression analysis of RMYB gene in transgenic cotton plants under drought, salt, and cold stress. *β-Actin* gene was used as internal control to normalize data and gene expression is indicated as a fold-increase relative to the control condition.

together. A single gene is regulated by multiple factors from the same gene family transmitting signals through the same element [19, 26, 27]. Epigenetic modifications and protein-protein interactions can measurably affect the regulatory potential of MYB transcription factors [4]. There is a monomeric DNA binding potential seen *in vitro* in the MYB proteins having two or more MYB repeats. They act as covalently attached dimers, while they make contact with DNA. Dimerization is responsible for the affinity and specificity with which these proteins bind DNA. This outcome diversified the overall functional characteristics, as well as sequence homology, environmental response, and cellular localization [28].

Modeling and interaction studies like metaDBSite analysis further authenticate the functional interaction of RMYB domain with the nucleic acid. About >76% of predicted DNA-protein active binding residues was found within the R2R3 DNA-binding domain. Aspartic acid, arginine, lysine, tryptophan, and serine covered the major portion of active residues within the domain and are potentially considered to be involved in the activities of MYB gene [29]. The relative binding affinities of CASTing, targets to the R2R3-*MYB* domain with kD values and error rates like ACCTAC, ACCAAT, ACCAAA, ACCATA. ACCAAC, ACCACA, and ACCACC, have been confirmed, and the preferred binding sequence was found to be ACCTAC [30]. The role of MYB transcription factor in various physiochemical processes, metabolic regulation, control of cell cycle, and involvement in various defense responses against biotic and abiotic environmental factors has been reported [31]. Variable but significant induction of expression was detected in the stem, root, and leaf tissues under salt stress which shows its critical role in the nucleus. The induction of *TaMYB1* by light under hypoxia, salt stress, and similarly by ABA and PEG application was observed in wheat [32]. Identification and characterization of full-length cDNA sequence of wheat MYB gene *TaPIMP1* against pathogen infection by using RT-PCR, RACE, and its overexpression in transgenic tobacco under drought and salt stress have also been studied [33]. Overexpression of *GhMYB7* in transgenic *Arabidopsis thaliana* enhanced the developing fibers, regulated the secondary cell wall biosynthesis, and acted as the potential transcriptional activator [34]. We also observed the response of RMYB gene toward multiple abiotic stresses and expression was induced in response to drought, salt, and low temperature. It indicates that the RMYB is a TF involved in regulating the response of roselle to low-temperature stress and desiccation regardless of salt stress in the arid and semiarid regions helping to relieve the plants' physiology while bringing it to the normal state. For further elucidation about the activity of MYB gene, overexpression of the RMYB transcript was observed in transgenic cotton plants in response to abiotic stresses. This study would help to cope with the changing external environmental conditions. Our results showed that the gene expression is consistent with the previously reported findings [8, 35]. Characterization of the R2R3 *SbMYB44* transcription factors from the haplotype *Salicornia brachiata* Roxb in response to salt stress has also been carried out which has shown its involvement in multiple abiotic stresses [24]. The stable expression of the drought-specific marker genes like *OsLEA3*, *OsRab16A*, and *OsDREB2A*, in *OsMYB2* overexpressing transgenic plants [36].

In conclusion, the RMYB transcript isolated from *Hibiscus sabdariffa* L. under salt stress is an R2R3 type MYB TF which is involved in stress tolerance mechanism. Overexpression of the RMYB gene in transgenic *Gossypium hirsutum* plants elucidates its potential role to cope with stressful environmental conditions. These transgenic plants could be a potential candidate for breeding and development of tolerant/resistant cultivars.

Abbreviations

DDRT-PCR:	Differential display reverse transcriptase polymerase chain reaction
RACE:	Random amplification of cDNA ends
CaMV:	Cauliflower mosaic virus
MYB:	Myeloblastosis
TFs:	Transcription factors
BLAST:	Basic Local Alignment Search Tool
UTRs:	Untranslated regions
Poly-A:	Polyadenylated A
ANOVA:	Analysis of variance
EST's:	Expression sequence tags.

Conflicts of Interest

The authors declare that the research was conducted in the absence of any commercial or financial relationships that could be construed as a potential conflict of interest.

Authors' Contributions

Bahaeldeen Babikar Mohamed and Muhammad Bilal Sarwar performed the wet lab experiments while Zarnab Ahmad and Bushra Rashid performed the bioinformatics analyses. Bushra Rashid and Sameera Hassan drafted the manuscript. Bushra Rashid and Tayyab Husnain designed and supervised the project. All authors read and approved the final manuscript.

Acknowledgments

The authors are also thankful to Dr. Muhammad Naveed Shahid and Mr. Mukhtar Ahmad for technical help and support. The authors are thankful to The World Academy of Sciences (TWAS) for financial support and to Dr. Bahaeldeen Babiker Mohamed in traveling to Pakistan to complete this study.

References

[1] M. Hasanuzzaman, K. Nahar, and M. Fujita, "Plant response to salt stress and role of exogenous protectants to mitigate salt-induced damages," in *Ecophysiology and Responses of Plants under Salt Stress*, pp. 25–87, Springer, New York, NY, USA, 2013.

[2] L. Hu, H. Li, L. Chen, Y. Lou, E. Amombo, and J. Fu, "RNA-seq for gene identification and transcript profiling in relation to root growth of bermudagrass (*Cynodon dactylon*) under salinity stress," *BMC Genomics*, vol. 16, p. 575, 2015.

[3] A. Maqbool, M. Zahur, M. Irfan et al., "Identification and expression of six drought responsive transcripts through differential display in desi cotton (*Gossypium arboreum*)," *Molekuliarnaia Biologiia*, vol. 42, pp. 559–565, 2007.

[4] C. Dubos, R. Stracke, E. Grotewold, B. Weisshaar, C. Martin, and L. Lepiniec, "MYB transcription factors in *Arabidopsis*," *Trends in Plant Science*, vol. 15, pp. 573–581, 2010.

[5] M. Cedroni, R. C. Cronn, K. Adams, T. Wilkins, and J. F. Wendel, "Evolution and expression of MYB genes in diploid and polyploid cotton," *Plant Molecular Biology*, vol. 51, pp. 313–325, 2003.

[6] H. Abe, T. Urao, T. Ito, M. Seki, K. Shinozaki, and K. Yamaguchi-Shinozaki, "Arabidopsis AtMYC2 (bHLH) and AtMYB2 (MYB) function as transcriptional activators in abscisic acid signaling," *The Plant Cell*, vol. 15, pp. 63–78, 2003.

[7] M. Denekamp and S. C. Smeekens, "Integration of wounding and osmotic stress signals determines the expression of the AtMYB102 transcription factor gene," *Plant Physiology*, vol. 132, pp. 1415–1423, 2003.

[8] C. Jung, J. S. Seo, S. W. Han et al., "Overexpression of *AtMYB44* enhances stomatal closure to confer abiotic stress tolerance in transgenic Arabidopsis," *Plant Physiology*, vol. 146, pp. 623–635, 2008.

[9] B. Mohamed, M. Sarwar, B. Rashid et al., "Physiological and biochemical responses of Roselle (*Hibiscus sabdariffa* L.) to NaCl stress," *Agrochimica*, vol. 57, pp. 248–263, 2013.

[10] P. Liang and A. B. Pardee, "Differential display of eukaryotic messenger RNA by means of the polymerase chain reaction," *Science*, vol. 257, pp. 967–971, 1992.

[11] P. Lang, C.-k. Zhang, R. C. Ebel, F. Dane, and W. A. Dozier, "Identification of cold acclimated genes in leaves of *Citrus unshiu* by mRNA differential display," *Gene*, vol. 359, pp. 111–118, 2005.

[12] A. Maqbool, M. Zahur, M. Irfan et al., "Identification and expression of six drought-responsive transcripts through differential display in desi cotton (*Gossypium arboreum*)," *Molecular Biology*, vol. 42, pp. 492–498, 2008.

[13] R. C. Muoki, A. Paul, A. Kumari, K. Singh, and S. Kumar, "An improved protocol for the isolation of RNA from roots of tea (*Camellia sinensis* (L.) O. Kuntze)," *Molecular Biotechnology*, vol. 52, pp. 82–88, 2012.

[14] J. Si, Z. Zhang, B. Lin, M. Schroeder, and B. Huang, "MetaDB-Site: a meta approach to improve protein DNA-binding sites prediction," *BMC Systems Biology*, vol. 5, p. S7, 2011.

[15] J. Gould, S. Banister, O. Hasegawa, M. Fahima, and R. H. Smith, "Regeneration of *Gossypium hirsutum* and *G. barbadense* from shoot apex tissues for transformation," *Plant Cell Reports*, vol. 10, pp. 12–16, 1991.

[16] P. K. Agarwal, P. S. Shukla, K. Gupta, and B. Jha, "Bioengineering for salinity tolerance in plants: state of the art," *Molecular Biotechnology*, vol. 54, pp. 102–123, 2013.

[17] C. Voelckel and I. T. Baldwin, "Detecting herbivore-specific transcriptional responses in plants with multiple DDRT-PCR and subtractive library procedures," *Physiologia Plantarum*, vol. 118, pp. 240–252, 2003.

[18] A. Feller, K. Machemer, E. L. Braun, and E. Grotewold, "Evolutionary and comparative analysis of MYB and bHLH plant transcription factors," *The Plant Journal*, vol. 66, pp. 94–116, 2011.

[19] C. Martin and J. Paz-Ares, "MYB transcription factors in plants," *Trends in Genetics*, vol. 13, pp. 67–73, 1997.

[20] I. Romero, A. Fuertes, M. Benito, J. Malpica, A. Leyva, and J. Paz-Ares, "More than *80 R2R3-MYB* regulatory genes in the genome of *Arabidopsis thaliana*," *Plant Journal*, vol. 14, pp. 273–284, 1998.

[21] J. Li, X. Li, L. Guo et al., "A subgroup of *MYB* transcription factor genes undergoes highly conserved alternative splicing in *Arabidopsis* and rice," *Journal of Experimental Botany*, vol. 57, pp. 1263–1273, 2006.

[22] I.-H. Oh and E. P. Reddy, "The myb gene family in cell growth, differentiation and apoptosis," *Oncogene*, vol. 18, pp. 3017–3033, 1999.

[23] S. Yang, J. P. Sweetman, S. Amirsadeghi et al., "Novel anther-specific myb genes from tobacco as putative regulators of phenylalanine ammonia-lyase expression," *Plant Physiology*, vol. 126, pp. 1738–1753, 2001.

[24] P. S. Shukla, P. Agarwal, K. Gupta, and P. K. Agarwal, "Molecular characterization of an MYB transcription factor from a succulent halophyte involved in stress tolerance," *AoB Plants*, vol. 7, article plv054, 2015.

[25] R. Stracke, M. Werber, and B. Weisshaar, "The *R2R3-MYB* gene family in *Arabidopsis thaliana*," *Current Opinion in Plant Biology*, vol. 4, pp. 447–456, 2001.

[26] P. D. Rabinowicz, E. L. Braun, A. D. Wolfe, B. Bowen, and E. Grotewold, "Maize *R2R3 Myb* genes: sequence analysis reveals amplification in the higher plants," *Genetics*, vol. 153, pp. 427–444, 1999.

[27] D. J. Gates, S. R. Strickler, L. A. Mueller, B. J. Olson, and S. D. Smith, "Diversification of R2R3-MYB transcription factors in the tomato family Solanaceae," *Journal of Molecular Evolution*, vol. 83, pp. 26–37, 2016.

[28] C. Yanhui, Y. Xiaoyuan, H. Kun et al., "The MYB transcription factor superfamily of Arabidopsis: expression analysis and phylogenetic comparison with the rice MYB family," *Plant Molecular Biology*, vol. 60, pp. 107–124, 2006.

[29] C. Kaya, S. Aydemir, O. Sonmez, M. Ashraf, and M. Dikilitas, "Regulation of growth and some key physiological processes in salt-stressed maize (Zea mays L.) plants by exogenous application of asparagine and glycerol," *Acta Botanica Croatica*, vol. 72, pp. 157–168, 2013.

[30] M. B. Prouse and M. M. Campbell, "Interactions between the R2R3-MYB transcription factor, *At*MYB61, and target DNA binding sites," *PLoS One*, vol. 8, article e65132, 2013.

[31] L. Zhang, G. Zhao, J. Jia, X. Liu, and X. Kong, "Molecular characterization of 60 isolated wheat MYB genes and analysis of their expression during abiotic stress," *Journal of Experimental Botany*, vol. 63, pp. 203–214, 2012.

[32] T. G. Lee, C. S. Jang, J. Y. Kim et al., "A Myb transcription factor (*TaMyb1*) from wheat roots is expressed during hypoxia: roles in response to the oxygen concentration in root environment and abiotic stresses," *Physiologia Plantarum*, vol. 129, pp. 375–385, 2007.

[33] H. Liu, X. Zhou, N. Dong, X. Liu, H. Zhang, and Z. Zhang, "Expression of a wheat MYB gene in transgenic tobacco enhances resistance to *Ralstonia solanacearum*, and to drought and salt stresses," *Functional & Integrative Genomics*, vol. 11, pp. 431–443, 2011.

[34] J. Huang, F. Chen, S. Wu, J. Li, and W. Xu, "Cotton *GhMYB7* is predominantly expressed in developing fibers and regulates secondary cell wall biosynthesis in transgenic *Arabidopsis*," *Science China Life Sciences*, vol. 59, pp. 194–205, 2016.

[35] Y.-K. Liang, C. Dubos, I. C. Dodd, G. H. Holroyd, A. M. Hetherington, and M. M. Campbell, "*At*MYB61, an R2R3-MYB transcription factor controlling stomatal aperture in *Arabidopsis thaliana*," *Current Biology*, vol. 15, pp. 1201–1206, 2005.

[36] A. Yang, X. Dai, and W.-H. Zhang, "A R2R3-type MYB gene, *OsMYB2*, is involved in salt, cold, and dehydration tolerance in rice," *Journal of Experimental Botany*, vol. 63, pp. 2541–2556, 2012.

15

Gene Coexpression Network Comparison via Persistent Homology

Ali Nabi Duman[1] **and Harun Pirim**[2]

[1]*Department of Mathematics and Statistics, King Fahd University of Petroleum and Minerals, Dhahran, Saudi Arabia*
[2]*Department of Systems Engineering, King Fahd University of Petroleum and Minerals, Dhahran, Saudi Arabia*

Correspondence should be addressed to Ali Nabi Duman; aliduman@kfupm.edu.sa

Academic Editor: Atsushi Kurabayashi

Persistent homology, a topological data analysis (TDA) method, is applied to microarray data sets. Although there are a few papers referring to TDA methods in microarray analysis, the usage of persistent homology in the comparison of several weighted gene coexpression networks (WGCN) was not employed before to the very best of our knowledge. We calculate the persistent homology of weighted networks constructed from 38 Arabidopsis microarray data sets to test the relevance and the success of this approach in distinguishing the stress factors. We quantify multiscale topological features of each network using persistent homology and apply a hierarchical clustering algorithm to the distance matrix whose entries are pairwise bottleneck distance between the networks. The immunoresponses to different stress factors are distinguishable by our method. The networks of similar immunoresponses are found to be close with respect to bottleneck distance indicating the similar topological features of WGCNs. This computationally efficient technique analyzing networks provides a quick test for advanced studies.

1. Introduction

Quantitative skills have become much more essential to distill meaning from the vast emerging and increasing diverse data sets since the technological advances in DNA sequencing that occurred at the end of the 20th century. Modern technological developments in high throughput data technologies such as microarrays and RNA-sequencing enable the generation of terabytes of data in a short amount of time. The type of the data generated comprises levels regarding the abundance of RNA, quantification of protein-protein interactions (PPI), and many other biological molecular interactions. The generated data is embraced for statistical inference and computational analysis including low-level data processing and high-level algorithmic analysis with computations and machine learning techniques. Making use of the data is a reverse engineering approach. Gene coexpression microarrays measure interactive activities of thousands of genes. In network terms, the nodes of the coexpression matrix represent the gene products and the edges of the matrix represent the relationship between the products (usually expressed by correlations). After a chip scanning and an image processing process, a matrix of coexpression values is obtained. The rows of the matrix refer to the gene products while the columns refer to the experiments/samples/tissues. The numeric values of the matrix are the expression values of genes across the experiments. The experiments may be "control versus treated" or "time course."

The networks constructed based on the gene expression similarity are called gene coexpression networks [1]. They can be named association, correlation, and influence networks [2] as well. Coexpression network analysis requires the selection of a similarity measure between genes and a clustering algorithm to decompose the network into functional clusters/modules following a meaningful experiment design [3, 4]. However, there exist clustering algorithms that do not require a distance matrix as an input, but rather they require the network itself (e.g., some community structure finding algorithms). The modules found by running a clustering algorithm require biological inference.

In order to have a high-level overview of a coexpression network construction and analysis, a few common processes can be summarized [5]:

(i) Obtaining a filtered data

(ii) Making use of network inference or guilt by association as in clustering

(iii) Enrichment analysis to see the biological relevance of computational outputs

(iv) Extension of the model(s) integrating multiple data types such as mRNA, miRNA data from RNA-req, TF, DNA-binding data from ChIP-seq, and protein interaction data from mass spectrometry

Ideally, network decomposition results in tight clusters/ modules with dense intracluster and sparse intercluster connections. Tight clusters are supposed to include biologically relevant genes in terms of functions or residing in the same pathway.

One of the widely used steps in constructing a gene coexpression network is trimming some of the edges based on a threshold [6–8]. Persistent homology, which is first developed to explore the topological features of point cloud data, is a topological invariant, and it addresses the problem of choosing a reasonable threshold. Our method employs persistent homology once the correlation similarity is calculated on the filtered networks.

Persistent homology is a new tool for studying the shape of a point cloud in application areas such as digital images [9, 10], dynamical systems [11], biomolecules [12], and high-dimensional data mining [13]. The persistent homological framework enables us to analyze multiscale networks in a consistent manner [14–16]. The output of the persistent homology of a network can be summarized visually using a *persistent diagram*, and the distance between two persistent diagrams can be measured via *bottleneck distance*. Here, we are using the persistent homology framework to do a threshold-free analysis of weighted gene coexpression networks constructed from 38 Arabidopsis microarray data sets. We list several advantages of our method:

(i) Our method does not require a choice of fixed threshold as it considers the networks at every possible threshold.

(ii) It gives a more robust result than an analysis of unweighted networks for which the results might depend on the choice of the threshold.

(iii) Persistent diagrams can be used for a standard data analysis method such as cluster analysis.

(iv) Our method eliminates the computational burden of analyzing many networks obtained for different thresholds.

Topological data analysis (TDA) has been applied to biological data before. Arsuaga et al. [17] associate a two-dimensional (2D) point cloud with each array comparative genomic hybridization (aCGH) profile and generate a sequence of simplicial complexes. They use these mathematical objects to identify DNA copy number aberrations by interrogating the topological properties. Camara et al. [18] use TDA for mapping meiotic recombination at fine scales. Comparing to standard linkage-based methods, they find that TDA can deal with a larger number of genomes in a computationally efficient way. Cang et al. [19] propose a support vector machine algorithm for protein classification. They choose the machine learning feature vectors from the persistent homology of the protein structure. Chan et al. [20] use persistent homology to capture both vertical and horizontal evolutions. They show that horizontal evolution exhibits nontrivial topology of dimension greater than zero. Nicolau et al. [21] introduce a topological method that identifies a unique subgroup of estrogen receptor-positive (ER+) breast cancers that express high levels of c-MYB and low levels of innate inflammatory genes. Perea et al. [22] present a novel method based on persistent homology to classify periodic or nonperiodic signals of microarray time series data. Their method successfully identifies the periodic genes in microarray data from the yeast cell cycle.

Here are the main results of the paper:

(i) We quantify topological features of WGCNs using persistent homology and apply the hierarchical clustering algorithm to the distance matrix whose entries are pairwise bottleneck distance values between the networks.

(ii) The immunoresponses to different stress factors are distinguishable by our method. The networks of similar immunoresponses are found to be close with respect to bottleneck distance indicating the similar topological features of WGCNs. Hence, persistent diagrams of the networks can be used to determine the topological and biological similarities.

2. Methods

Topological methods address several problems that arise in biological data analysis [13]. We now summarize three of these which are related to the analysis of our microarray data sets. First of these problems is to extract qualitative information from a given set of data prior to quantitative methods. This might include studying the characteristics of the data space such as determining the connected components, loops, and higher dimensional surfaces. In the biological context, these methods have already been used, for example, in identifying a novel subgroup of a certain disease [17, 21], cataloguing the type of exchange of genomic material [20], classifying protein domain [23], and discovering periodicity in gene expression time series data [22].

The second issue in biological data analysis is the choice of a natural coordinate system. A particular choice of a coordinate system might not have an essential meaning during the analysis. Topological methods, which are coordinate-free and depend only on the chosen metric, enable us to compare the data sets given in different coordinate systems

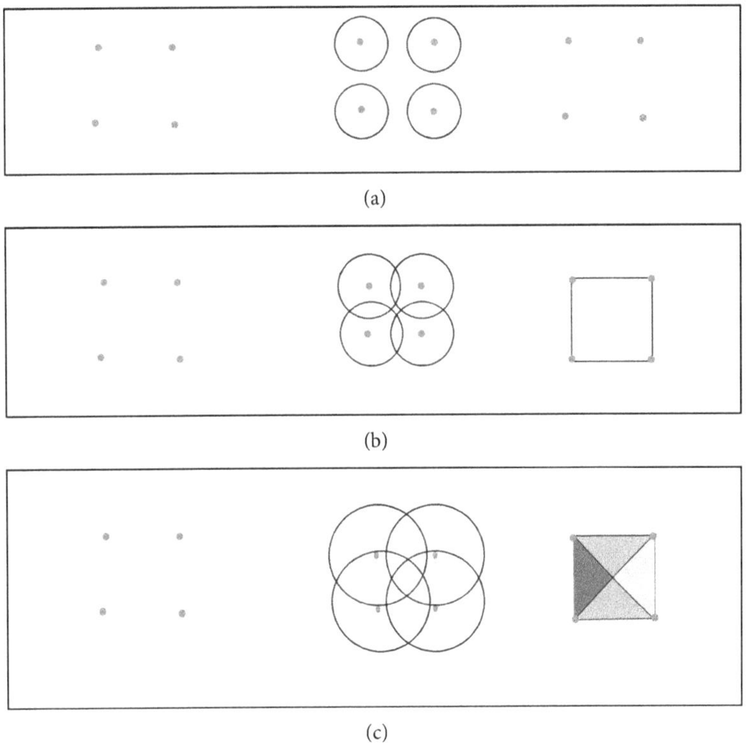

Figure 1: Construction of Rips complex. The leftmost figures show point cloud data, the middle the covering of the balls centered at each point, and the rightmost Rips complex for the corresponding balls of each radius.

where there is a concept of similarity in general, not only the Euclidean metric.

Thirdly, fixing an optimal parameter in conventional clustering algorithms might not reveal sufficient information about the data set under consideration. Hence, it is preferred to consider the entire set of parameters at once. This raises the question of what the relationship between the information obtained from different parameters is. Topology deals with this problem via the concept of *functoriality* which is used to compute the topological invariants from discrete approximations.

Topology ideally aims to find the homeomorphism type of a topological space. Roughly speaking, we would like to classify the spaces up to stretching and bending but not tearing and gluing. However, in most of the cases, it is very hard to get the homeomorphism type of a space. Hence, we need to consider other invariants: homotopy, homology, cohomology, and so on. In order to find these topological invariants of a data space, we need to construct a combinatorial approximation of the space called *simplicial complex*:

Definition 1. A *simplicial complex* K consists of a set of objects, $V(K)$, called *vertices* and a set, $S(K)$, of finite nonempty subsets of $V(K)$, called simplices such that (i) any nonempty subset of a simplex is also a simplex, (ii) every one element set $\{v\}$, where $v \in V(K)$, is a simplex, and (iii) the intersection of any two simplices is also a simplex.

The *dimension of a simplex* $\sigma \in S(K)$ is defined as dim $(\sigma) = |\sigma| - 1$: simplices consisting of a single element are zero-dimensional, and simplices consisting of two elements are one-dimensional and so on. The *dimension of the complex* dim (K) is defined as the largest dimension of any of its simplices.

One can construct a simplicial complex from a data set in $\mathbb{R}^n$ using a *Rips complex* (see Figure 1).

Definition 2. Let $X = \{x_\alpha\}$ be a set of points in $\mathbb{R}^n$. The *Rips complex* $R(X, \epsilon)$ (also called Vitoris-Rips complex) is the simplicial complex whose k-simplices correspond to $k+1$-tuples of points $\{x_a\}_0^k$ which are pairwise within distance ϵ.

The Rips complex can be considered a variant of Čech complex construction. The problem with the Čech construction is that it is computationally expensive as it requires the storage of simplices of various dimensions. On the other hand, the Rips complex is computationally more efficient as the edges in the complex completely determine the complex. Although there are other efficient complexes (e.g., witness complex and alpha complex), in this work, it is sufficient for us to use the Rips complex because of the moderate size of the filtered microarray coexpression data under study.

Once we obtain the simplicial complex K from our data set, we assign a vector space via homology, which is a classical invariant within algebraic topology. We now summarize the construction of the homology. One can refer to [24] for a more detailed description and properties.

(i) The first step is to construct the chain group C_p, which consists of the formal sums of the

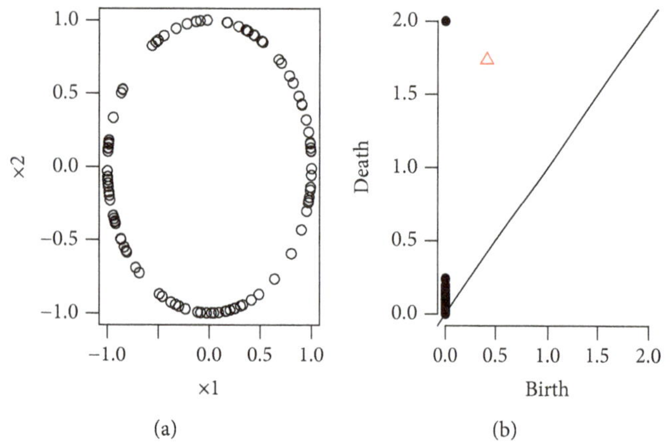

FIGURE 2: (b) is the persistent diagram of point cloud data on (a).

p-dimensional simplices $c = \sum \gamma_i \sigma_i$, where γ_i are in a field F. These formal sums are called *p-chains.*

(ii) The next step is to define the *boundary map*, $\partial : C_p \longrightarrow C_{p-1}$ which maps each p-chain to the sum of $p-1$-dimensional subsets, *faces*, of its p-simplices. One can easily check that the square of the boundary map is zero; that is, $\partial \circ \partial = 0$.

(iii) Hence, one obtains the following *chain complex*:

$$\cdots \longrightarrow C_{p+1} \longrightarrow C_p \longrightarrow C_{p-1} \longrightarrow \cdots \quad (1)$$

(iv) As a result of $\partial \circ \partial = 0$, one can define the *pth homology group* $H_p(K)$ as the quotient of the *p-cycles* Z_p, elements of C_p which are mapped to 0 by ∂, and *p-boundaries* B_p, which is the image of C_{p+1} under ∂.

As the coefficients of the p-chains are chosen from a field F, the resulting abelian group $H_p(K)$ is torsion free. Hence, $H_p(K) = Z_p/B_p = F^{\beta_p}$ is a vector space over F, with rank β_p. This number β_p is called *pth Betti number* and is equal to 0 if p is greater than the dimension of K.

Given a *simplicial map* $f : K \longrightarrow L$ between simplicial complexes K and L, where f takes simplices to simplices and it is linear, and f induces a homomorphism on homology $f_p : H_p(K) \longrightarrow H_p(L)$. This functoriality is crucial, when we next introduce the concept of persistence.

We now review the persistent homology which is the main mathematical tool that we use in our study. One can refer to the survey articles [13–16] for more detailed discussions. Given a fixed possibly obtained from the distribution of a data set X, the Rips complex $R(X, \epsilon)$ is in most cases not sufficient to extract details about the shape of X. Hence, one needs to check $R(X, \epsilon)$ for different values to get a satisfactory level of homological information. In order to determine the topological features that persist in a parametrized family of spaces, we construct the family of spaces using the Rips complexes $R(X, \epsilon_i)$ of a data set for increasing

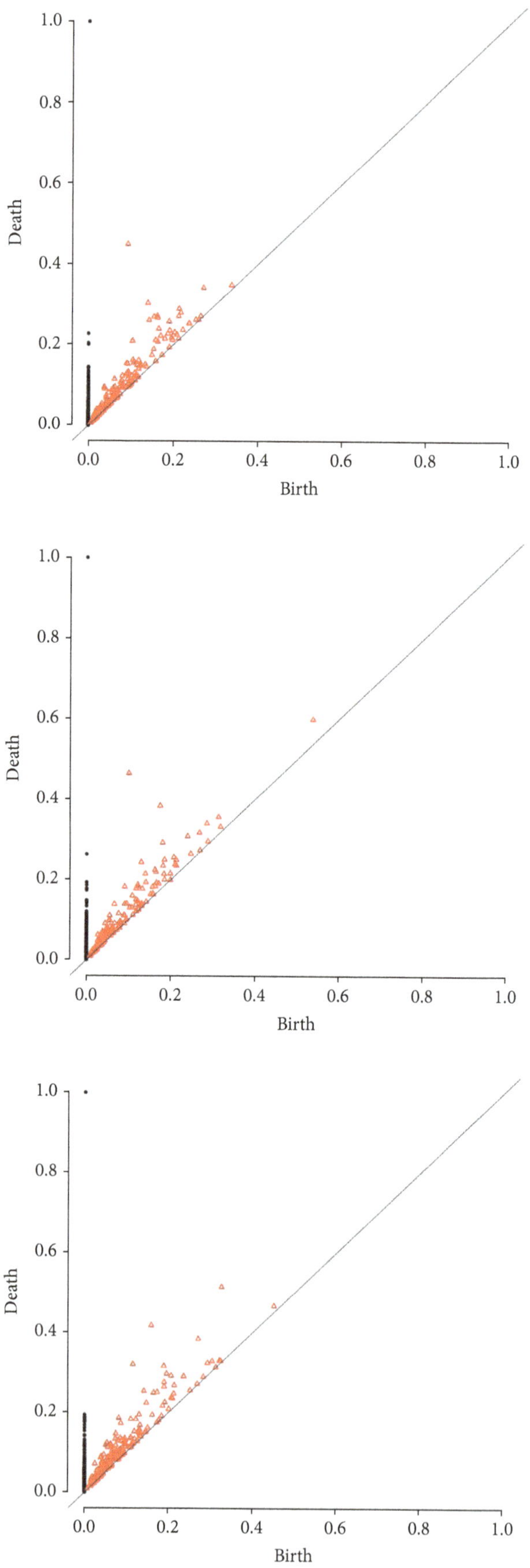

FIGURE 3: Persistent diagrams of experiments 5, 6, and 8.

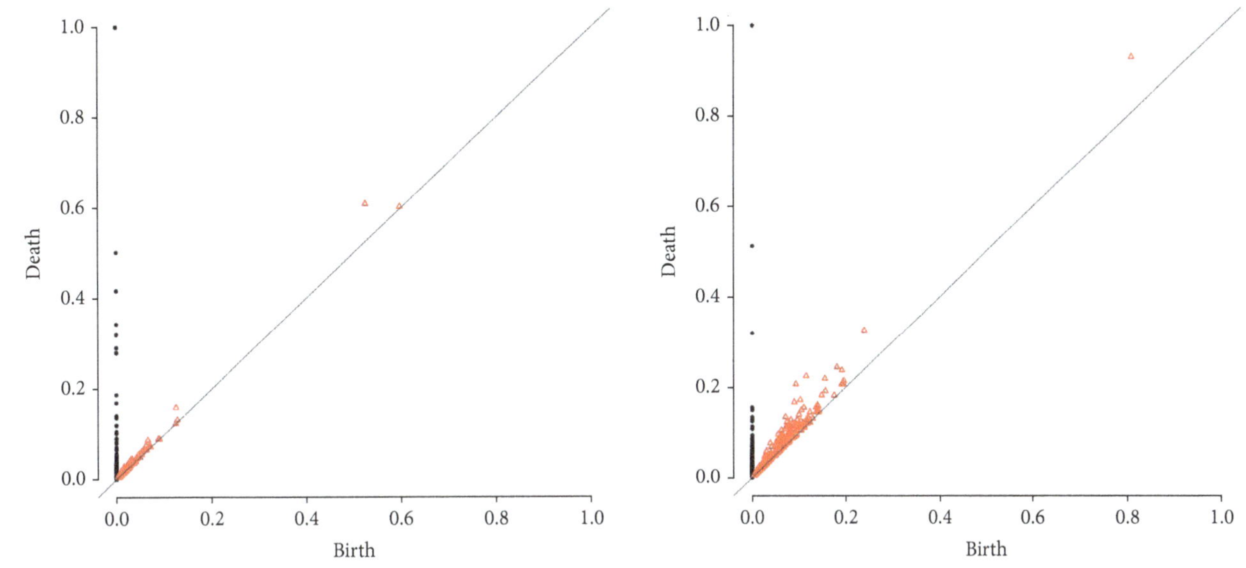

FIGURE 4: Persistent diagrams of experiments 31 and 32.

parameter values $\{\epsilon_i\}_i^n$. The inclusion maps give us a filtration of spaces:

$$R(X, \epsilon_1) \hookrightarrow R(X, \epsilon_2) \hookrightarrow \cdots \hookrightarrow R(X, \epsilon_N). \tag{2}$$

We next obtain the following sequence of vector spaces and linear maps using the functoriality of homology:

$$H_k(R(X, \epsilon_1)) \longrightarrow H_k(R(X, \epsilon_2)) \longrightarrow \cdots \longrightarrow H_k(R(X, \epsilon_N)). \tag{3}$$

One decomposes this algebraic structure into summands of the form $0 \longrightarrow F \longrightarrow \cdots \longrightarrow F \longrightarrow 0$, which points out the spaces $R(X, \epsilon_i)$, where a certain homological feature (i.e., connected components, loops, and voids) manifests. The set of these summands corresponds with finite subsets of $P = \{(\epsilon_{\text{birth}}, \epsilon_{\text{death}}) \mid \epsilon_{\text{birth}} \in [0, \infty), \epsilon_{\text{death}} \in [0, \infty), \epsilon_{\text{birth}} < \epsilon_{\text{death}}\}$, where ϵ_{birth} and ϵ_{death} are the birth and the death times, respectively. The *persistence* of a feature is defined as the difference between its death and its birth time. The features with larger persistence provide the most important information about the topology, while the features with short persistence are most of the time considered noise [19, 20, 23, 25].

The set P can visually be represented as a collection of points of the first quadrant of the coordinate plane. This visual representation is called *persistence diagram* and can be considered a replacement for Betti numbers in the case of a collection of thresholds $\{\epsilon_i\}$. With the help of the persistence diagram, one can identify the significant homological features, which are located far from the main diagonal and filter the noisy features, which can be found near the main diagonal. For example, for $k = 0$, we obtain the persistence of the connected components, and for $k = 1$, we obtain the persistence of the loops.

In Figure 2, a point cloud data and corresponding persistent diagram are shown. The zero-dimensional features are represented by black dots. The persistent one-dimensional feature represented by a red triangle reveals the existence of the big loop. In this work, we are mainly interested in the evolution of the connected component of the networks. Hence, the birth time of each component is fixed as 0. As the threshold increases, the number of component decreases and no new component is born (Figures 3 and 4).

In order to compare two networks, we need to find the distance between the persistent diagrams of these two networks. The most common metric to measure the distance between two persistence diagrams is the *bottleneck distance* which is known to be stable under small perturbations due to the bottleneck stability theorem [26].

Definition 3. The *bottleneck* distance between two persistent diagrams PD and PD$'$ is defined as

$$d_B\left(\text{PD}, \text{PD}'\right) = \inf_{\lambda} \sup_{p \in \text{PD}} \|p - \lambda(p)\|_\infty, \tag{4}$$

where λ ranges over all bijections from P to P' and $\|\cdot\|_\infty$ is the L^∞-norm on the plane.

As the diagonals are also considered in the diagram with infinite multiplicity, any bijection λ can map an extra point in the diagram with a point on the diagonal. Hence, it is not necessary to have the same number of points in the diagrams. The stability of the bottleneck distance is given by the following inequality:

$$d_B\left(\text{PD}, \text{PD}'\right) \le 2d_{\text{GH}}(X, Y) \le 2d_{\text{H}}(X, Y), \tag{5}$$

where PD and PD$'$ are persistence diagrams for the data sets X and Y in a Euclidean space, respectively, and

Table 1: Pathogen resistance Arabidopsis microarray data collected. This is the part of the table that appeared in [34], and it is under CC BY 4.0 license.

ID	GEO data set	Plant	Stress group	Stress
1	GSE12856	Arabidopsis	PTI	Nonhost
2	GSE13739	Arabidopsis	Induced resistance	Induced resistance (SA)
3	GSE14961	Arabidopsis	Induced resistance	Induced resistance (SA)
4	GSE15236	Arabidopsis	Fungi	*Fusarium oxysporum*
5	GSE16471	Arabidopsis	PTI	PTI
6	GSE16472	Arabidopsis	PTI	PTI
7	GSE16497	Arabidopsis	Induced resistance	Induced resistance (Aphid)
8	GSE17382	Arabidopsis	PTI	PTI
9	GSE17875	Arabidopsis	Fungi	*Botrytis cinerea*
10	GSE19273	Arabidopsis	Bacteria	*Ralstonia solanacearum*
11	GSE20188	Arabidopsis	Induced resistance	Induced resistance (insecticides)
12	GSE21762	Arabidopsis	Induced resistance	Induced resistance (JA)
13	GSE21920	Arabidopsis	Bacteria	*Pseudomonas syringae*
14	GSE26679	Arabidopsis	Fungi	Golovinomyces cichoracearum
15	GSE26973	Arabidopsis	Induced resistance	Induced resistance (exudates)
16	GSE28800	Arabidopsis	Induced resistance	Induced resistance (chemistry)
17	GSE431	Arabidopsis	Fungi	*Erysiphe cichoracearum*
18	GSE5513	Arabidopsis	Induced resistance	Induced resistance (PTI)
19	GSE5752	Arabidopsis	Induced resistance	Induced resistance (SA)
20	GSE5753	Arabidopsis	Induced resistance	Induced resistance (SA)
21	GSE5754	Arabidopsis	Induced resistance	Induced resistance (SA)
22	GSE5755	Arabidopsis	Induced resistance	Induced resistance (SA)
23	GSE5756	Arabidopsis	Induced resistance	Induced resistance (SA)
24	GSE5757	Arabidopsis	Induced resistance	Induced resistance (SA)
25	GSE5758	Arabidopsis	Induced resistance	Induced resistance (SA)
26	GSE6831	Arabidopsis	Induced resistance	SAR(JA)
27	GSE8319	Arabidopsis	PTI	PTI
28	GSE10426	Arabidopsis	Fungi	*Plasmodiophora brassicae*
29	GSE10713	Arabidopsis	Fungi	*Fusarium oxysporum* pv. raphani
30	GSE13390	Arabidopsis	Bacteria	*Pseudomonas syringae* pv. tomato
31	GSE15880	Arabidopsis	Fungi	*Botrytis cinerea*
32	GSE15881	Arabidopsis	Fungi	*Botrytis cinerea*
33	GSE18757	Arabidopsis	Bacteria	*Ralstonia solanacearum*
34	GSE25838	Arabidopsis	Fungi	*Botrytis cinerea*
35	GSE34081	Arabidopsis	Bacteria	*Pseudomonas syringae* pv. tomato
36	GSE7990	Arabidopsis	Induced resistance	Induced resistance (ISR, *Bradyrhizobium*)
37	GSE8877	Arabidopsis	Fungi	*Plasmodiophora brassicae*
38	GSE31230	Arabidopsis	Bacteria	*Ralstonia solanacearum*

d_H and d_{GH} are Hausdorff and Gromov-Hausdorff distances (Equation (1) in [22]). The bottleneck distance is the $q = \infty$ version of the Wasserstein distance defined as

$$d_q\left(\mathrm{PD}, \mathrm{PD}'\right) = \left[\inf_{\lambda} \sum_{p \in \mathrm{PD}} \|p - \lambda(p)\|_{\infty}^{q}\right]^{1/q}. \quad (6)$$

The Wasserstein distance d_q includes all the differences between the diagrams while the bottleneck distance d_B measures the single largest distance between the persistence diagrams. Thus, we have

$$d_B < d_q. \quad (7)$$

The Wasserstein distance is also more sensitive to the small differences due to noise as the sensitivity of d_q decreases when q increases.

The distance between networks can be used to determine the topological and biological similarities. In this paper, we

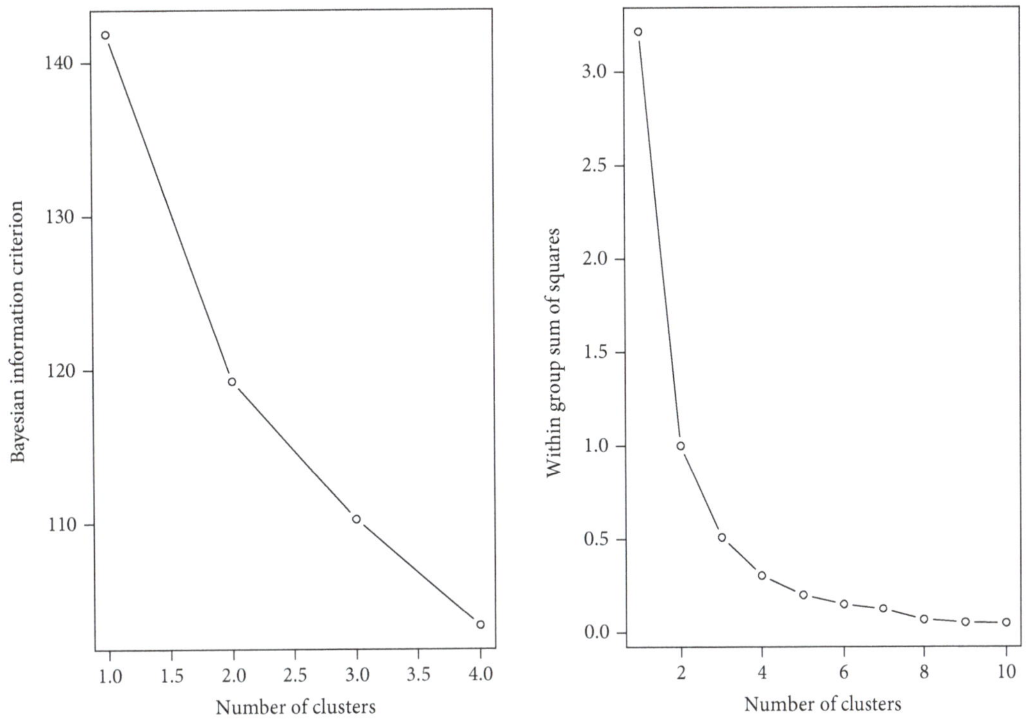

FIGURE 5: Determination of the number of clusters based on the Bayesian information criterion and the within sum of squares.

use the bottleneck distance d_B to construct a distance matrix whose entries are the pairwise distance between the experiments represented as networks. The networks are then clustered using Ward's hierarchical clustering algorithm applied in *R*. The optimum number of clusters is predicted using both the Bayesian information criterion (BIC) and the within sum of squares (WSS).

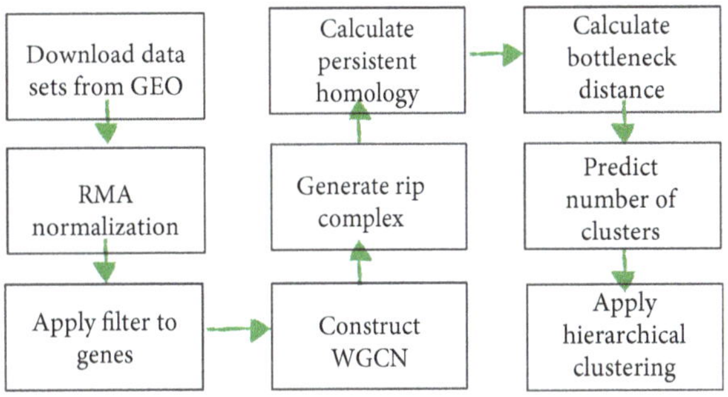

FIGURE 6: Work flow.

3. Results

The pathogen resistance microarrays for *Arabidopsis thaliana* data sets are downloaded from the Gene Expression Omnibus (GEO) database. The data sets can be accessed entering the GEO ID into the query field in the link https://www.ncbi.nlm.nih.gov/geo/query/acc.cgi. Table 1 summarizes the data sets based on stress groups and stresses. The existing four stress groups: PTI, bacteria, induced resistance, and fungi, belong to 6, 6, 16, and 10 experiments, respectively. The studies with the pathogen *Botrytis cinerea* are the most abundant.

Persistent homology application workflow starts with the robust multiarray average (RMA) normalization for the data sets. R Bioconductor affy package is used for the normalization of Affymetrix data. The two-color microarray data sets are preprocessed using the marray and Agi4x44PreProcess libraries. The expression matrices are constructed for each microarray experiment independently. Each experiment comprises several microarrays.

After the normalization, highly coexpressed genes are retained through a gene filtering procedure so that only the significantly expressed genes are retained in the data matrix. R genefilter package is used, while any other filtering method could be employed. Around 400 genes are remained after filtering each data [21, 27].

The filtered genes are used to construct the weighted coexpression network using the absolute value of the Pearson correlation as the similarity measure, which is commonly used in a gene expression cluster analysis [28]. We use this measure to calculate the dependence between the expression levels of the genes across samples. These similarity measures take values in the same interval [0, 1], where 0 indicates nondependence and 1 indicates total dependence or maximum similarity. Unlike the standard approach in WGCN analyses [29, 30], we do not need to specify a threshold to punish weak correlations and emphasize strong ones because weak correlation is regarded as noise by persistent homology as explained in the previous section.

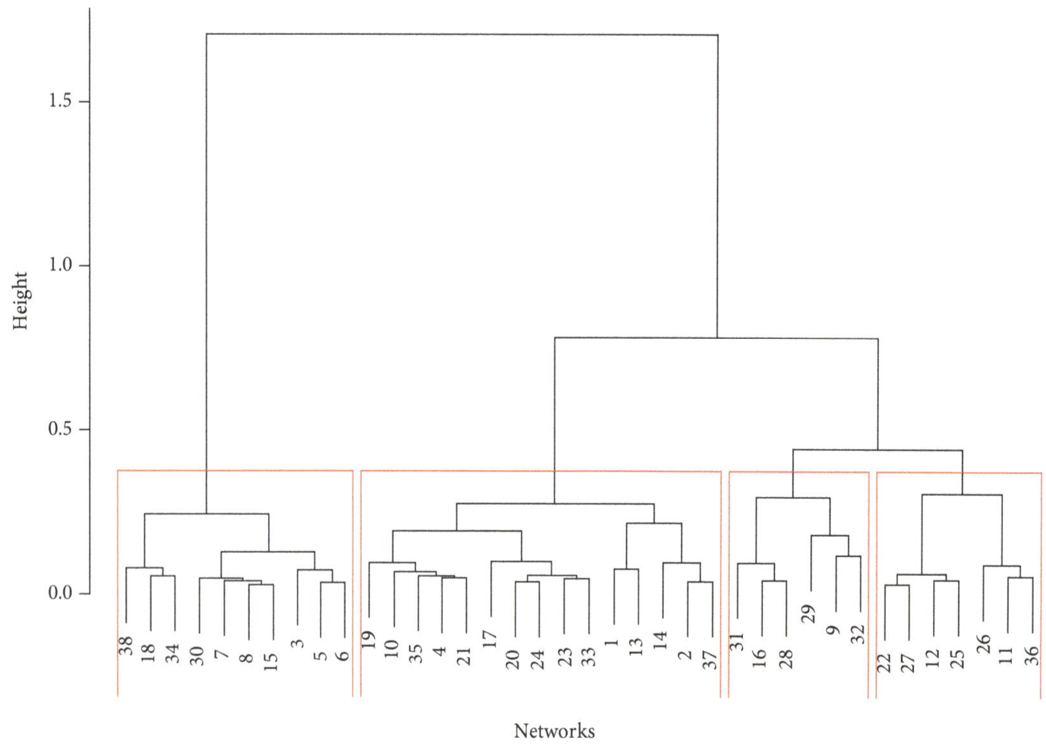

FIGURE 7: The clustering of zero-dimensional persistent diagrams.

In the next step, Rips complexes are generated for each weighted network in order to determine the persistent topological features (connected components, loops, voids, etc.). We then calculate the persistent homology of the Rips complexes. In this study, we only consider 0 and 1 dimensional homologies [17, 18, 20, 22]. According to our knowledge, higher dimensional homology has not been used to analyze microarray data. Higher dimensional features such as voids might not have biological significance in weighted gene coexpression networks.

The results for each network are summarized in persistent diagrams. Afterwards, the distance matrix including the distance values between persistent diagrams is obtained employing the bottleneck distance, which is the most commonly used metric to measure the distance between two persistent diagrams [26]. The persistent homology and bottleneck distance are calculated using R TDA package [31].

The final distance matrix is used as an input to a hierarchical clustering method in order to compare 38 WGCNs. Hierarchical clustering is frequently used in WGCN analyses since it works well when there are many singleton clusters and when cluster sizes vary greatly [29]. The optimum number of clusters is predicted to be four using both the Bayesian information criterion (BIC) and the *K-means* within the sum of squares method (Figure 5). They are mainly utilized to choose the number of clusters according to the intrinsic complexity present in a particular data set. BIC is employed using the R adegenet package [32]. The workflow is illustrated in Figure 6.

The hierarchical clustering using the bottleneck distance with four cluster results in the partition as seen on the dendrogram (Figure 7). The clustering here is obtained from zero dimensional persistent diagrams. Figure 8 shows hierarchical clustering of one-dimensional persistent diagrams.

We have four different stress groups in our data set: PTI, bacteria, fungi, and induced resistance. The networks from the same stress group were found to be in the same clusters. Here are some examples from the clustering of zero-dimensional persistent diagrams (Figure 7):

(i) The networks 5, 6, 8, and 18 associated with the stress group PTI are close to each other; hence, they are all in the first cluster. This means that they show similar topological features as a result of the same immune responses to PTI.

(ii) Four out of 6 bacteria experiments (IDs 10, 13, 33, and 35; *Ralstonia solanacearum* and *Pseudomonas syringae*) are located in the second cluster. Thus, we can conclude that the immunity processes of Arabidopsis that are represented in these networks derived from the responses to pathogens *Ralstonia solanacearum* and *Pseudomonas syringae* could share some similarities.

(iii) Five out of 6 experiments in cluster 3 (IDs 9, 28, 29, 31, and 32; *Botrytis cinerea*, *Plasmodiophora brassicae*, and *Fusarium oxysporum* pv. raphani) are related to fungal experiments. Thus, we could say that Arabidopsis has similar immunity responses to *Botrytis cinerea*, *Plasmodiophora brassicae*, and *Fusarium oxysporum* pv. raphani. Figure 4 shows the persistent diagrams of experiments 31 and 32

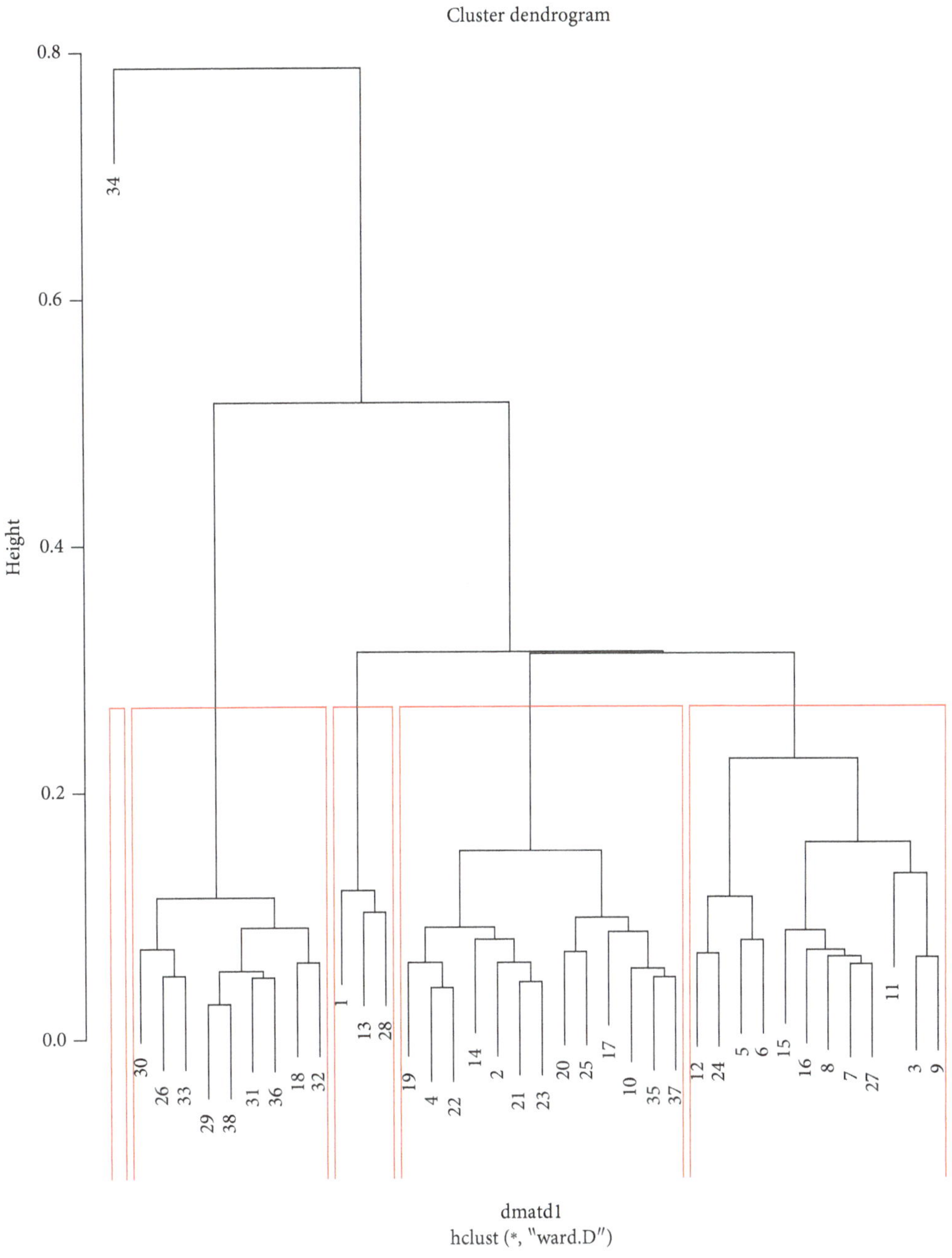

FIGURE 8: The clustering of one-dimensional persistent diagrams.

of the same study for which the one-dimensional features also resemble each other.

(iv) A similar result can be observed in cluster 4, where 6 out of 7 experiments are identified as induced resistance experiments (IDs 11, 12, 22, 25, 26, and 36; *insecticides*, *SA*, *JA*, and *SR*; *Bradyrhizobium*).

(v) Networks 19, 20, 21, 23, and 24 from the same study of natural genetic variation for plant disease resistance responses to salicylic acid (SA) are found in the second cluster. The networks 20, 23, and 24 are particularly close to each other revealing similar responses to SA.

Specifically, the similarity between persistent diagrams of networks 5, 6, and 8, which are relatively close with respect to bottleneck distance (Figures 7 and 8), indicates that the coexpression networks of these data sets have similar topological features (see Figure 3). This represents the comparable immunity process in each of these experiments. Indeed, this result agrees with the original study [33]. The goal of these experiments was to identify CPK5ac and CPK11ac early target genes (network 5) and early 22 amino acid peptide of bacterial flagellin (flg22) responsive genes (networks 6 and 8) in Arabidopsis mesophyll cells (networks 5 and 6) and Arabidopsis seedlings (network 8). They discover that CPK5ac and CPK11ac activate the same genes, mimicking flg22.

Consequently, a small bottleneck distance between two persistent diagrams of two networks could have a biological meaning in correspondence with the experiment as some clusters mainly contain specific stress groups. The clustering of the bottleneck distance matrix with the hierarchical algorithm identifies theoretically similar networks based on topological and biological aspects.

On the other hand, some distinct stress groups can have similar WGCNs. For example, induced resistance experiment 16 is in the same group with fungal experiments 9, 28, 29, 31, and 32 of cluster 3. Similarly, PTI experiment 27 is in cluster 4 along with 6 induced resistance experiments (IDs 11, 12, 22, 25, 26, and 36; *insecticides*, *SA*, *JA*, and *SR*; *Bradyrhizobium*).

One can also notice that the topological features of the networks from the same stress group can also be different. For instance, networks 28 and 34 (*Plasmodiophora brassicae* and *Botrytis cinerea*) from the fungal experiments are not close to each other.

4. Comparison with the Principal Component Analysis Method

Principal component analysis can also be used to provide a framework to construct and compare weighted gene coexpression networks (WGCNs) employing the same data sets [34]. Unlike in persistent homology, a threshold needs to be determined to construct the network. In this case, the threshold is chosen using mutual information for similarity measurement and a clustering coefficient-based method. PCA is applied to compare WGCNs. Similar WGCNs are projected, that is, clustered, closely on the principal component space using the K-means algorithm. Here is the comparison between PCA and persistent homology methods:

(i) According to PCA results, 6 bacteria experiments on Arabidopsis are associated to different clusters (see Figure 7 in [34]) while 4 out of 6 experiments are located in the same cluster in the persistent homology method. Similar results can be observed for fungi experiments as well. The persistent homology method returns a much better result (i.e., 5 out of 6 fungi experiments are in the same cluster) unlike the PCA approach where fungi groups are spread over distinct clusters.

(ii) According to both methods, PTI experiments 5, 6, 8, and 18 are close to each other (see Figure 6 in [29]).

(iii) Similarly, in both clustering results, induced resistance experiments appear in the same cluster (see Figure 7 in [34]). However, persistence homology is able to detect that networks 20, 23, and 24 revealing similar responses to SA are closer to each other.

5. Conclusion

In this paper, a topological data analysis procedure is applied to weighted gene coexpression networks (WGCN) which are constructed from microarray data sets of pathogen-infected *Arabidopsis thaliana*. We quantify topological features of WGCNs using persistent diagrams and apply the hierarchical clustering algorithm to the distance matrix whose entries are pairwise bottleneck distance between the networks. The closeness under the bottleneck distance between the zero-dimensional persistent diagrams of WGCNs indicates similar plant immune responses and conserved signaling pathways. The cluster analysis shows that a small bottleneck distance between two persistent diagrams of two networks could have a biological meaning in correspondence with the experiment as certain clusters dominantly contain specific stress groups.

Our method has recovered what principal component analysis (PCA) approach in the previous studies has identified from the same data set [34]. For certain stress groups, the persistent homology approach outperforms PCA-based clustering.

By using persistent homology along with other methods that complement its strengths and lessen its weaknesses, this method can be used as a tool in WGCN analyses. Indeed, persistent homology can identify the sets of different stress factors from the high-throughput biological data sets such as microarrays. Hence, we provide another evidence that the shape of the data can provide useful information in the study of translational functional genomics.

The proposed method does not utilize any computationally expensive algorithms and can be applied to any similar biological networks such as RNA-seq and protein-protein interaction (PPI) networks.

Conflicts of Interest

The authors declare that there is no conflict of interest regarding the publication of this paper.

Acknowledgments

The authors would like to thank the King Fahd University of Petroleum and Minerals for the support during the course of the work.

References

[1] E. A. R. Serin, H. Nijveen, H. W. M. Hilhorst, and W. Ligterink, "Learning from co-expression networks: possibilities and challenges," *Frontiers in Plant Science*, vol. 7, p. 444, 2016.

[2] T. Fuller, P. Langfelder, A. Presson et al.H. S. Lu, B. Schölkopf, H. Zhao et al., "Review of weighted gene coexpression network analysis," in *Handbook of Statistical Bioinformatics*, Springer Handbooks of Computational Statistics, pp. 369–388, Springer, Berlin, Heidelberg, 2011.

[3] M. B. Hossen and M. Siraj-Ud-Doulah, "Identification of robust clustering methods in gene expression data analysis," *Current Bioinformatics*, vol. 12, no. 6, 2017.

[4] P. Dafas and A. d'Avila Garcez, "Discovering meaningful rules from gene expression data," *Current Bioinformatics*, vol. 2, no. 3, pp. 157–164, 2007.

[5] H. Bolouri, "Modeling genomic regulatory networks with big data," *Trends in Genetics*, vol. 30, no. 5, pp. 182–191, 2014.

[6] B. R. Borate, E. J. Chesler, M. A. Langston, A. M. Saxton, and B. H. Voy, "Comparison of threshold selection methods for microarray gene co-expression matrices," *BMC Research Notes*, vol. 2, no. 1, p. 240, 2009.

[7] H. Lee, H. Kang, M. K. Chung, B.-N. Kim, and D. S. Lee, "Persistent brain network homology from the perspective of dendrogram," *IEEE Transactions on Medical Imaging*, vol. 31, no. 12, pp. 2267–2277, 2012.

[8] A. D. Perkins and M. A. Langston, "Threshold selection in gene co-expression networks using spectral graph theory techniques," *BMC Bioinformatics*, vol. 10, Supplement 11, p. S4, 2009.

[9] P. Bendich, H. Edelsbrunner, and M. Kerber, "Computing robustness and persistence for images," *IEEE Transactions on Visualization and Computer Graphics*, vol. 16, no. 6, pp. 1251–1260, 2010.

[10] O. Delgado-Friedrichs, V. Robins, and A. Sheppard, "Skeletonization and partitioning of digital images using discrete morse theory," *IEEE Transactions on Pattern Analysis and Machine Intelligence*, vol. 37, no. 3, pp. 654–666, 2015.

[11] H. Kurtuldu, K. Mischaikow, and M. F. Schatz, "Measuring the departures from the Boussinesq approximation in Rayleigh–Bénard convection experiments," *Journal of Fluid Mechanics*, vol. 682, pp. 543–557, 2011.

[12] Z. Cang, L. Mu, and G. W. Wei, "Representability of algebraic topology for biomolecules in machine learning based scoring and virtual screening," *PLOS Computational Biology*, vol. 14, no. 1, article e1005929, 2018.

[13] G. Carlsson, "Topology and data," *Bulletin of the American Mathematical Society*, vol. 46, no. 2, pp. 255–308, 2009.

[14] H. Edelsbrunner and J. Harer, "Persistent homology—a survey," in *Surveys on Discrete and Computational Geometry: Twenty Years Later*, vol. 453 of Contemporary Mathematics, pp. 257–282, American Mathematical Society, Providence, Rhode Island, 2008.

[15] H. Edelsbrunner and D. Morozov, "Persistent homology: theory and practice," in *European Congress of Mathematics Kraków, 2 - 7 July, 2012*, pp. 31–50, European Mathematics Society, 2013.

[16] R. Ghrist, "Barcodes: the persistent topology of data," *Bulletin of the American Mathematical Society*, vol. 45, no. 1, pp. 61–76, 2008.

[17] J. Arsuaga, T. Borrman, R. Cavalcante, G. Gonzalez, and C. Park, "Identification of copy number aberrations in breast cancer subtypes using persistence topology," *Microarrays*, vol. 4, no. 3, pp. 339–369, 2015.

[18] P. G. Camara, D. I. S. Rosenbloom, K. J. Emmett, A. J. Levine, and R. Rabadan, "Topological data analysis generates high-resolution, genome-wide maps of human recombination," *Cell Systems*, vol. 3, no. 1, pp. 83–94, 2016.

[19] Z. Cang, L. Mu, K. Wu, K. Opron, K. Xia, and G. W. Wei, "A topological approach for protein classification," *Computational and Mathematical Biophysics*, vol. 3, no. 1, 2015.

[20] J. M. Chan, G. Carlsson, and R. Rabadan, "Topology of viral evolution," *Proceedings of the National Academy of Sciences of the United States of America*, vol. 110, no. 46, pp. 18566–18571, 2013.

[21] M. Nicolau, A. J. Levine, and G. Carlsson, "Topology based data analysis identifies a subgroup of breast cancers with a unique mutational profile and excellent survival," *Proceedings of the National Academy of Sciences of the United States of America*, vol. 108, no. 17, pp. 7265–7270, 2011.

[22] J. A. Perea, A. Deckard, S. B. Haase, and J. Harer, "SW1PerS: sliding windows and 1-persistence scoring; discovering periodicity in gene expression time series data," *BMC Bioinformatics*, vol. 16, no. 1, p. 257, 2015.

[23] K. Xia, Z. Zhao, and G. W. Wei, "Multiresolution topological simplification," *Journal of Computational Biology*, vol. 22, no. 9, pp. 887–891, 2015.

[24] A. Hatcher, *Algebraic Topology*, Cambridge University Press, 2002.

[25] A. Collins, A. Zomorodian, G. Carlsson, and L. J. Guibas, "A barcode shape descriptor for curve point cloud data," *Computers & Graphics*, vol. 28, no. 6, pp. 881–894, 2004.

[26] D. Cohen-Steiner, H. Edelsbrunner, and J. Harer, "Stability of persistence diagrams," *Discrete & Computational Geometry*, vol. 37, no. 1, pp. 103–120, 2007.

[27] T. D. Pham, C. Wells, and D. I. Crane, "Analysis of microarray gene expression data," *Current Bioinformatics*, vol. 1, no. 1, pp. 37–53, 2006.

[28] B. Zhang and S. Horvath, "A general framework for weighted gene co-expression network analysis," *Statistical Applications in Genetics and Molecular Biology*, vol. 4, no. 1, 2005.

[29] S. Horvath, *Weighted Network Analysis*, Springer, 2011.

[30] D. Lim, N. K. Kim, S. H. Lee et al., "Characterization of genes for beef marbling based on applying gene coexpression network," *International Journal of Genomics*, vol. 2014, Article ID 708562, 10 pages, 2014.

[31] B. T. Fasy, J. Kim, F. Lecci, C. Maria, D. L. Millman, and V. Rouvreau, "R package "TDA"," 2016, https://CRAN.R-project.org/package=TDA.

[32] R Development Core Team, *R: A Language and Environment for Statistical Computing*, R Foundation for Statistical Computing, Viena, 2011.

[33] M. Boudsocq, M. R. Willmann, M. McCormack et al., "Differential innate immune signalling via Ca (2+) sensor protein kinases," *Nature*, vol. 464, no. 7287, pp. 418–422, 2010.

[34] L. G. Leal, C. López, and L. López-Kleine, "Construction and comparison of gene co-expression networks shows complex plant immune responses," *PeerJ*, vol. 2, article e610, 2014.

16

Data Mining *Mycobacterium tuberculosis* Pathogenic Gene Transcription Factors and their Regulatory Network Nodes

Guangxin Yuan[1] **Yu Bai**[2] **Yuhang Zhang**[1] **and Guangyu Xu**[1]

[1]*College of Pharmacy, Beihua University, Jilin, Jilin 132013, China*
[2]*Pharmaceutical College, Jilin Medical University, Jilin, Jilin 132011, China*

Correspondence should be addressed to Guangyu Xu; xuguangyu2005@163.com

Academic Editor: Jinfa Zhang

Tuberculosis (TB) is one of the deadliest infectious diseases worldwide. In *Mycobacterium tuberculosis*, changes in gene expression are highly variable and involve many genes, so traditional single-gene screening of *M. tuberculosis* targets has been unable to meet the needs of clinical diagnosis. In this study, using the National Center for Biotechnology Information (NCBI) GEO Datasets, whole blood gene expression profile data were obtained in patients with active pulmonary tuberculosis. Linear model-experience Bayesian statistics using the Limma package in R combined with *t*-tests were applied for nonspecific filtration of the expression profile data, and the differentially expressed human genes were determined. Using DAVID and KEGG, the functional analysis of differentially expressed genes (GO analysis) and the analysis of signaling pathways were performed. Based on the differentially expressed gene, the transcriptional regulatory element databases (TRED) were integrated to construct the *M. tuberculosis* pathogenic gene regulatory network, and the correlation of the network genes with disease was analyzed with the DAVID online annotation tool. It was predicted that IL-6, JUN, and TP53, along with transcription factors SRC, TNF, and MAPK14, could regulate the immune response, with their function being extracellular region activity and protein binding during infection with *M. tuberculosis*.

1. Introduction

Tuberculosis (TB) is one of the deadliest infectious diseases in the world [1, 2]. One-third of the global population is estimated to be infected with *Mycobacterium tuberculosis*, and 5–10% of these patients have active disease [3]. Since the World Health Organization (WHO) announced TB as a global emergency, incidence and mortality of tuberculosis remain high, despite the large cost invested during this period [4]. A dominant reason behind this is the long-term latency of *M. tuberculosis* and lack of development of new, effective antituberculosis drugs. *M. tuberculosis* is an intracellular pathogen, and it is difficult for human antibodies to enter the cell from serum, and the cell-mediated immune response plays a very important role in either development of infection to active disease or containment of the pathogen. Therefore, the study of *M. tuberculosis* genes related to immunity and their regulatory network has become a focus of the current research.

Traditional single-gene screening of *M. tuberculosis* targets has been insufficient towards the development of new therapies [5]. Tuberculosis infectious disease is a complex disease involving the body's immune system, and traditional single-gene screening patterns do not fully and accurately reflect the occurrence, development, and pathogenesis of infectious diseases [6]. Therefore, integrating multidisciplinary approaches to construct an active pulmonary TB immune response gene regulatory network and then utilizing high-throughput screening to look for diagnostic targets in body fluids would add significant understanding of the regulatory network involved in the immune response to *M. tuberculosis* [7]. This strategy may lay a foundation for the early clinical diagnosis of TB and the development of new effective drugs for TB treatment [8].

In this study, using the National Center for Biotechnology Information (NCBI) GEO Datasets [9], the fluid (whole blood) gene expression profile data were obtained in patients with active pulmonary tuberculosis. Linear model-experience

Bayesian statistics using the Limma package in R combined with *t*-tests were applied for nonspecific filtration of the expression profile data [10], and the differentially expressed genes were screened out. Using DAVID (Database for Annotation Visualization and Integrated Discovery) [11] and KEGG (Kyoto Encyclopedia of Genes and Genomes) [12], the functional analysis of differentially expressed genes (GO analysis) [13] and the analysis of signaling pathways (pathway analysis) [14] were performed. Based on the differentially expressed gene set, the transcriptional regulatory element databases (TRED) [15] were integrated to construct the *M. tuberculosis* pathogenic gene regulatory network, and the correlation of the network genes with disease was analyzed with the DAVID online annotation tool.

2. Method

2.1. Gene Chip Array Data. The gene chip array data were retrieved from GEO Datasets of the National Center for Biotechnology Information (NCBI, USA) database using keywords "tuberculosis" and "*Homo sapiens*" [porgn:_txid9606] [9]. Data from Affymetrix gene chips from human subjects infected with *M. tuberculosis* and with the original CEL documents available were selected.

We chose gene expression profile of GSE16250 from GEO database, which was a public and freely available database. GSE20050 was based on agilent GPL1352 platform. The GSE52819 dataset included 12 samples, containing 6 TB samples and 6 healthy samples. We also downloaded the Series Matrix File of GSE54992 from GEO database (Table 1).

2.2. Chip Data Processing. The Affymetrix Expression Console software tool was applied for the background correction of the collected chip data and the conversion of probe fluorescence values into gene expression values [16]. The Affymetrix Transcriptome Analysis Console was used for the logarithmic calculation and standardization of the chip data. The difference in mRNA expression between the chips in healthy people and those infected with *M. tuberculosis* was compared by SAM, and the genes with fold change > 2.0 or <−2.0 and p value < 0.05 were selected as differentially expressed genes. The differentially expressed genes were further screened by a Venn diagram; considering the differences between the chip platforms, the genes that overlapped with three or more than three platforms were selected.

2.3. Transcription Factor and Target Gene Screening. Transcription factor and target gene screening analysis was carried out at the TRED website http://rulai.cshl.edu/TRED [15].

2.4. Regulatory Network Diagram. Eighteen transcription factors (TF) and 887 target genes were predicted to a combined total of 887 TF-to-target pairs. The relation obtained from the analysis on the differential coexpression was mapped to the human transcription factors and target gene pairs to obtain transcription regulation pairs. Finally, Cytoscape software was used for plotting [17].

2.5. Gene Function Annotation Analysis. This analysis was carried out using DAVID database at website. 1075 genes were included in the analysis, and the human genome was used as the source of background genes [11].

2.6. Analysis of Differentially Expressed Gene Functions. Based on the NCBI Gene Ontology database, the GO annotation of these genes was performed to obtain all GO terms which genes were involved [13]. Fisher's exact test and χ^2 test were used to calculate the significance level and misdiagnosis rate of each GO, and the p value was calibrated with the misdiagnosis rate, thereby screening out the GO significance reflected by the differentially expressed genes ($p < 0.05$). Using the European Bioinformatics Institute (EBI) database, the experimental results were manually analyzed.

2.7. Analysis of KEGG Gene Pathways. All pathways used in this study were downloaded from KEGG databases. In total, 78 relevant KEGG pathways were downloaded from http://www.genome.jp/kegg/ in September 2016 [12].

3. Results

3.1. Gene Chip Data. All the chip data in this study were taken from GEO Datasets of the National Center for Biotechnology Information (NCBI) database. The gene expression data of humans infected with *M. tuberculosis* based on Affymetrix gene chips with the original CEL document were selected, and 4 groups of gene chips were determined, as shown in Table 1.

These 4 groups of chips included those whose chip number of chips was 6, 7, 12, and 21, and each chip group was divided into two subgroups, a healthy subgroup and TB patient subgroup.

3.2. Chip Data Processing. The logarithm and standardization of chip data were conducted using Affymetrix's Transcriptome Analysis Console, and the differentially expressed mRNA between the chips of healthy people (healthy group) and those infected with *M. tuberculosis* (TB patient group) was compared by the SAM method to determine differentially expressed genes. The number of differentially expressed genes of each platforms was 5569, 2442, 616, and 578, respectively. These genes were cross-screened again, and the results showed that there were 130 differentially expressed genes on the 4 platforms, 224, 63, 504, and 154 on three platforms, respectively, with a total of 1075 genes (Figure 1).

3.3. TB Transcription Factor Regulatory Network Diagram. Using TRED (Transcriptional Regulatory Element Database) database (http://rulai.cshl.edu/TRED), transcription factors interacting among the 1075 expressed genes were predicted and analyzed, and 887 genes were found to correspond to 18 transcription factors (Table 2).

The regulatory network diagram of *M. tuberculosis* pathogenic gene transcription factors of the 18 transcription factors and their corresponding target genes was mapped with Cytoscape software (Figure 2).

Sixteen target genes were regulated by more than 1 transcription factor (Table 3), of which the target genes regulated

Table 1: TB chip data.

Dataset ID	Sample ID	Sample number	Control sample number	Disease sample number	Platforms	Organism	Submission date	Manufacturer
GSE16250 [33]	GSM409134–GSM409139	6	3	3	GPL570	*Homo sapiens*	May 27, 2009	Affymetrix
GSE20050 [34]	GSM501249–GSM501255	7	2	5	GPL1352	*Homo sapiens*	July 19, 2004	Affymetrix
GSE52819 [35]	GSM1276660–GSM1276662 GSM1276666–GSM1276668	12	6	6	GPL6244	*Homo sapiens*	Dec. 05, 2007	Affymetrix
GSE54992 [36]	GSM1327526–GSM1327531 GSM1327533, GSM1327535, GSM1327540–GSM1327542, GSM1327544, GSM1327546, GSM1327549, GSM1327550	21	11	10	GPL570	*Homo sapiens*	Nov. 07, 2003	Affymetrix

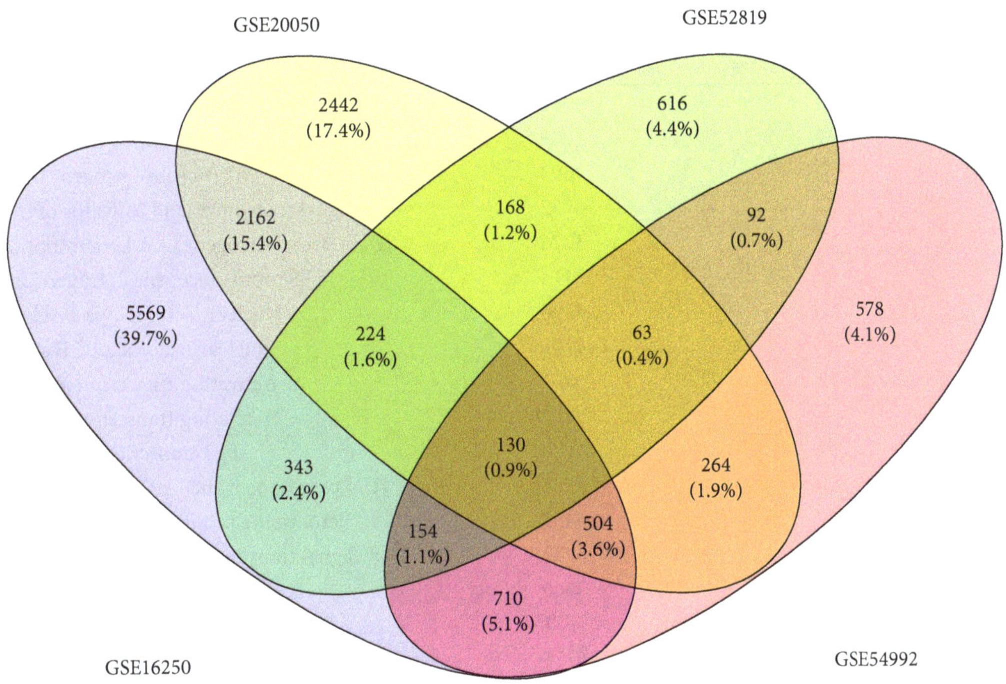

FIGURE 1: Screening results of genes overlapping ≥3 platforms. The blue part is GSE16250, the yellow part is GSE20050, the green part is GSE52819, and the pink part is GSE54992.

by the most transcription factors were IL-6, MAPK14, RELA, and FOS, regulated by 4, 3, 3, and 3 TFs, respectively.

3.4. Network Nodes. As shown in Figure 2, statistical analysis of network nodes in the transcription factor regulatory network diagram found more than 10 nodes that had 44 genes, of which there were 20 nodes that had more than 4 genes, including IL-6, ETS2, TNF, and OPN1MW (Table 4). Furthermore, IL-6 was regulated by the most transcription factors, and genes such as JUN and IL-5 were also closely related to the regulation of *M. tuberculosis* pathogenic gene transcription factors.

3.5. GO Function Annotation Analysis. GO function annotation for 1075 different genes was carried out and sorted based on the *p* values, and 10 pathways were determined (Table 5). These 10 pathways were closely related to the disease, and the TB pathway was ranked in the top five of the pathways found.

4. Discussion

Tuberculosis caused by the infection of *M. tuberculosis* remains a threat to human health throughout the world. Currently, about 2 billion (1/3) people worldwide are infected with *M. tuberculosis*, but only 10% will develop active tuberculosis and the remaining 90% will make the disease to curb due to the body's own well-developed immune system [18]. Our previous study found that the immune response mechanism of host cells plays a crucial role in the process of microbial infection [19]. The disease process of active tuberculosis is transformed into the active stage and depends not only on the adaptive changes of mycelium caused by genes regulation but also on the transcription factor regulation network of the host's immune system [20]. To understand the defense mechanism of the host immune system against *M. tuberculosis* infection will provide a theoretical basis for the progress and prevention of active tuberculosis.

In this study, utilizing GEO Datasets from the NCBI database and using the Affymetrix Transcriptome Analysis Console, the logarithm transformation, and standardization of chip data were carried out, and 1075 differentially expressed *M. tuberculosis* pathogenic genes were found. The TRED database was used to predict and analyze these 1075 genes, and as a result, 18 transcription factors and 887 genes were found, and an *M. tuberculosis* pathogenic gene transcription regulation network diagram was constructed (Figure 2). Through a series of statistical analysis based on Figure 2, we found that many of these key factors are closely related to the human immune system.

As shown in Figure 2, there were 2 transcription factors that corresponded highly to the target genes, namely, JUN (236 target genes) and NFKB1 (221 target genes), whereas the other transcription factors corresponded to less than 50 target genes. JUN is closely linked with SLE (systemic lupus erythematosus), and SLE is a typical autoimmune disease involving multiple organs and systems. Doniz-Padilla et al. [21] found that JUN expression level in peripheral blood mononuclear cells (PBMCs) of patients with SLE was significantly higher than in individuals in the control group. JUN might play an important role in the transcriptional regulation of FCGR2B promoter activity, while FCGR2B has been shown to be closely associated with the pathogenesis of SLE [22]. These results suggest that c-JUN may be involved in the pathogenesis of SLE. NFKB1 gene is one of members in

Table 2: 18 transcription factors and their target genes.

TF	Target gene number	Gene ID	Description
SMAD1	30	4086	SMAD family member 1
SMAD4	39	4089	SMAD family member 4
MYBL1	7	4603	MYB proto-oncogene like 1
STAT1	57	6772	Signal transducer and activator of transcription 1
JUNb	10	3726	JUNb proto-oncogene, AP-1 transcription factor subunit
ETS2ETS2	64	2114	ETS proto-oncogene 2, transcription factor
NF-κB1	211	4790	Nuclear factor kappa B subunit 1
NF-κB2	8	4791	Nuclear factor kappa B subunit
ATF3	12	467	Activating transcription factor 3
ATF5	1	22809	Activating transcription factor 5
SP3	95	6670	Sp3 transcription factor
JUN	236	3725	JUN proto-oncogene, AP-1 transcription factor subunit
PPARD	27	5467	Peroxisome proliferator activated receptor delta
STAT2	2	6773	Signal transducer and activator of transcription 2
REL	26	5966	REL proto-oncogene, NF-κB subunit
RELB	7	5971	RELB proto-oncogene, NF-κB subunit
STAT4	8	6775	Signal transducer and activator of transcription 4
SMAD3	47	4088	SMAD family member 3

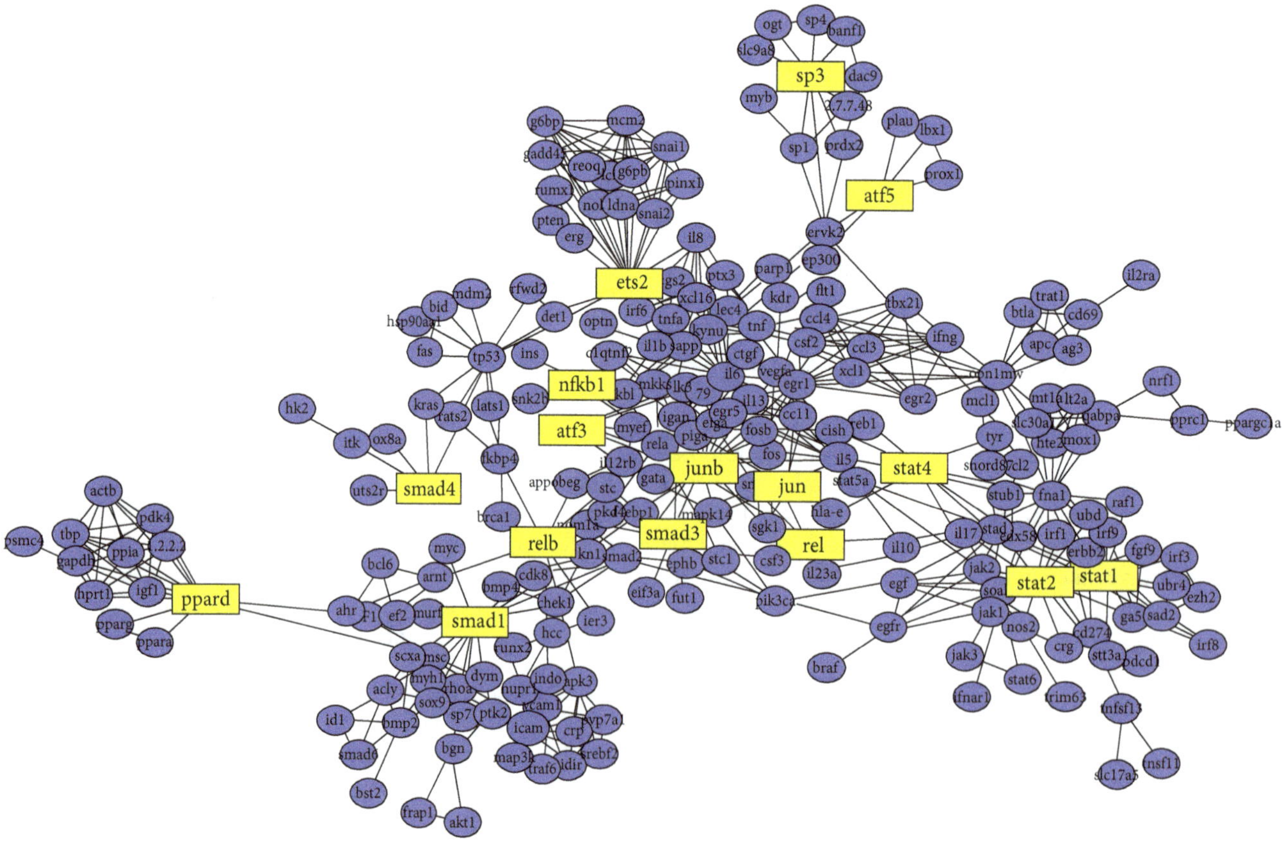

Figure 2: Regulatory network diagram of the 18 transcription factors found in TB patients. The yellow parts represent 18 transcription factors, and the blue parts are their corresponding target genes.

TABLE 3: Genes regulated by transcription factors.

Genes regulated by transcription factors	Number of transcription factors regulating the target genes	Network node (Figure 2)
IL-6	4	30
MAPK14	3	—
RELA	3	11
FOS	3	13
JUN	2	13
NF-kappa 1	2	14
IL-1b	2	16
COL11	2	—
EGR1	2	18
EGR3	2	—
FOSb	2	—
IL-13	2	—
IL-5	2	10
SGK1	2	—
SMAD2	2	10
SRC	2	—

NF-κB family of widespread transcription factors. It plays an important role in the immune response, inflammation, and cell growth and development.

The statistics of genes regulated by the transcription factors in Figure 2 also showed that the target genes regulated by more than 3 transcription factors were IL-6, MAPK14, RELA, and FOS; JUN was also regulated by the 2 transcription factors. Studies have shown that 4 genes, including IL-6, MAPK14, RELA, and FOS, are closely related to the immune system. IL-6 is a B cell stimulating factor and can also activate macrophages. It is produced not only by T cells but also by macrophages, fibroblasts, epithelial cells, and other cells outside the lymphatic system. IL-6 can provide a non-T cell-dependent response to acute and chronic infection with *M. tuberculosis*. IL-6 can act on multiple cells, such as B cells, liver cells, hybridoma cells, and plasma cells, and enhance immune resistance through its proinflammatory activity and the effect on the production of the other cytokines. Different from other type-2 cytokines, IL-6 not only is essential for the activation of T cells to secret IFN-γ but also presents as the main molecule to induce and protect T cells during *M. tuberculosis* infection and strengthen activities of IFN-γ [23]. Ladel et al. [24] found that after infection with *M. tuberculosis*, the production of IFN-γ in IL-6-deficient mice was reduced and their lifetime was shortened compared with wild-type mice. Although many studies have shown that IL-6 could induce a protective immune response against *M. tuberculosis* infection, some reports have demonstrated that IL-6 also has an adverse effect to inhibit the responsiveness of macrophages to IFN-γ [25]. MAPK14 gene is a member of mitogen-activated protein kinase (MAPK) family, and the MAPK cascade is a ubiquitous intracellular serine/threonine protein kinase super family. Thus, an important substance that can transmit the cytoplasmic signaling into the nucleus and cause the changes in the nucleus closely linked with the cell apoptosis and proliferation, tumor genesis, and oxidative stress injury of intestinal epithelial cells [26].

TABLE 4: Network node statistics.

Gene	Network node
IL-6	30
ETS2	29
TNF	25
OPN1MW	21
IFNAL	19
STAT1	18
SMAD1	18
EGR1	18
JUNb	17
IL-1b	16
STAT3	14
NF-kappa B	14
STAT2	13
JUN	13
FOS	13
TP53	12
TNFα	12
PTX3	12
PTGS2	12
PPARD	12
KYNU	12
IRF6	12
IL-8	12
CXCL16	12
CLEC4E	12
ATF3	12
STAT4	11
SNAI2	11
SNAIL	11
SMAD3	11
SLC2A1	11
RELA	11
REC8	11
PINX1	11
NOL3	11
MCM2	11
lLDHA	11
IGFBP3	11
GADD45G	11
G6PD	11
SP3	10
SMAD2	10
RELB	10
IL-5	10

Table 5: GO function annotation of differentially expressed genes.

Pathway	Gene number	p value	Benjamini
Influenza A	40	$2.6E-10$	$7.2E-8$
Osteoclast differentiation	30	$4.8E-9$	$6.6E-7$
Legionellosis	19	$3.1E-8$	$2.8E-6$
NF-kappa B signaling pathway	24	$5.4E-8$	$3.7E-6$
Tuberculosis	36	$6.9E-8$	$3.8E-6$
TNF signaling pathway	26	$1.6E-7$	$7.5E-6$
Hepatitis B	30	$7.4E-7$	$2.9E-5$
Herpes simplex infection	34	$1.5E-6$	$5.2E-5$
Cytokine-cytokine receptor interaction	39	$2.3E-6$	$7.0E-5$
RIG-I-like receptor signaling pathway	19	$2.4E-6$	$6.5E-5$

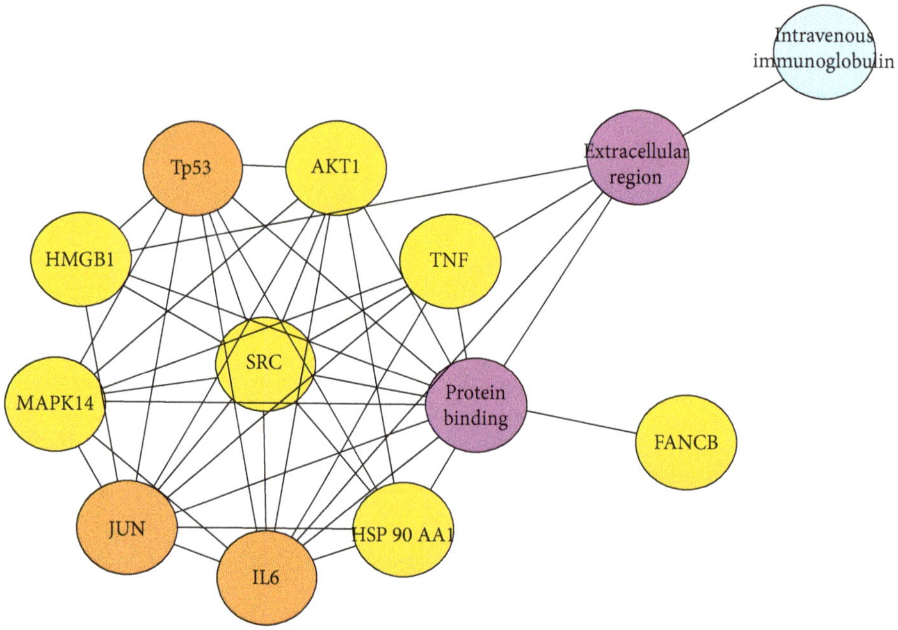

Figure 3: Network diagram of the relationship between network nodes. The blue part is the drug, the two purple parts are Gene Ontology, the yellow parts are the transcription factors, and the brown parts are the network nodes.

M. tuberculosis can exist in a nonproliferative state in the body of its host, thus resulting in a delayed course of the disease and a persistent recurrence of infection. The persistence of *M. tuberculosis* in macrophages may exert a huge survival pressure from bactericidal components in the cells. During its long evolutionary process and coevolution with humans, *M. tuberculosis* has developed a unique signal transmission system to ensure that it can start different response elements in different infection cycles to regulate the expression of genes for the adaptation to the environment. Among the numerous response (stringent response) regulatory elements, RELA plays a key role [27–29]. Studies have showed that the resistance and tolerance of bacteria to environmental pressure are dependent on RELA and the resulting ppGpp (P) signal can mediate bacterial resistance to antibiotics, UV-resistant survival, and DNA damage repair. *M. tuberculosis* with inactivated RELA cannot survive for a long time in an environment with the abovementioned types of pressure [30]. There are some differences in the morphology and chromaticity between long-term persisters and *M. tuberculosis* active propagules; persisters isolated from the lungs show extremely close manifestations to those under conditions of pressure, especially lack of nutrition, which is completely consistent with the starting condition of RELA. Another study showed that RELA is closely related to the bacterial morphology and colony formation caused by changes in the cell wall of mycobacterium, which is important for the formation of *M. tuberculosis* L-forms, namely, cell wall-deficient *M. tuberculosis*. RELA is thus a good target against persistent tuberculosis infection [31, 32].

FOS gene is a member of FOS family, along with JUN family members and other activated transcription factor protein family members including activator protein 1 (AP-1). AP-1 is an important transcription regulation factor in the nucleus, playing an important role in the transduction process of many signals in the body, and is the intersection of a series of nuclear cellular signal transduction pathways [8]. FOS can transmit extracellular stimulatory signals through the upstream signaling pathway to AP-1 and then activate the AP-1 to regulate the expression of a series of downstream target genes with AP-1 DNA recognition sites, that is, binding the specific AP-1 DNA recognition site 5′-TGA (G/C TCA-3′), also known as the TPA response element (TRE), to allow the body to make an adaptive response to external stimuli. In addition, AP-1 plays an important regulation role in SLE.

The statistics of network nodes is shown in Figure 2. The results showed that there were 4 genes with more than 20 network nodes, respectively, IL-6, ETS2, TNF, and OPN1MW (Table 4). As discussed previously, IL-6 is a gene regulated by transcription factors and JUN and IL-5 are also the genes regulated by transcription factors and the network node. Based on analysis of previous publications and analysis of all the network nodes, we suggest a certain portion between most of the networks (Figure 3), also closely related to the transcription factor SRC. Many of these nodes were targets of intravenous immunoglobulin drugs (Figure 3), and it is well known that intravenous immunoglobulin can rapidly raise the level of the IgG in the blood of recipients to enhance the body's ability to fight infections and regulate immune function. We also found that the major GO terms of these network nodes were in the extracellular region and were protein binding. Therefore, we presume that the network nodes IL-6, JUN, and TP53, along with transcription factors SRC, TNF, and MAPK14, can regulate the body's immune function, and their gene function should be mainly acting in the extracellular region and be protein binding.

5. Conclusions

In summary, 1075 genes were screened out through GEO database and a transcription factor regulation network diagram was constructed. Through the statistical analysis of the transcription factors and network nodes of their regulatory network diagram, we speculate that IL-6, JUN, and TP53, along with transcription factors SRC, TNF, and MAPK14, could regulate the human immune function, evading the attack on the host immune system and ultimately leading to the existence of *M. tuberculosis* in a nonproliferative state in the body of the host, thereby resulting in a delayed course of the disease and a recurrent persistent infection.

Conflicts of Interest

The authors declare that they have no competing interests.

Authors' Contributions

Guangxin Yuan and Yu Bai contributed equally to this study.

Acknowledgments

This work was supported by the National Natural Science Foundation of China (81401712), Beihua University Youth Research Innovation Team Training Program, and Research and Development Innovation Team of Animal and Plant Resources in Changbai Mountain and also by the National and Local United Engineering Research and Development Center for Research on Active Peptides of Medicinal Plants and Animals in Changbai Mountains, Jilin Province Chinese Medicine Science and Technology Projects (2017249 and 2017086), Jilin Province Health Planning Science and Technology Innovation Projects (2017ZC003).

References

[1] N. K. Dutta, S. Mehra, P. J. Didier et al., "Genetic requirements for the survival of tubercle bacilli in primates," *The Journal of Infectious Diseases*, vol. 201, no. 11, pp. 1743–1752, 2010.

[2] M. I. Saramba and D. Zhao, "A perspective of the diagnosis and management of congenital tuberculosis," *Journal of Pathogens*, vol. 2016, Article ID 8623825, 8 pages, 2016.

[3] H. Getahun, A. Matteelli, I. Abubakar et al., "Management of latent *Mycobacterium tuberculosis* infection: WHO guidelines for low tuberculosis burden countries," *European Respiratory Journal*, vol. 46, no. 6, pp. 1563–1576, 2015.

[4] J. Weiner and S. H. Kaufmann, "High-throughput and computational approaches for diagnostic and prognostic host tuberculosis biomarkers," *International Journal of Infectious Diseases*, vol. 56, pp. 258–262, 2016.

[5] E. Hong-Geller and S. N. Micheva-Viteva, "Functional gene discovery using RNA interference-based genomic screens to combat pathogen infection," *Current Drug Discovery Technologies*, vol. 7, no. 2, pp. 86–94, 2010.

[6] M. S. Vasava, M. N. Bhoi, S. K. Rathwa, M. A. Borad, S. G. Nair, and H. D. Patel, "Drug development against tuberculosis: past, present and future," *The Indian Journal of Tuberculosis*, vol. 64, no. 4, pp. 252–275, 2017.

[7] T. T. Zhou, "Network systems biology for targeted cancer therapies," *Chinese Journal of Cancer*, vol. 31, no. 3, pp. 134–141, 2011.

[8] Z. Hmama, S. Pena-Diaz, S. Joseph, and Y. Av-Gay, "Immunoevasion and immunosuppression of the macrophage by *Mycobacterium tuberculosis*," *Immunological Reviews*, vol. 264, no. 1, pp. 220–232, 2015.

[9] S. Davis and P. S. Meltzer, "GEOquery: a bridge between the Gene Expression Omnibus (GEO) and BioConductor," *Bioinformatics*, vol. 23, no. 14, pp. 1846-1847, 2007.

[10] E. R. Morrissey, M. A. Juarez, K. J. Denby, and N. J. Burroughs, "Inferring the time-invariant topology of a nonlinear sparse gene regulatory network using fully Bayesian spline autoregression," *Biostatistics*, vol. 12, no. 4, pp. 682–694, 2011.

[11] W. Huang Da, B. T. Sherman, and R. A. Lempicki, "Systematic and integrative analysis of large gene lists using DAVID bioinformatics resources," *Nature Protocols*, vol. 4, no. 1, pp. 44–57, 2009.

[12] H. Ogata, S. Goto, K. Sato, W. Fujibuchi, H. Bono, and M. Kanehisa, "KEGG: Kyoto Encyclopedia of Genes and Genomes," *Nucleic Acids Research*, vol. 27, no. 1, pp. 29–34, 1999.

[13] M. Ashburner, C. A. Ball, J. A. Blake et al., "Gene ontology: tool for the unification of biology. The Gene Ontology Consortium," *Nature Genetics*, vol. 25, no. 1, pp. 25–29, 2000.

[14] G. Xu, Z. Ni, Y. Shi et al., "Screening essential genes of *Mycobacterium tuberculosis* with the pathway enrichment method," *Molecular Biology Reports*, vol. 41, no. 11, pp. 7639–7644, 2014.

[15] F. Zhao, Z. Xuan, L. Liu, and M. Q. Zhang, "TRED: a Transcriptional Regulatory Element Database and a platform for *in silico* gene regulation studies," *Nucleic Acids Research*, vol. 33, no. Database issue, pp. D103–D107, 2005.

[16] J. Lu, J. C. Lee, M. L. Salit, and M. C. Cam, "Transcript-based redefinition of grouped oligonucleotide probe sets using AceView: high-resolution annotation for microarrays," *BMC Bioinformatics*, vol. 8, no. 1, p. 108, 2007.

[17] P. Shannon, A. Markiel, O. Ozier et al., "Cytoscape: a software environment for integrated models of biomolecular interaction networks," *Genome Research*, vol. 13, no. 11, pp. 2498–2504, 2003.

[18] K. D. Mayer-Barber, B. B. Andrade, S. D. Oland et al., "Host-directed therapy of tuberculosis based on interleukin-1 and type I interferon crosstalk," *Nature*, vol. 511, no. 7507, pp. 99–103, 2014.

[19] A. L. Banuls, A. Sanou, N. T. Anh, and S. Godreuil, "*Mycobacterium tuberculosis*: ecology and evolution of a human bacterium," *Journal of Medical Microbiology*, vol. 64, no. 11, pp. 1261–1269, 2015.

[20] J. D. Blischak, L. Tailleux, A. Mitrano, L. B. Barreiro, and Y. Gilad, "Mycobacterial infection induces a specific human innate immune response," *Scientific Reports*, vol. 5, no. 1, p. 16882, 2015.

[21] L. Doniz-Padilla, V. Martinez-Jimenez, P. Nino-Moreno et al., "Expression and function of Cbl-b in T cells from patients with systemic lupus erythematosus, and detection of the 2126 A/G Cblb gene polymorphism in the Mexican mestizo population," *Lupus*, vol. 20, no. 6, pp. 628–635, 2011.

[22] M. Olferiev, E. Masuda, S. Tanaka, M. C. Blank, and L. Pricop, "The role of activating protein 1 in the transcriptional regulation of the human FCGR2B promoter mediated by the -343 G → C polymorphism associated with systemic lupus erythematosus," *Journal of Biological Chemistry*, vol. 282, no. 3, pp. 1738–1746, 2007.

[23] I. S. Leal, B. Smedegard, P. Andersen, and R. Appelberg, "Interleukin-6 and interleukin-12 participate in induction of a type 1 protective T-cell response during vaccination with a tuberculosis subunit vaccine," *Infection and Immunity*, vol. 67, no. 11, pp. 5747–5754, 1999.

[24] C. H. Ladel, C. Blum, A. Dreher, K. Reifenberg, M. Kopf, and S. H. Kaufmann, "Lethal tuberculosis in interleukin-6-deficient mutant mice," *Infection and Immunity*, vol. 65, no. 11, pp. 4843–4849, 1997.

[25] B. M. Saunders, A. A. Frank, I. M. Orme, and A. M. Cooper, "Interleukin-6 induces early gamma interferon production in the infected lung but is not required for generation of specific immunity to *Mycobacterium tuberculosis* infection," *Infection and Immunity*, vol. 68, no. 6, pp. 3322–3326, 2000.

[26] Y. Zhou, Q. Wang, B. Mark Evers, and D. H. Chung, "Oxidative stress-induced intestinal epithelial cell apoptosis is mediated by p38 MAPK," *Biochemical and Biophysical Research Communications*, vol. 350, no. 4, pp. 860–865, 2006.

[27] M. Sajish, S. Kalayil, S. K. Verma, V. K. Nandicoori, and B. Prakash, "The significance of EXDD and RXKD motif conservation in Rel proteins," *Journal of Biological Chemistry*, vol. 284, no. 14, pp. 9115–9123, 2009.

[28] D. Rifat, W. R. Bishai, and P. C. Karakousis, "Phosphate depletion: a novel trigger for *Mycobacterium tuberculosis* persistence," *The Journal of Infectious Diseases*, vol. 200, no. 7, pp. 1126–1135, 2009.

[29] L. D. Lu, Q. Sun, X. Y. Fan, Y. Zhong, Y. F. Yao, and G. P. Zhao, "Mycobacterial MazG is a novel NTP pyrophosphohydrolase involved in oxidative stress response," *Journal of Biological Chemistry*, vol. 285, no. 36, pp. 28076–28085, 2010.

[30] R. Dhiman, M. Raje, and S. Majumdar, "Differential expression of NF-κB in mycobacteria infected THP-1 affects apoptosis," *Biochimica et Biophysica Acta*, vol. 1770, no. 4, pp. 649–658, 2007.

[31] D. J. Murphy and J. R. Brown, "Identification of gene targets against dormant phase *Mycobacterium tuberculosis* infections," *BMC Infect ious Diseases*, vol. 7, no. 1, p. 84, 2007.

[32] K. T. Park, J. L. Dahl, J. P. Bannantine et al., "Demonstration of allelic exchange in the slow-growing bacterium *Mycobacterium avium* subsp. paratuberculosis, and generation of mutants with deletions at the *pknG*, *relA*, and *lsr2* loci," *Applied and Environmental Microbiology*, vol. 74, no. 6, pp. 1687–1695, 2008.

[33] N. Reyes, A. Bettin, I. Reyes, and J. Geliebter, "Microarray analysis of the in vitro granulomatous response to *Mycobacterium tuberculosis* H37Ra," *Colombia Médica*, vol. 46, no. 1, pp. 26–32, 2015.

[34] M. J. Kim, H. C. Wainwright, M. Locketz et al., "Caseation of human tuberculosis granulomas correlates with elevated host lipid metabolism," *EMBO Molecular Medicine*, vol. 2, no. 7, pp. 258–274, 2010.

[35] M. Verway, M. Bouttier, T. T. Wang et al., "Vitamin D induces interleukin-1β expression: paracrine macrophage epithelial signaling controls *M. tuberculosis* infection," *PLoS Pathogens*, vol. 9, no. 6, article e1003407, 2013.

[36] Y. Cai, Q. Yang, Y. Tang et al., "Increased complement C1q level marks active disease in human tuberculosis," *PLoS One*, vol. 9, no. 3, article e92340, 2014.

17

Transcriptome Analysis Reveals Dynamic Fat Accumulation in the Walnut Kernel

Shuang Zhao,[1,2] Xuemei Zhang,[1,3] Yanping Su,[4] Yilan Chen,[5] Yang Liu,[1] Meng Sun,[1] and Guohui Qi[1,3]

[1]*Department of Forestry, Agricultural University of Hebei, Baoding 071000, China*
[2]*Cotton Research Institute, Hebei Academy of Agriculture and Forestry Sciences, Shijiazhuang 050000, China*
[3]*Research Center for Walnut Engineering and Technology of Hebei, Lincheng 054304, China*
[4]*Department of Life Science, Normal University of Langfang, Langfang 065000, China*
[5]*Baoding Forest Seedling Management Station, Baoding 071000, China*

Correspondence should be addressed to Guohui Qi; bdqgh@sina.com

Academic Editor: Antonio Ferrante

Walnut (*Juglans regia* L.) is an important woody oilseed species cultivated throughout the world. In this study, comparative transcript profiling was performed using high-throughput RNA sequencing technology at the following three stages of walnut fat synthesis in the "Lvling" walnut cultivar: the initial developmental stage (L1), the fast developing stage (L2), and the last developing stage (L3). A total of 68.18 GB of data were obtained on the three developmental stages, and 92% to 94% of clean data were able to be located to the reference genome. Further comparisons of the transcripts in the three libraries revealed that 724, 2027, and 4817 genes were differentially expressed between the L2 and L1 (L2vsL1), L3 and L2 (L3vsL2), and L3 and L1 (L3vsL1) samples, respectively. Through the GO gene enrichment analysis, differentially expressed genes (DEGs) in L2vsL1, L3vsL2, and L3vsL1 were enriched into 3, 0, and 2 functional categories, respectively. According to the KEGG enrichment analysis, DEGs in L2vsL1, L3vsL2, and L3vsL1 were annotated into 77, 110, and 3717 taxonomic metabolic pathways in the KEGG database, respectively. Next, we analyzed expression levels of genes related to fat synthesis. Our results indicated that ACCase, LACS, and FAD7 were the key genes related to fat synthesis. The high-throughput transcriptome sequencing of walnut in different developmental stages has greatly enriched the current genomic available resources. The comparison of DEGs under different developmental stages identified a wealth of candidate genes involved in fat synthesis, which will facilitate further genetic improvement and molecular studies of the walnut.

1. Introduction

Edible oil, an important source of fat for humans, has received increasing attention as a necessary dietary ingredient. However, China edible oil is highly dependent on foreign imports. In 2013, a total of 11.694 million tons of oil were extruded from Chinese-produced oilseed crops, but the annual total demand for edible oil was 30.408 million tons; therefore, the self-sufficiency rate for edible oil in China was 38.5% [1]. The dependence on foreign edible oil increased to 61.5%, far exceeding the international security warning line. It is projected that by 2030, the demand for vegetable oil will double. This demand can be met only by increasing the oil content of current oilseed crops or introducing new high-yield oil crops [2].

Walnut (*Juglans regia* L.) is an important woody oilseed species. Walnut oil, a high-quality cooking oil, is comparable to olive oil [3]. The fat content of walnut can be as high as 65.21 g/100 g, and polyunsaturated fatty acids account for 47.17 g/100 g; the polyunsaturated fatty acids are primarily linoleic acid (38.90 g/100 g) and linolenic acid (9.08 g/100 g). The ratio of oleic acid (ω-6) to α-linolenic acid (ω-3) in walnut kernels is close to (4–6) : 1, and this nutritional component is beneficial for human health. This component can effectively prevent cardiovascular disease and is beneficial for type 2 diabetes patients [4, 5]. Given that walnut oil is an important functional plant fat, this oil is being widely studied in various countries. Increasing walnut yields and selecting high-oil walnut varieties are essential strategies for increasing the production of high-grade edible oil and reducing China's dependence on imported edible oil. Understanding the genetic mechanism of walnut oil metabolism regulation is important to improve the oil content of walnut and provide an important theoretical basis for research.

The transcriptome is the total expressed RNA of a specific tissue or cell at a certain stage of development or in a specific state. Transcriptome analysis can reveal differences in the expression level of the same gene in different states and can explore the interaction between different genes and the functional characteristics of each gene [6]. Transcriptome analysis can obtain information on gene expression levels, structural characteristics, function, and intergene interactions; therefore, this technique is an effective means to study specific biological processes and molecular mechanisms under biological conditions [7]. In recent years, many scholars have used transcriptome sequencing to research fat metabolism in many species, such as oil palm [8, 9], castor [10], *Jatropha curcas* [11, 12], *Camellia oleifera* [13], *Idesia polycarpa* [14], and *Symplocos paniculata* [15]. The sequencing and transcriptome analyses of oil palm revealed that *WRINKLED1 (WRI1)* regulates the accumulation of pericarp fat in oil palm, but the expression of *WRI1* is 75 times higher in the pericarp than in the seed; *WRI1* plays a central role in nut fat biosynthesis. *LEAFY COTYLEDON1 (LEC1), LEC2, ABSCISIC ACID INSENSITIVE3 (ABI3)*, and *FUSCA3* are expressed in nuts but not in the mesocarp. *PICKLE (PKL)*, a transcriptional regulator, is expressed in both the mesocarp and the kernel. Research on *Idesia polycarpa* by Li et al. [14] also revealed that the expression of the oil-related transcription factor WRI1 is higher in the pulp than in the seeds, whereas other transcription factors, *LEC1, LEC2, ABI4, ABI3*, and *FUS3*, are expressed only in the seeds. An analysis of acetyl-CoA activity in castor by Brown et al. [10] revealed that ricinoleic acid acyl-CoA does not dominate during endosperm oil formation. This finding suggested that the expression of *RcDGAT2* and *RcPDAT1* increased oleate-12 hydroxylase activity, thereby increasing the ricinoleic acid content of the organ. Liu et al. [15] analyzed the transcriptome of *S. paniculata* at different development stages; the result founded that the key regulatory enzymes involved in lipid metabolism *(ACC, KASII, KASIII, FATA, FATB, ACSL, SAD, FAD2, GPAT9, LPCAT1, DGAT1*, and *PDAT1)* were determined and they play vital roles in the oil accumulation in *S. paniculata* fruit and fatty acid composition of the fruit oil.

In this study, transcriptome sequencing technology was used to analyze differentially expressed genes of the walnut kernels at various developmental stages and to explore the mechanism underlying walnut fat synthesis at the gene transcriptional level. The transcriptional expression information of specific genes during the development of the walnut was analyzed. This information provides a theoretical reference for future research on the mechanism underlying the synthesis of the fat in the walnut kernel, laying a foundation for the cloning and functional analysis of genes related to oil synthesis.

2. Materials and Methods

2.1. Plant Materials. The walnut cultivar Lvling was used in this study. Three 7-year-old trees that were disease-free and exhibited consistent, vigorous production were selected. Beginning at 50 days after fertilization (DAF), samples were taken every 10 days, ending at 130 DAF. A total of 5 fruits with normal growth, no disease and consistent size were collected in each direction (north, south, west, and east) on each tree canopy, that is, 20 fruits were collected per plant. The samples used for the transcriptome sequencing were the fruits collected at 60, 90, and 120 DAF. The kernels were removed and put into self-sealing bags, labeled as samples L1 (60 DAF), L2 (90 DAF), and L3 (120 DAF), with three biological replicates each. The samples were rapidly frozen with liquid nitrogen and stored in a −80°C freezer. The remaining fruits, which were used for the analysis of fat and fatty acid content, were removed from the husk and placed in an oven and dried at 40°C until their weight stabilized.

2.2. Determination of Fat and Fatty Acid Content. The crude fat content was measured by Soxhlet extraction [16], and then the fat was methyl esterified. The methyl esterified fatty acid mixture was analyzed using gas chromatography [17, 18], and the peak area normalization method was used to calculate the components in a unified way. There were three replicates in each treatment, and each experiment was repeated three times. Data were analyzed using the SPSS software (version 17). Significant differences were assessed by using Duncan's multiple range test at the 5% probability level. All the data were expressed as mean ± standard error (SE).

The formula $(M_{n+1} - M_n)$ represents the increase of kernel fat in two consecutive samplings, the formula $(T_{n+1} - T_n)$ represents the number of days between two consecutive samplings, and the formula for calculating the walnut kernel fat accumulation rate is as follows: KW $(\text{g d}^{-1}) = (M_{n+1} - M_n)/(T_{n+1} - T_n)$.

2.3. RNA Extraction and Quality Verification. RNA extraction from samples and transcriptome sequencing was completed by Beijing Novo Gene Technology Co. Ltd. The total RNA of the walnut kernel samples was extracted using the Tiangen DP441 Polysaccharide Polyphenol Kit. The purity, concentration, and integrity of the RNA samples were accurately detected using 1% agarose gel electrophoresis, NanoDrop 2000, Qubit 2.0, and Agilent 2100.

2.4. cDNA Library Construction and Sequencing. After the samples were qualified, the eukaryotic mRNA was enriched using magnetic beads with oligo (dT) bound to the polyA tail of the mRNA by A-T complementary pairing. The mRNA was then broken into short fragments, and single-stranded cDNA was synthesized with a six-base random primer using mRNA as a template. Then, double-stranded cDNA was synthesized by adding buffer solution, dNTPs, and DNA polymerase I, followed by the purification of the double-stranded cDNA using AMPure XP beads. The purified double-stranded cDNA was then subjected to end repair, an A tail was added, and the sequencing adapter was ligated. Next, the fragment size was selected by AMPure XP beads, and PCR enrichment was performed to obtain the final cDNA library. After the construction of the library, the length of the inserted fragments in the library was detected using the Agilent 2100, and the effective concentration of the library was accurately quantified (effective concentration of the library: >2 nM) by qPCR to ensure library quality. After passing the library check, Illumina HiSeq sequencing was performed after pooling the libraries according to the requirements of effective concentration and target data volume.

2.5. Raw Sequencing Data Processing and Sequence Assembly. The raw image data files obtained by high-throughput sequencing were converted into the sequenced reads by raw CASAVA (base calling) analysis, namely, raw reads. To obtain clean reads, the raw reads were filtered to remove the reads with adapters, the reads of *N*, which accounted for more than 10% of the reads (*N* indicates that the base information cannot be determined) and the low-quality reads (Q phred ≤ 20 bases for the entire length of the read of the 50% above reads). The follow-up analyses were all based on the resulting clean reads. The HISAT software was selected to analyze the genome localization on the filtered sequencing sequence; the reference genome path is https://treegenesdb.org/Drupal/FTP/Genomes/Reju/v1.0/.

2.6. Screening and Analysis of Differentially Expressed Genes. After DESeq standardization, differential expression analysis was performed. The standard of differential gene screening was padj < 0.05. A gene ontology (GO) functional analysis and a KEGG pathway enrichment analysis were used to determine differentially expressed genes.

2.7. Real-Time qRT-PCR Analysis. Eight unigenes involved in walnut fat accumulation were chosen for validation by qRT-PCR. In this experiment, the reference gene was β-actin. The primers were designed with the Primer Premier 6.0 software (Table 1). RNA was extracted from the fruit according to the previously detailed method. The qRT-PCR was performed using an Ultra SYBR Mixture (With ROX) (CWbio Co. Ltd, cat no. CW0956) on Line Gene 9600 Plus Fluorescence quantitative PCR apparatus (Bioer, Hangzhou, China). Three replicates of each sample were conducted to calculate the average Ct values. The relative expression level was calculated by the comparative $2^{-\Delta\Delta Ct}$ method [19, 20].

3. Results

3.1. Changes of Fat Content in the Walnut Fruit at Various Developmental Stages. The walnut embryos were in the sol state at the early stage of development and began to solidify at 50 DAF, at which time the walnut kernel fat content was 8.3%. The rate of kernel fat at 50 to 60 DAF increased was low, whereas the period at 60 to 90 DAF exhibited the fastest increase. At 90 to 120 DAF, the increase of fat content slowed down. At 120 to 130 DAF, the walnut was completely mature, and the fat content tended to stop increasing. The fat content at 130 DAF was 69.8%.

Table 1: qRT-PCR validation primers.

Gene	Primer sequence (5′ to 3′)	Tm (°C)
β-Actin	CTCTTCCAGCCATCCATGATCG	57.63
	CCACTGAGGACAATATTGCCAT	55.16
LACS7	CAGGAGTTGAGGTTGTGACATATTC	56.05
	AGAGCAGCCATATCATCCATTAGTT	56.28
LACS1	CAGGTTAGGTGGTCGGATTAGACT	57.86
	TGATGGCTCCTCAAGTGGATTGT	58.43
ACC1	GAGAATTGCTGAAGAGTCGCTGAT	57.73
	TGTTGGCTCCACCTTGCTTAGA	58.56
FAD7	ACTTCAACTCTGTGCTGGTCTCT	58.07
	CCTCCTCGTAGATAACTCCATTCCT	57.29
LACS9	CGACTTACCACTTCCAGCCGATA	58.66
	ACTTCCAACAGCAGCAATCACATT	58.22
ACX5	TGATGAGGGTTTCACAGGTTACAAG	57.32
	AAGCATAAGCAGAAGCCAGCAA	57.86
PKT3	GCAGGATGGCGGCATTCTATG	58.74
	AGAGGCAGTAGCAGCAGCAG	59.29
PAS2	CGTCTTCTTCGGATGGCTTCA	57.33
	CAGTGTTGCTGAGATCGGAGAT	56.92

The accumulation rate of walnut fat displays a single peak curve (Figure 1). The fat slowly accumulated between 0 and 60 DAF and accumulated rapidly at 60 to 80 DAF, then reaching a peak at 90 DAF. Thereafter, the fat accumulated slowly until 130 DAF, at which point the fruits were fully mature and the fat content tended to stop increasing.

Based on the analysis of the kernel fat content at various developmental stages of Lvling walnuts, the optimal periods for analyzing transcriptomics were determined to be the early stage of fat synthesis (60 DAF), the peak stage of fat synthesis (90 DAF), and the late stage of fat synthesis (120 DAF).

3.2. Changes of Fatty Acid Content in the Walnut Kernel at Various Developmental Stages. A total of 14 kinds fatty acids have been detected in the walnut kernel, including 6 unsaturated fatty acids and 8 saturated fatty acids (Table 2). Walnut fat was composed primarily of the unsaturated fatty acids (linoleic acid, oleic acid, and α-linolenic acid) and the saturated fatty acids (palmitic acid and stearic acid). The total relative content of these five fatty acids was found to be greater than 98% at each period. The linoleic acid content gradually increased from 60 days to 90 DAF, reaching the highest value of 69.66% at 90 DAF and then gradually decreasing, reaching 63.29% at 120 DAF. Oleic acid showed an upward trend from 60 to 120 DAF, reaching 20.15% at 120 DAF and then dropping by 0.87% at 130 DAF. α-Linolenic acid showed an upward trend at 60 to 70 DAF, reaching a maximum of 12.22% at 70 DAF and then declining. At 90 DAF, the α-linolenic acid content slightly increased and then exhibited a downward trend. The content of palmitic acid was 17.45% at 60 DAF, reaching the lowest value at 100 DAF and then gradually increasing over the subsequent month to reach 6.46% at 130 DAF. The content of stearic acid was relatively

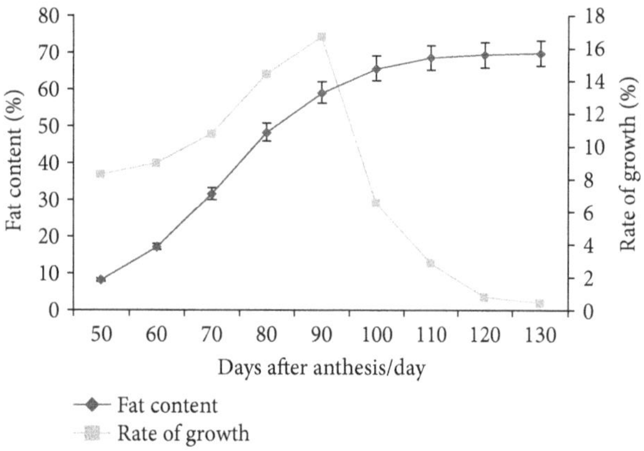

Figure 1: Dynamic changes of fat content in walnut kernel.

smaller, reached the highest value, 2.87%, at 100 DAF. The lowest palmitic acid content observed was 2.55% at 130 DAF. Based on the above analysis, the relative content of saturated fatty acids and unsaturated fatty acids in walnut kernel fats displayed opposing trends: the saturated fatty acids gradually declined, whereas the unsaturated fatty acids showed a gradual upward trend.

3.2.1. Transcriptome Sequencing Data Analysis

3.2.2. Transcriptome Sequencing Quality. A total of 4.73×10^8 raw reads were obtained after sequencing the transcriptome from three developmental stages, and 4.54×10^8 clean reads remained after filtration (Table 3). A total of 68.18 GB of data were obtained. The Q20 of clean reads were all above 96.75%, the percentage of Q30 bases was between 91.88% and 92.63%, and the GC content was approximately 46.49% to 49.64%. The square of the Pearson's correlation coefficient (R^2) between each sample repetition was greater than 0.8 in every case, indicating that the repeatability among the samples was satisfactory and could be used for subsequent bioinformatics analysis.

3.2.3. Comparison of Transcriptome Data with Reference Sequences. The alignment results showed that between 92% and 94% of the clean data could be located on the genomes, and the percentage of unique alignment positions was above 74% (Figure 2). Over 85% of the sequences aligned to the genome were located in exon regions. A small percentage of the sequences were found to be located in intergenic regions, likely due to the incomplete annotation of the genome, and a small percentage of the sequences were located in intron regions, which may be due to the retention of mRNA precursors and the inactivation of variable cleavage.

3.2.4. Analysis of Differentially Expressed Genes

(1) Distribution of Differentially Expressed Genes in the Walnut Kernel at Various Developmental Stages. The results of the analysis of differentially expressed genes in the walnut kernel at various developmental stages showed that the number of differentially expressed genes in L2vsL1, L3vsL2, and L3vsL1 was 724, 2027, and 4817, respectively (Figure 3(a)). The corresponding number of upregulated and downregulated genes was 400, 724, and 2637 and 324, 1303, and 4817, respectively. The number of upregulated genes was higher than that of downregulated genes in the case of L2vsL1, but the number of downregulated genes in the other two comparisons was significantly higher than that of upregulated genes.

As shown in the Wien diagram of differentially expressed genes in different combinations (Figure 3(b)), the specific DEGs of L2vsL1, L3vsL1, and L3vsL2 were 119, 5246, and 189, respectively. There were 424, 1657, and 54 genes that were differentially expressed simultaneously in two combinations, and 127 genes were differentially expressed across all three combinations. Among them, 54 genes were related to metabolism, and 33 were related to the synthesis of secondary metabolites (Table 4).

(2) GO Enrichment Analysis of Differentially Expressed Genes. A GO functional enrichment analysis of differentially expressed genes in the walnut kernel at various developmental stages was conducted. As shown in Figure 4, the genes expressed in L3vsL2 were not significantly enriched, whereas the genes differentially expressed in L2vsL1 were significantly enriched in the following functional categories (Figure 4(a)): nucleic acid-binding transcription factor activity, transcription factor activity, sequence-specific DNA binding, and negative regulation of cell proliferation. The genes differentially expressed in L3vsL1 were enriched predominantly in the following two major categories: biological processes and molecular functions (Figure 4(b)). In the biological processes group, biochemical processes and metabolic processes accounted for the highest proportion, followed by the biosynthesis of each substance, nucleic acid template transcription, DNA template transcription, macromolecular metabolic process regulation, and gene expression regulation. In the molecular functional group, sequence-specific DNA synthesis transcription factor activity and nucleic acid-binding transcription factor activity accounted for the highest proportion, followed by sequence-specific DNA synthesis, cytoskeletal protein synthesis, polymer complex synthesis, and protein complex body synthesis. In the genes differentially expressed in L3vsL1, there were more downregulated genes than upregulated genes in most enriched items.

(3) KEGG Metabolic Pathway Enrichment Analysis. KEGG, an online database that contains information on genomes, enzymatic pathways, and biochemical substances, can be used to further study the complex behavior of genes [21]. The primary metabolic pathways and signal transduction pathways related to the DEGs of the walnut kernel during different developmental stages were identified by a KEGG enrichment analysis (Figure 5). The results showed that 307 genes that were differentially expressed in L2vsL1 were annotated to 77 metabolic pathways in the KEGG database, 1094 genes that were differentially expressed in L3vsL2 were annotated to 110 metabolic pathways, and 3717 genes that were differentially expressed in L3vsL1 were annotated to 121 metabolic pathways. The pathways enriched by more DEGs were primarily starch and sucrose metabolism, glycolysis/gluconeogenesis, amino acid metabolism, fatty

Table 2: Dynamic changes of fatty acid content in walnut kernel fat (%).

Types of fatty acids		Days after fertilization/day							
		60	70	80	90	100	110	120	130
Unsaturated fatty acid	Linoleic acid (C18 : 2)	60.86 ± 2.01f	66.87 ± 2.10c	68.36 ± 2.56b	69.66 ± 2.03a	64.99 ± 2.34d	64.53 ± 2.01d	63.29 ± 2.35e	64.34 ± 1.32d
	Oleic acid (C18 : 1)	6.67 ± 0.13 g	8.37 ± 0.03f	12.29 ± 0.12de	12.87 ± 0.15d	18.92 ± 0.21bc	19.53 ± 0.45ab	20.15A ± 0.20a	19.28 ± 0.13b
	α-Linolenic acid (C18 : 3)	12.02 ± 0.08ab	12.22 ± 0.12a	9.3 ± 0.05c	8.54 ± 0.03d	7.38 ± 0.04e	6.88 ± 0.10 g	7.36 ± 0.06ef	6.93 ± 0.04 fg
	Palmitoleic acid (C16 : 1)	0.25 ± 0.02a	0.15 ± 0.03b	0.13 ± 0.001b	0.06 ± 0.001c	0.06 ± 0.001c	0.06 ± 0.002c	0.07 ± 0.001c	0.06 ± 0.002c
	all-*cis*-11,14-Eicosadienoic acid (C20 : 2)	0.07 ± 0.01a	0.08 ± 0.001a	0.03 ± 0.001b	0.04 ± 0.001b	0.02 ± 0.001bc	0.03 ± 0.001b	0.03 ± 0.001b	0.02 ± 0.001bc
	cis-11-Eicosenoic acid (C20 : 1)	0.40 ± 0.003a	0.26 ± 0.002b	0.19 ± 0.001c	0.19 ± 0.001c	0.19 ± 0.001c	0.17 ± 0.001c	0.17 ± 0.001c	0.18 ± 0.001c
Saturated fatty acid	Margaric acid (17 : 0)	0.12 ± 0.01	0.06 ± 0.001	0.07 ± 0.001	0.07 ± 0.001	0.06 ± 0.001	0.07 ± 0.001	0.07 ± 0.001	0.07 ± 0.001
	Behenic acid (22 : 0)	0.34 ± 0.003a	0.10 ± 0.001b	0.02 ± 0.001b	0.02 ± 0.001b	0.03 ± 0.001b	0.01 ± 0.001b	0.03 ± 0.001b	0.02 ± 0.001b
	Lauric acid (12 : 0)	0.06 ± 0.001a	0.01 ± 0.001b	0b	0.02 ± 0.001b	0.02 ± 0.001b	0.02 ± 0.001b	0.01 ± 0.001b	0b
	Pentadecylic acid (15 : 0)	0.25 ± 0.002a	0.08 ± 0.001b	0.04 ± 0.001c	0.02 ± 0.001c	0.02 ± 0.001c	0.01 ± 0.001c	0.02 ± 0.001c	0.02 ± 0.001c
	Myristic acid (C14 : 0)	0.13 ± 0.005a	0.07 ± 0.001b	0.06 ± 0.001b	0.03 ± 0.001c	0.02 ± 0.001c	0.02 ± 0.001c	0.02 ± 0.001c	0.01 ± 0.001c
	Palmitic acid (C16 : 0)	17.45 ± 0.01a	10.06 ± 0.001b	7.57 ± 0.02c	5.93 ± 0.04ef	5.35 ± 0.03 fg	5.85 ± 0.01ef	6.06 ± 0.02de	6.46 ± 0.03d
	Stearic acid (C18 : 0)	1.29 ± 0.003e	1.59 ± 0.001d	1.90 ± 0.001 cd	2.5 ± 0.001bc	2.87 ± 0.005a	2.75 ± 0.007a	2.65 ± 0.002ab	2.55 ± 0.008bc
	Arachidic acid (20 : 0)	0.09 ± 0.002a	0.08 ± 0.001a	0.05 ± 0.001b	0.05 ± 0.001b	0.07 ± 0.001a	0.07 ± 0.001a	0.07 ± 0.001a	0.06 ± 0.001ab

Note: values represent the mean ± SE. Means within each line followed by the same letter are not significantly different at $P < 0.05$ according to Duncan's multiple range test.

Table 3: Data statistics of transcriptome sequencing.

Sample name	Number of clean reads	Clean bases (G)	Unique mapped reads	Q20 (%)	Q30 (%)	GC content
L1-1	56,786,156	8.52	46,399,939	97.06	92.63	47.34
L1-2	47,864,764	7.18	39,778,196	97.04	92.56	46.74
L1-3	48,659,022	7.3	40,863,104	96.93	92.35	46.49
L2-1	51,398,780	7.71	39,559,096	96.91	92.24	49.13
L2-2	52,897,882	7.93	43,079,514	96.94	92.34	47.63
L2-3	50,390,996	7.56	39,187,890	96.75	91.88	49.64
L3-1	50,680,344	7.6	41,211,813	96.78	92.01	48.31
L3-2	46,300,696	6.95	37,672,468	97.01	92.46	48.35
L3-3	49,505,362	7.43	40,171,554	96.97	92.44	48.01
Total	454,484,002	68.18				

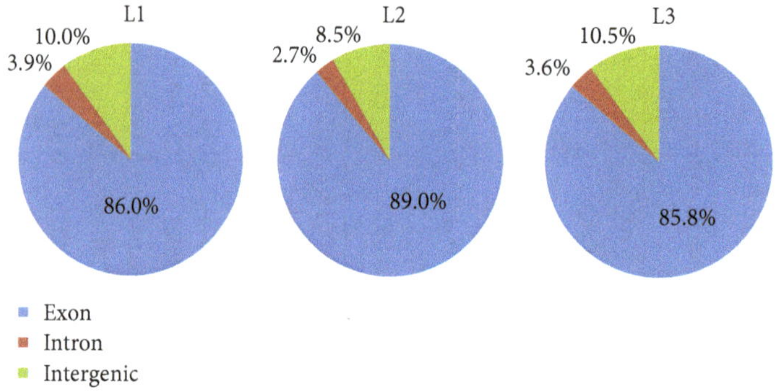

Figure 2: Distribution of reads in different regions of the genome in three periods.

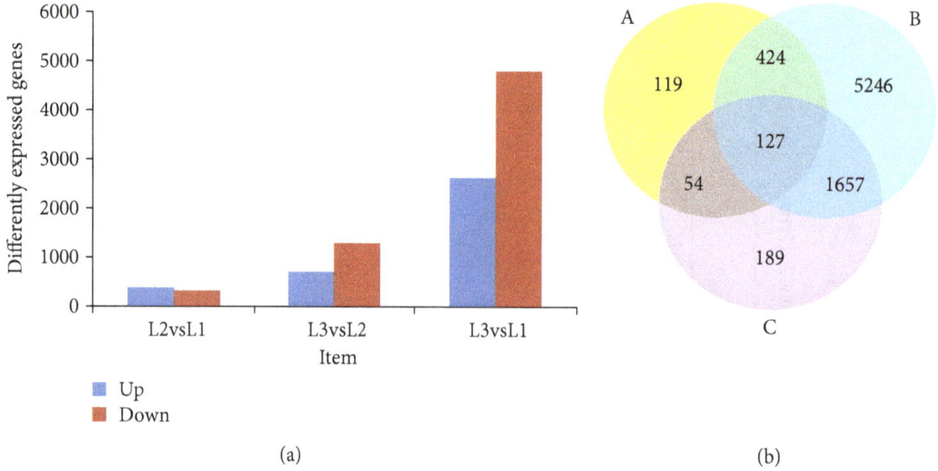

Figure 3: Distribution of differentially expressed genes in walnut kernel at different developmental stages. (a) Number of differentially expressed genes in different combinations. (b) Venn diagram of differentially expressed genes in different combinations: A: L2vsL1; B: L3vsL1; and C: L3vsL2.

acid metabolism, and secondary metabolite and amino acid synthesis. Among these pathways, the number of different genes expressed in different pathways varied, and the difference in genes of L3vsL1 in all pathways was greater than L3vsL2 and L2vsL1, indicating that the physiological processes also changed greatly with the development of the fruit.

(4) Identification of Fat Metabolism-Related Genes. The primary purpose of this study was to discover the mechanisms underlying walnut fat synthesis and to identify the key genes that control walnut fat synthesis. This analysis of the expression levels genes related to kernel fat biosynthesis at the 60 DAF, 90 DAF, and 120 DAF revealed certain genes that regulate the synthesis of walnut fat.

Among the differentially expressed genes annotated by KEGG, there were many genes related to fatty acid extension, unsaturated fatty acid biosynthesis, fatty acid metabolism, fatty acid degradation, fatty acid biosynthesis, and glycerolipid metabolism (Table 5).

Table 4: KEGG analysis of common differentially expressed genes.

KEGG pathway	Pathway ID	Input number
Protein processing in endoplasmic reticulum	ath04141	15
Taurine and hypotaurine metabolism	ath00430	3
Starch and sucrose metabolism	ath00500	12
Endocytosis	ath04144	8
Photosynthesis-antenna proteins	ath00196	3
Plant-pathogen interaction	ath04626	9
Stilbenoid, diarylheptanoid, and gingerol biosynthesis	ath00945	4
Flavonoid biosynthesis	ath00941	2
Carbon fixation in photosynthetic organisms	ath00710	4
SNARE interactions in vesicular transport	ath04130	3
Phenylpropanoid biosynthesis	ath00940	7
Biosynthesis of secondary metabolites	ath01110	33
RNA degradation	ath03018	4
Phenylalanine metabolism	ath00360	4
Spliceosome	ath03040	6
Metabolic pathways	ath01100	54
Glycolysis/gluconeogenesis	ath00010	3
Carbon metabolism	ath01200	6
Biosynthesis of amino acids	ath01230	6

Changes in the expression of these genes were likely the basis for the observed differences in fat content and fatty acid composition of the walnut kernel at various developmental stages.

The site of the initial synthesis of plant fatty acids is the plastid, and the synthesis is catalyzed by fatty acid synthase. Pyruvate dehydrogenase (PHD) generates acetyl coenzyme A (acetyl-CoA), which is the raw material for fatty acid synthesis. The first step of fatty acid biosynthesis is the conversion of acetyl-CoA to malonyl-CoA, catalyzed by acetyl coenzyme A carboxylase (ACC). Malonyl groups are condensed with acetyl-CoA, catalyzed by acyltransferase (ACP), followed by the sequential addition of 2 carbon units to extend the fatty acid chain [22]. A total of 23 differentially expressed genes involved in fatty acid synthesis were identified in this study, including 7 long-chain acyl-CoA synthetases (LACS), ACC1, 4 beta-ketoacyl-ACP synthases (KASs), FAB1, SSI2, FATB, FTM1, and MOD1.

Two important enzymes in the fat synthesis pathway are acetyl-CoA carboxylase and LACS. The expression of ACC1 was upregulated from L1 to L3 (Table 6). LACS plays an important role in lipid metabolism. Whether the reaction is anabolic or catabolic, the essential substrate, fatty acyl-CoA, is catalyzed by LACS. In this study, the expression of LACS1 was found to be upregulated from L1 to L3; the expression of LACS7 was not significantly different in L1 and L2 and was upregulated from L2 to L3. The expression of LACS9 was upregulated from L1 to L3 and was not significantly different between L2 and L1 or L3. KAS, which is an acyl carrier, catalyzes the condensation reaction of fatty acid synthesis [22]. In this study, a total of 4 KAS-related genes were detected, and their expression levels were downregulated from L1 to L3.

The content of unsaturated fatty acids in walnut fat is greater than 60%. In this experiment, 14 differentially expressed genes related to unsaturated fatty acid synthesis were found, including 5 desaturases, 3 thiolases, 2 dehydratases, 1 reductase, 1 oxidase, 1 protease, and 1 phosphatase. The expression levels of 2 sulfur lyases were upregulated, and 1 sulfur lyase was downregulated. Of the 5 desaturases, only FAD7 was upregulated. The expression levels of 2 PAS2 genes were upregulated, and the expression of Rossmann-fold NAD(P)-binding domain-containing protein was upregulated from L1 to L 2.

3.3. Validation of DEG Expression by qRT-PCR. To validate the expression of the DEGs obtained from RNA-Seq, 8 DEGs were selected for qRT-PCR, including LACS1, LACS7, LACS9, FAD7, ACX5, PKT3, PAS2, and ACC1. The expression trends of 8 DEGs were different; LACS1, LACS7, PKT3, and ACC1 genes were upregulated from 60 to 120 days after fertilization, while ACX5 and LACS9 were gradually downregulated (Figure 6). FAD7 upregulated rapidly at 90 days after fertilization and gradually downregulated after 120 DAF. Correlation analysis (Table 7) showed that the correlation coefficients of 8 genes related to the walnut lipid synthesis between expression and transcriptome sequencing were above 0.76, suggesting that our transcriptome analysis was accurate and reliable.

4. Discussion

The walnut kernel was found to be rich in unsaturated fatty acids. In this experiment, the fat content and fatty acid composition of the precocious Lvling walnut kernel were determined at various developmental stages. The results showed that the walnut kernel began to solidify at 50 DAF and that the fat content rapidly accumulated between 60 and 90 DAF, reaching a peak at 90 DAF and then decreasing. At 130 DAF, the fruit is mature and tends to stop growing. However, the trend of fatty acid composition was inconsistent with fat content, and this result was consistent with that of the study reported by Hu et al. [23]. The content of saturated fatty acids in the walnuts gradually decreased with the development of the fruit, whereas the content of unsaturated fatty acids exhibited an upward trend. This accumulation of fatty acid was similar to that of woody oil plants such as *Symplocos paniculata* [15] and *Swida wilsoniana* [24, 25]. Based upon the analysis of the fruit fat, the three developmental stages selected for further transcriptome analysis were 60, 90, and 120 DAF.

In this study, the transcriptome of the fruit at various developmental stages was sequenced using the Illumina technique, and the functions of the assembled unigenes were annotated. In addition, the differentially expressed genes related to fat synthesis were screened using the DESeq method. Certain key genes were also analyzed, and these genes included genes involved in the synthesis of acetyl-CoA and genes related to fatty acid synthesis. This study provides a theoretical basis for walnut fat synthesis. From the

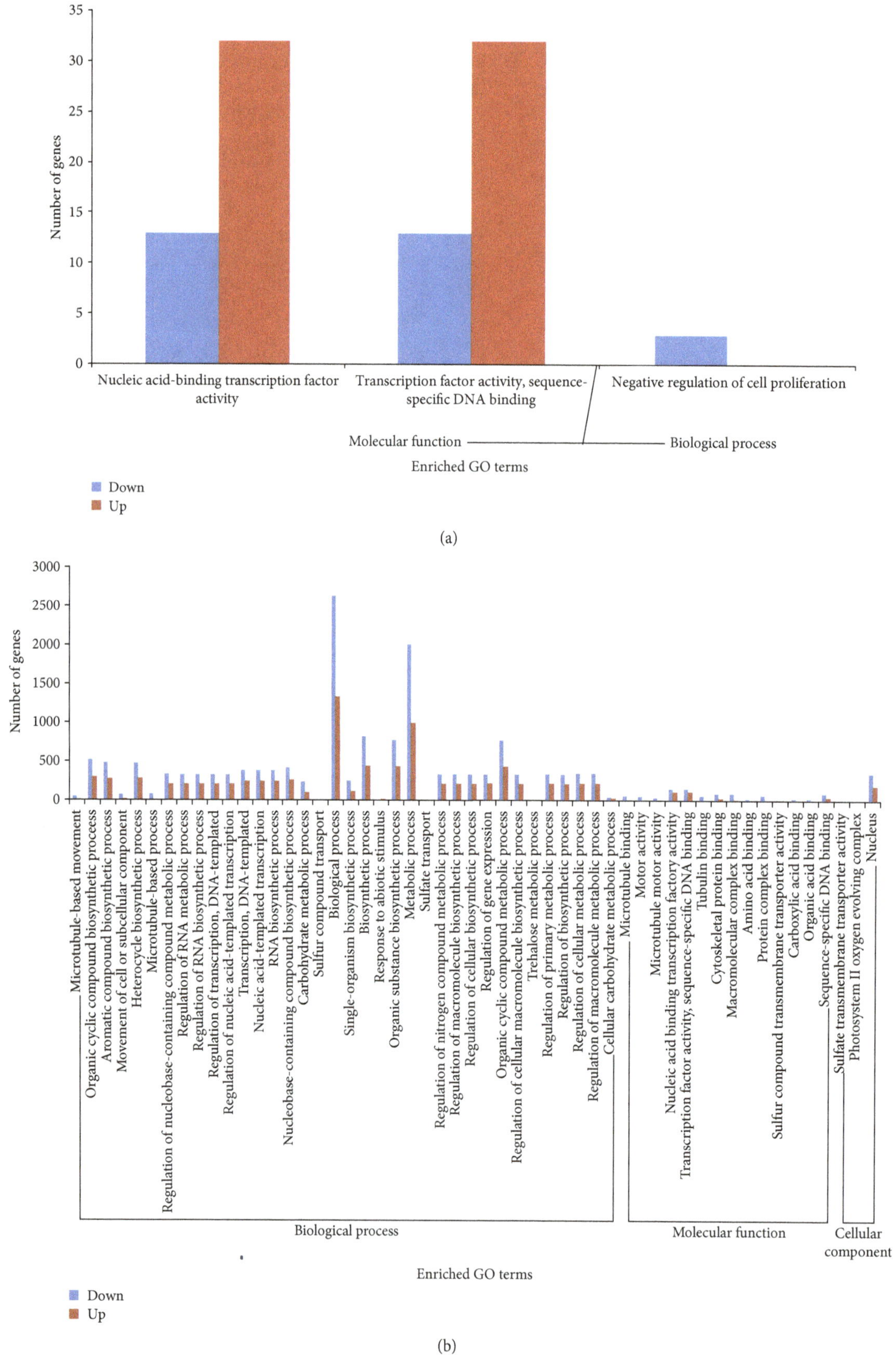

FIGURE 4: Clustering analysis diagram of significantly enriched GO terms from differentially expressed genes between different samples. (a) Clustering analysis diagram of significantly enriched GO terms from differentially expressed genes in L2vsL1. (b) Clustering analysis diagram of significantly enriched GO terms from differentially expressed genes in L3vsL1.

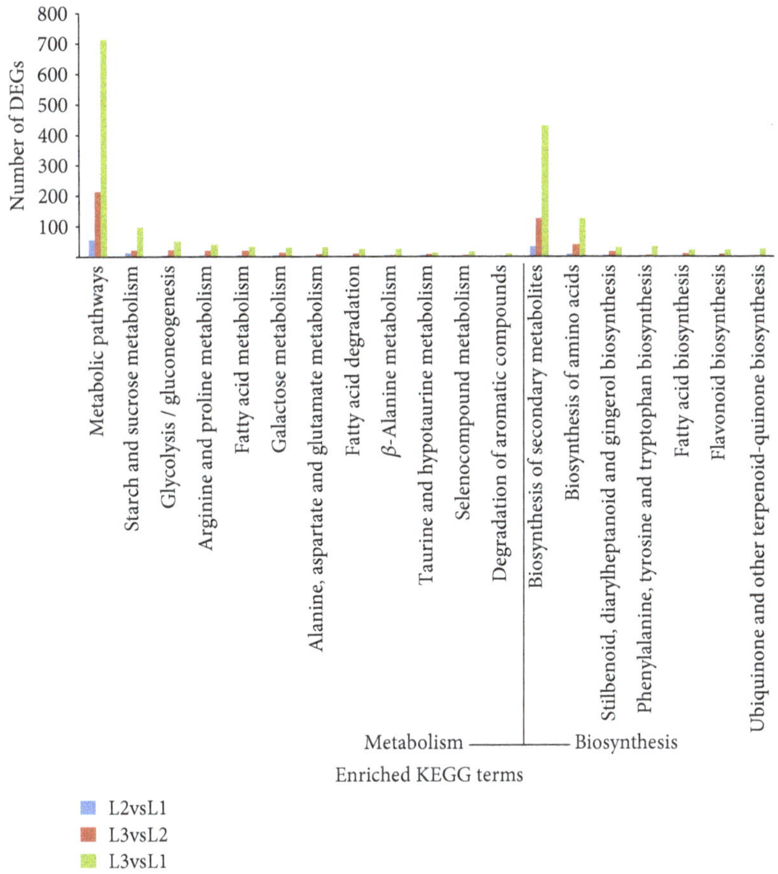

FIGURE 5: KEGG enrichment analysis of differentially expressed genes among different samples.

TABLE 5: KEGG analysis of differentially expressed genes in lipid metabolism.

Term	Database ID	Background number	Input number L2vsL1	Input number L3vsL2	Input number L3vsL1
Fatty acid metabolism	ath01212	73	4	17	32
Biosynthesis of unsaturated fatty acids	ath01040	36	2	7	11
Fatty acid biosynthesis	ath00061	40	2	9	20
Fatty acid degradation	ath00071	40	2	10	22
Fatty acid elongation	ath00062	31	1	4	13
Glycerophospholipid metabolism	ath00564	83	1	13	31
Glycerolipid metabolism	ath00561	52	0	5	22

abovementioned analysis of genes encoding key enzymes and genes involved in the regulation process in the walnut fatty acid biosynthetic pathway, it was found that with gradual maturation, the fat content of the walnut kernel increased. The high expression of LACS and ACCase accelerates fatty acid biosynthesis. Thus, it can be concluded that LACS and ACCase are key targets for the gene regulation of the fat content in walnuts.

Acetyl-CoA carboxylase catalyzes the conversion of acetyl-CoA into malonyl-CoA, which constitutes the initial reaction in fatty acid biosynthesis [26]. Therefore, ACCase is a key enzyme in the process of fat synthesis and is a major rate-limiting enzyme [27, 28]. In many organisms, ACCases in plastids and in the cytosol catalyze the conversion of acetyl-CoA conversion into malonyl-CoA, but the function of ACCase in plants is not exactly the same. Malonyl-CoA generated by the ACCase in plastids is used for the de novo biosynthesis of fatty acids, whereas malonyl-CoA generated by ACCase in the cytosol is used for fatty acid extension, flavonoid formation, and the synthesis of many other secondary metabolites. The seed-specific overexpression or constitutive overexpression of ACC1, an *Arabidopsis* homozygous ACCase gene, increases the fatty acid content in *Arabidopsis* and the potato [29, 30]. In this study, ACC1, a differentially expressed ACCase gene, was found to be upregulated from stage 1 to stage 2, indicating that ACC1 is a key enzyme involved in walnut fat synthesis. Long-chain acyl-CoA synthase (LACS, EC 6.2.1) plays an important role in lipid

Table 6: Differentially expressed genes in fatty acid synthesis.

Gene ID	Gene symbol	Annotation	L2vsL1	L3vsL2	L3vsL1
WALNUT_00023412	LACS1	Long-chain acyl-CoA synthetase 1		D	D
WALNUT_00006224	LACS1	Long-chain acyl-CoA synthetase 1			U
WALNUT_00028238	LACS2	Long-chain acyl-CoA synthetase 2		U	D
WALNUT_00012724	LACS4	Long-chain acyl-CoA synthetase 4	D		D
WALNUT_00003419	LACS7	Long-chain acyl-CoA synthetase 7		U	U
WALNUT_00018437	LACS7	Long-chain acyl-CoA synthetase 7		D	D
WALNUT_00022757	LACS9	Long-chain acyl-CoA synthetase 9			U
WALNUT_00011609	FAB1	3-Oxoacyl-[acyl-carrier-protein] synthase		D	
WALNUT_00012795	SSI2	Acyl-[acyl-carrier-protein] desaturase		D	D
WALNUT_00014166	FATB	Fatty acyl-ACP thioesterases		D	
WALNUT_00019717	FTM1	Stearoyl-ACP desaturase		D	U
WALNUT_00021234	CAC3	Acetyl coenzyme A carboxylase carboxyl transferase subunit alpha		D	D
WALNUT_00022861	FATA	FatA acyl-ACP thioesterase		D	D
WALNUT_00002636	MOD1	Enoyl-[acyl-carrier protein] reductase I		D	D
WALNUT_00009311	MOD1	Enoyl-[acyl-carrier-protein] reductase (NADH)		D	D
WALNUT_00012185	ACC1	Homomeric acetyl-CoA carboxylase (Hom-ACCase)	U		U
WALNUT_00024655	KAS_III	3-Ketoacyl-[acyl-carrier-protein] synthase III			D
WALNUT_00025417	KAS_I	Beta-ketoacyl-[acyl-carrier-protein] synthase I		D	D
WALNUT_00028064	KAS_I	Beta-ketoacyl-[acyl-carrier-protein] synthase I		D	D
Novel01692	KAS_I	Beta-ketoacyl-[acyl-carrier-protein] synthase I		D	D
WALNUT_00011323		3-Hydroxyacyl-[acyl-carrier-protein] dehydratase		D	D
WALNUT_00003181		NAD(P)-binding Rossmann-fold superfamily protein		D	D
WALNUT_00018045		Short-chain type dehydrogenase		D	
WALNUT_00021629	PKT3	3-Ketoacyl-CoA thiolase 2			U
WALNUT_00026180	PKT3	3-Ketoacyl-CoA thiolase 2		U	
WALNUT_00009879	PKT3	3-Ketoacyl-CoA thiolase 2			D
WALNUT_00003181		3-Oxoacyl-[acyl-carrier-protein] reductase		D	D
WALNUT_00011902	FAD2	Omega-6 fatty acid desaturase ω-6	D		D
WALNUT_00002632	FAD2	Omega-6 fatty acid desaturase ω-6		D	
WALNUT_00015387	FAD7	Fatty acid desaturase 7	U		U
WALNUT_00018114	PAS2	Very-long-chain (3R)-3-hydroxyacyl-[acyl-carrier protein] dehydratase			U
WALNUT_00026237	PAS2	Very-long-chain (3R)-3-hydroxyacyl-[acyl-carrier protein] dehydratase		U	U
WALNUT_00018045		Rossmann-fold NAD(P)-binding domain-containing protein	U	D	
WALNUT_00019717		Acyl-[acyl-carrier-protein] desaturase		D	D
WALNUT_00020859		Protein-tyrosine phosphatase			D
WALNUT_00008997	ACX5	Acyl-CoA oxidase 5		D	D
WALNUT_00012795	SSI2	Acyl-[acyl-carrier-protein] desaturase	D		D

Note: U: significantly upregulated; D: significantly downregulated.

metabolism; whether the reaction is anabolic or catabolic, the necessary substrate, fatty acyl-CoA, is catalyzed by LACS. Physiological processes in plants, such as fatty acid chain extension and desaturation modification, transport of fatty acids, and the generation of other lipid derivatives of fatty acids, are all dependent on LACS [31]. The LACS protein family in *Arabidopsis thaliana* contains LACS1–LACS9. LACS2 and LACS9 have been reported to play important roles in lipid synthesis and accumulation [32, 33]. In this study, LACS1, LACS2, LACS4, LACS7, and LACS9 were found to be differentially expressed in the various stages of walnut development. For example, LACS1, LACS7, and LACS9 were upregulated from stage 1 to stage 3, indicating that LACS played a key role in walnut kernel fatty acid synthesis.

In this study, the differentially expressed genes related to unsaturated fatty acid synthesis were also analyzed, and 14 differentially expressed genes were found. Fatty acid desaturases exist in all plants and can be classified into the following three groups according to their catalytic substrates [34]: (1) acyl-carrier protein desaturases, which are present in the plant cytoplasmic matrix, introduce double bonds into

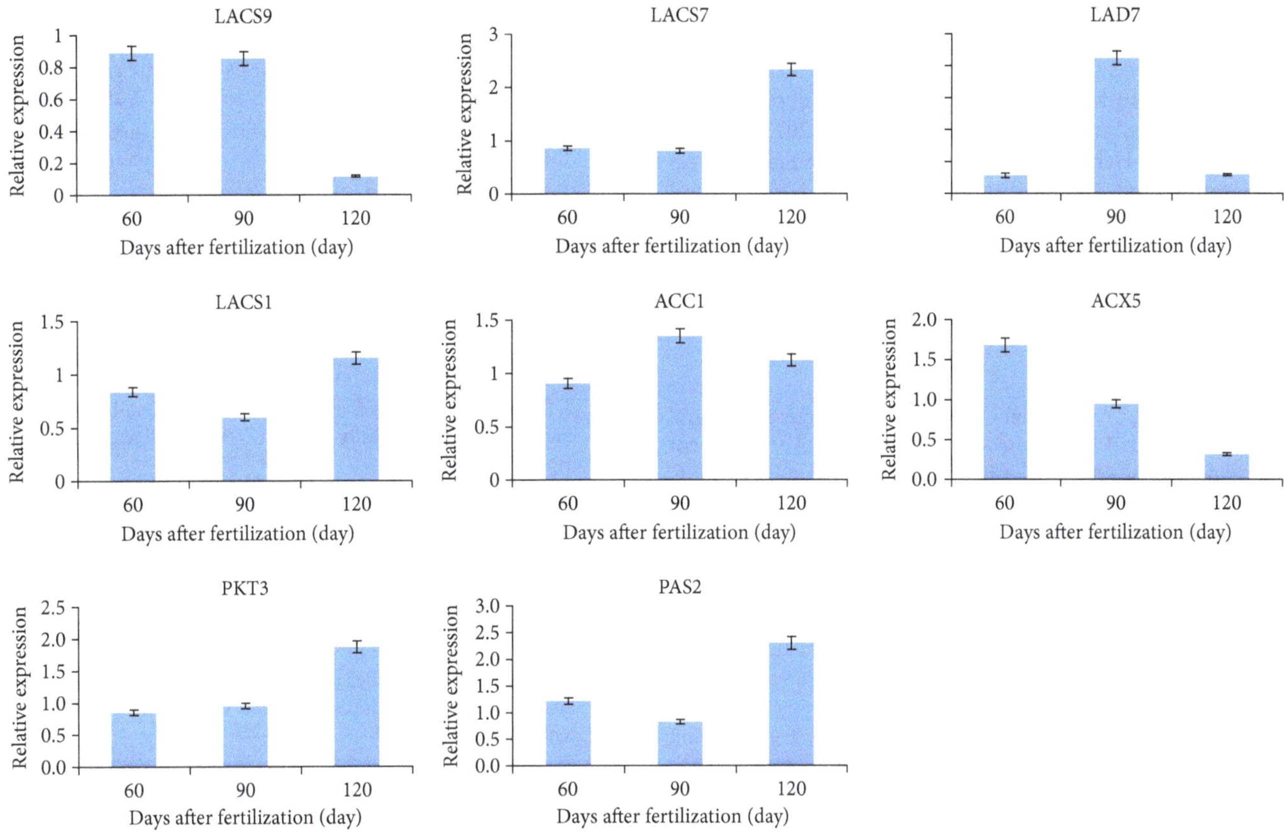

Figure 6: Expression analysis of fatty acid synthesis candidate genes in walnut kernels at different developmental stages.

Table 7: Correlation coefficients of candidate genes related to the walnut lipid synthesis between expression and transcriptome sequencing.

Gene ID	Enzyme names	Correlation coefficient
WALNUT_00003419	LACS7	0.92
WALNUT_00006224	LACS1	0.96
WALNUT_00008997	ACX5	0.99
WALNUT_00012185	ACC1	0.88
WALNUT_00015387	FAD7	0.96
WALNUT_00021629	PKT3	0.98
WALNUT_00022757	LACS9	0.76
WALNUT_00026237	PAS2	0.97

acyl carbon chains that bind to ACP; (2) acyl-CoA desaturases, which are present in the endoplasmic reticulum of plant cells, introduce double bonds into the acyl of acyl-CoA; and (3) acyl-lipid desaturases, which are present in the plasma membrane of plant cells, introduce double bonds into the acyl carbon chains of glycerolipids and glycolipids. In this study, 5 fatty acid desaturase-encoding genes were found, including 3 acyl-lipid desaturase genes (FAD2, FAD7) and 2 acyl-ACP desaturase genes.

The acyl-lipid desaturase catalyzes the fatty acid carbon chains binding with glycerol to form double bonds. The acyl-lipid desaturases screened in this study include 2 ω-6 fatty acid desaturases (FAD2) and 1 ω-3 fatty acid desaturase (FAD7). The ω-6 fatty acid desaturases introduce a double bond at the ω-6 position of the mono-olefm fatty acid carbon chain of glycerol to form the diene fatty acid carbon chain [35], whereas the ω-3 fatty acid desaturase catalyzes the biosynthesis of octadecatrienoic acid and hexadecatrienoic acid, forming the third double bond in a polyunsaturated fatty acid. α-Linolenic acid, which is present in walnut fat, is a polyunsaturated fatty acids with 3 double bonds, and FAD7 is most likely the key enzyme that catalyzes the synthesis of α-linolenic acid.

Many studies indicate that differences in the gene sequence could also influence the fat accumulation in walnut fruits. In this study, the genes related to walnut kernel fatty acid synthesis were screened, and their expression patterns during the lipid synthesis phase were analyzed. However, the differences in the sequence of genes affecting walnut fat content were not discussed in this paper. So, this study was the foundation for uncovering the mechanism underlying walnut kernel fatty acid formation. The mechanism of walnut oil synthesis still needs further study. We anticipate that future research based upon this study will reveal the entire mechanism underlying the synthesis of walnut fat.

Conflicts of Interest

The authors declare that they have no conflicts of interest.

Acknowledgments

This research was financially supported by the National Key Technology Research and Development Program of the Ministry of Science and Technology of China (2013BAD14B 0103) and the Hebei Science and Technology Plan Project "Innovation and Demonstration of Walnut Near-Natural Production Technology in Mountainous Areas of Hebei Province" (no. 16236810D).

References

[1] R. Y. Wang, "Analysis on China edible oil market supply and demand in 2013 and opinion on accelerating the development of woody oil industry," *China Oils and Fats*, vol. 39, no. 6, pp. 1–5, 2014.

[2] K. D. Chapman and J. B. Ohlrogge, "Compartmentation of triacylglycerol accumulation in plants," *Journal of Biological Chemistry*, vol. 287, no. 4, pp. 2288–2294, 2012.

[3] F. M. You, K. R. Deal, J. Wang et al., "Genome-wide SNP discovery in walnut with an AGSNP pipeline updated for SNP discovery in allogamous organisms," *BMC Genomics*, vol. 13, no. 1, p. 354, 2012.

[4] A. Magi, M. Benelli, A. Gozzini, F. Girolami, F. Torricelli, and M. L. Brandi, "Bioinformatics for next generation sequencing data," *Genes*, vol. 1, no. 2, pp. 294–307, 2010.

[5] M. Li, Y. Liu, C. Sun et al., "Research advance about nutrients and medicinal value of walnut," *Journal of the Chinese Cereals and Oils Association*, vol. 6, pp. 166–170, 2009.

[6] V. E. Velculescu, L. Zhang, W. Zhou et al., "Characterization of the yeast transcriptome," *Cell*, vol. 88, no. 2, pp. 243–251, 1997.

[7] U. Nagalakshmi, Z. Wang, K. Waern et al., "The transcriptional landscape of the yeast genome defined by RNA sequencing," *Science*, vol. 320, no. 5881, pp. 1344–1349, 2008.

[8] F. Bourgis, A. Kilaru, X. Cao et al., "Comparative transcriptome and metabolite analysis of oil palm and date palm mesocarp that differ dramatically in carbon partitioning," *Proceedings of the National Academy of Sciences of the United States of America*, vol. 108, no. 30, pp. 12527–12532, 2011.

[9] R. Singh, M. Ong-Abdullah, E. T. L. Low et al., "Oil palm genome sequence reveals divergence of interfertile species in old and new worlds," *Nature*, vol. 500, no. 7462, pp. 335–339, 2013.

[10] A. P. Brown, J. T. M. Kroon, D. Swarbreck et al., "Tissue-specific whole transcriptome sequencing in castor, directed at understanding triacylglycerol lipid biosynthetic pathways," *PLoS One*, vol. 7, no. 2, article e30100, 2012.

[11] P. Natarajan and M. Parani, "De novo assembly and transcriptome analysis of five major tissues of Jatropha curcas L. using GS FLX titanium platform of 454 pyrosequencing," *BMC Genomics*, vol. 12, no. 1, p. 191, 2011.

[12] S. Sato, H. Hirakawa, S. Isobe et al., "Sequence analysis of the genome of an oil-bearing tree, Jatropha curcas L," *DNA Research*, vol. 18, no. 1, pp. 65–76, 2011.

[13] X. Tan, Y. Jang, B. Wang, and L. Zhang, "Cloning and sequence analysis of BC gene from Camellia oleifera," *Journal of Central South University of Forestry & Technology*, vol. 2, pp. 1-9, 2010.

[14] R.-J. Li, X. Gao, L.-M. Li, X.-L. Liu, Z.-Y. Wang, and S.-y. Lü, "De novo assembly and characterization of the fruit transcriptome of Idesia polycarpa reveals candidate genes for lipid biosynthesis," *Frontiers in Plant Science*, vol. 7, 2016.

[15] Q. Liu, Y. Sun, J. Chen et al., "Transcriptome analysis revealed the dynamic oil accumulation in Symplocos paniculata fruit," *BMC Genomics*, vol. 17, no. 1, p. 929, 2016.

[16] "GB/T 5512–2008: Inspect of grain and oilseeds-determination of crude fat content in grain".

[17] "GB/T17376-2008; Animal and vegetable fats and oils - preparation of fatty acid methyl esters".

[18] "GB/T17377-2008; Analysis of animal and vegetable fats and fatty acid methyl esters by gas chromatography".

[19] J. Bennett, D. Hondred, and J. C. Register, "Keeping qRT-PCR rigorous and biologically relevant," *Plant Cell Reports*, vol. 34, no. 1, pp. 1–3, 2015.

[20] F. Wang, L. Li, H. Li et al., "Transcriptome analysis of rosette and folding leaves in Chinese cabbage using high-throughput RNA sequencing," *Genomics*, vol. 99, no. 5, pp. 299–307, 2012.

[21] M. Kanehisa, M. Araki, S. Goto et al., "KEGG for linking genomes to life and the environment," *Nucleic Acids Research*, vol. 36, no. Database, pp. D480–D484, 2007.

[22] P. D. Bates, S. Stymne, and J. Ohlrogge, "Biochemical pathways in seed oil synthesis," *Current Opinion in Plant Biology*, vol. 16, no. 3, pp. 358–364, 2013.

[23] Z. Y. Hu, W. Hua, L. Zhang et al., "Seed structure characteristics to form ultrahigh oil content in rapeseed," *PLoS One*, vol. 8, no. 4, article e62099, 2013.

[24] Z. Xiao, C. Li, J. Chen, R. Liu, and X. Wu, "Analysis of oil composition in different parts of Swida wilsoniana Wanger. fruit," *China Oil and Fats*, vol. 17, no. 1, p. 929, 2016.

[25] J. Z. Chen, C. Z. Li, and P. W. Li, "Fruit development and fatty acid accumulation of Swida wilsoniana," *Hunan Forestry Science & Technology*, vol. 40, no. 2, pp. 9–14, 2013.

[26] T. Konishi, K. Shinohara, K. Yamada, and Y. Sasaki, "Acetyl-CoA carboxylase in higher plants: most plants other than gramineae have both the prokaryotic and the eukaryotic forms of this enzyme," *Plant & Cell Physiology*, vol. 37, no. 2, pp. 117–122, 1996.

[27] J. E. Cronan Jr and G. L. Waldrop, "Multi-subunit acetyl-CoA carboxylases," *Progress in Lipid Research*, vol. 41, no. 5, pp. 407–435, 2002.

[28] J. Mu, H. Tan, Q. Zheng et al., "LEAFY COTYLEDON1 is a key regulator of fatty acid biosynthesis in Arabidopsis," *Plant Physiology*, vol. 148, no. 2, pp. 1042–1054, 2008.

[29] D. Klaus, J. B. Ohlrogge, H. E. Neuhaus, and P. Dörmann, "Increased fatty acid production in potato by engineering of acetyl-CoA carboxylase," *Planta*, vol. 219, no. 3, pp. 389–396, 2004.

[30] K. Roesler, D. Shintani, L. Savage, S. Boddupalli, and J. Ohlrogge, "Targeting of the Arabidopsis homomeric acetyl-coenzyme A carboxylase to plastids of rapeseeds," *Plant Physiology*, vol. 113, no. 1, pp. 75–81, 1997.

[31] S. Lü, T. Song, D. K. Kosma, E. P. Parsons, O. Rowland, and M. A. Jenks, "Arabidopsis CER8 encodes LONG-CHAIN ACYL-COA SYNTHETASE 1 (LACS1) that has overlapping functions with LACS2 in plant wax and cutin synthesis," *Plant Journal*, vol. 59, no. 4, pp. 553–564, 2009.

[32] J. Schnurr, J. Shockey, and J. Browse, "The acyl-CoA synthetase encoded by LACS2 is essential for normal cuticle development in Arabidopsis," *Plant Cell*, vol. 16, no. 3, pp. 629–642, 2004.

[33] J. A. Schnurr, J. M. Shockey, G. J. de Boer, and J. A. Browse, "Fatty acid export from the chloroplast. Molecular characterization of a major plastidial acyl-coenzyme a synthetase from Arabidopsis," *Plant Physiology*, vol. 129, no. 4, pp. 1700–1709, 2002.

[34] S. Mekhedov, O. M. de Ilarduya, and J. Ohlrogge, "Toward a functional catalog of the plant genome. A survey of genes for lipid biosynthesis," *Plant Physiology*, vol. 122, no. 2, pp. 389–402, 2000.

[35] D. Zhang, I. L. Pirtle, S. J. Park et al., "Identification and expression of a new delta-12 fatty acid desaturase (FAD2-4) gene in upland cotton and its functional expression in yeast and Arabidopsis thaliana plants," *Plant Physiology and Biochemistry*, vol. 47, no. 6, pp. 462–471, 2009.

18

Multiple Regression Analysis Reveals MicroRNA Regulatory Networks in *Oryza sativa* under Drought Stress

Jiajia Chen and Liangzhi Li

School of Chemistry, Biology and Materials Engineering, Suzhou University of Science and Technology, No. 1 Kerui Road, Suzhou 215011, China

Correspondence should be addressed to Jiajia Chen; njucjj@126.com

Academic Editor: Giandomenico Corrado

Drought is a major abiotic stress that reduces rice development and yield. miRNAs (microRNAs) are known to mediate posttranscriptional regulation under drought stress. Although the importance of individual miRNAs has been established, the crosstalks between miRNAs and mRNAs remain unearthed. Here we performed microarray analysis of miRNAs and matched mRNA expression profiles of drought-treated rice cultivar Nipponbare. Drought-responsive miRNA-mRNA regulations were identified by a combination of a partial least square (PLS) regression approach and sequence-based target prediction. A drought-induced network with 13 miRNAs and 58 target mRNAs was constructed, and four miRNA coregulatory modules were revealed. Functional analysis suggested that drought-response miRNA targets are enriched in hormone signaling, lipid and carbohydrate metabolism, and antioxidant defense. 13 candidate miRNAs and target genes were validated by RT-qPCR, hierarchical clustering, and ROC analysis. Two target genes (DWARF-3 and P0651G05.2) of miRNA coregulatory modules were further verified by RLM-5′ RACE. Together, our integrative study of miRNA-mRNA interaction provided attractive candidates that will help elucidate the drought-response mechanisms in *Oryza sativa*.

1. Introduction

Rice (*Oryza sativa*) is one of the main staple crops that feed more than half of the global population [1]. Rice is acclimated to rain-fed or fully irrigated fields and is susceptible to drought stress [2]. The yield of rice is frequently challenged by dwindling water resources. There is a growing need to find novel drought-responsive molecules to improve the rice cultivars [3].

miRNAs (microRNAs) are short noncoding RNAs that mediate gene expression by mRNA degradation or translational repression [4]. Currently, 757 mature rice miRNA sequences have been identified in miRBase release 22 [5].

miRNAs have emerged as important players in a wide range of biological processes in plants [6]. They are involved in the resistance to environmental stresses by regulating the expression of stress-responsive genes [7–12]. Over the past decade, high-throughput technologies, e.g., microarray hybridization and RNA-seq, have accelerated the transcriptomic studies and shed light into the miRNA repertoires under drought stress conditions [13]. Drought-specific miRNAs have been investigated in different crops, including wheat [14–16], barley [17], maize [18], and Brachypodium [19], and dehydration-responsive miRNA expression patterns were detected. Several drought-related miRNA families have been identified in rice [20–24].

Drought-responsive miRNAs regulate their target mRNAs, most of which are functional proteins involved in drought stress [25]. Therefore, the function of a miRNA is often inferred via its target genes [26]. However, the identification of miRNA target genes represents a combinatorial challenge. Early miRNA target prediction methods rely on sequence complementarities, conservation, free energy, or target accessibility. However, these tools usually predict hundreds of target genes for each miRNA. Therefore, the number of false-positives is large, making it difficult to identify targets of real interest. More recent expression-based methods incorporate mRNA and miRNA expression data

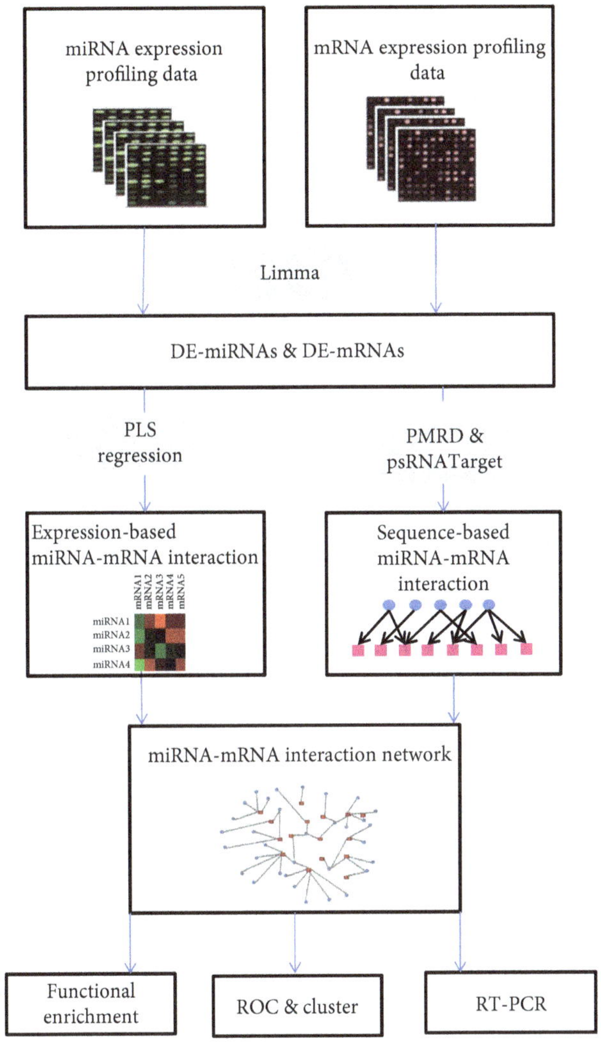

FIGURE 1: Flow chart of the general procedure of this study.

to predict functional miRNA-mRNA interactions. Although useful, most of the algorithms utilize the pairwise negative correlation coefficient to measure the interaction. Focusing exclusively on linear inverse correlations, they may undermine the many to many miRNA-mRNA relationships.

The aim of the present study was to detect accurate miRNA-mRNA associations in rice under drought stress. We proposed an integrative approach that combines PLS regression and sequence-based target prediction to identify potential miRNA and their targets that are important in drought response. We constructed a miRNA-mRNA regulatory network, and identified 4 miRNA-mRNA regulatory modules in the network that were associated with drought response. Furthermore, we evaluated the expression level and anticorrelation of the putative miRNA and targets. Functional analysis uncovered some drought-responsive canonical pathways that are specifically responsive to drought stress. This report demonstrates the utility of our approach to obtain new insights into posttranscriptional gene regulation at the miRNA level in drought-stressed rice. The schematic diagram of the procedures is given in Figure 1.

2. Materials and Methods

2.1. Plant Materials and Stress Treatment. Seeds of the rice cultivar Nipponbare were sterilized with 3% (*v*/*v*) NaClO, rinsed with distilled water, and pregerminated in moist filter paper for 48 h at 28°C. Thereafter, the seeds were transferred to sandy loam soil and grown under controlled conditions (16/8 h photoperiod, day/night air temperature of 25°C/20°C, 60% RH, and light intensity of 250 μmol m^{-2} s^{-1}). Four-week-old plants were dehydrated by 20% polyethylene glycol 6000 for 48 h to simulate moderate drought stress of −0.5 MPa osmotic potential [27]. The control group was maintained under well irrigation conditions. Fifteen plants were included in each treatment and control group. After the treatment, the leaf disks (6 mmϕ) were excised from the uppermost rice leaf blades from both stress and control plants and stored in liquid nitrogen. Total RNA was isolated from 400 mg leave tissues by using the mirVana™ miRNA Isolation Kit according to the manufacturer's instructions. Three biological replicates were prepared for microarray analysis.

2.2. Microarray Analysis. miRNA microarray (LC Sciences, Houston, USA) contains 4 triplicates of 509 unique miRNA probes of 29 species in Sanger miRBase Release 19.0. *Arabidopsis thaliana* 5S rRNA and PUC2-20B were used as positive controls. Approximately 5 μg of the total RNA was size-fractionated, and the isolated small RNAs (~300 nt) were 3′-polyadenylated using polyA polymerase (Takara, Dalian, China) and subsequently ligated to Cy5 for fluorescent staining. Hybridization was performed at 34°C for 20 h. The mRNA expression was analyzed on Agilent-015241 Rice Gene Expression 4x44K Microarray (Feature Number version). The microarray raw data were submitted to the NCBI/GEO database and can be found under the accession number GSE99083 (https://www.ncbi.nlm.nih.gov/geo/query/acc.cgi?acc=GSE99083).

2.3. Statistical Analysis. Data was processed in R programming environment v3.3.1. Raw expression files were background subtracted with the Normex method. Normalization was performed using the quantile method. Differentially expressed probe sets were identified using empirical Bayes method using the Limma v3.10. Signals with adjusted P value < 0.05 (Benjamini-Hochberg) were listed as significant.

2.4. miRNA Target Prediction. PMRD (http://bioinformatics.cau.edu.cn/PMRD/) and psRNATarget (http://plantgrn.noble.org/psRNATarget/) were searched for potential target sites. PMRD was searched with default parameters. The maximum expectation of psRNATarget was set as 2.5 to minimize the number of nonauthentic targets, and other settings remained default. Only the targets predicted by both databases were chosen to increase the prediction accuracy.

2.5. miRNA-mRNA Interaction Prediction. PLS regression can be used to measure the relationship between a given gene and other genes using their expression profiles [28, 29]. Briefly, consider a dataset of p miRNAs and m mRNAs for which N measurements are recorded. Let Y_i be the expression profile for the ith mRNA, $1 < i < p$. PLS first computes a number $v < N$ of latent factors $t_i^{(l)}$, called components for each mRNA, which represents the lth latent variable out of all orthogonal latent variables ($l << p$), and then a separate regression model is fitted:

$$Y_i = \sum_{l=1}^{v} \beta_{il} t_i^{(l)} + e_i, \quad (1)$$

where e_i is an n-dimensional vector of residuals.

Ordinary least squares calculate as follows:

$$\hat{\beta}_{il} = \frac{t_i^{(l)^T} Y_i}{t_i^{(l)^T} t_i^{(l)}}, \quad 1 \leq l \leq v. \quad (2)$$

The overall association score is then given by

$$\hat{S}_{ij} = \frac{\sum_{l=1}^{v} \beta_{il}}{\hat{\sigma}(x)} \hat{\sigma}(y_i). \quad (3)$$

PLS regression was run with 30 deregulated miRNAs and 329 mRNAs to calculate the association scores for each gene pair and recorded in a 329×30 matrix.

2.6. Significance Evaluation. Repeated bootstrapping (1000 times) was performed to assess the statistical significance of the miRNA-mRNA relationship (b_{ij}). The interactions with P value < 0.01 and FDR < 0.05 were regarded as significant.

2.7. Network Construction. miRNA-mRNA interaction network was built and visualized with Cytoscape v3.4.0.

2.8. Functional Enrichment Analysis. The targets of DE-miRNAs under drought were mapped to GO and KEGG for functional enrichment using the DAVID Bioinformatics tool [30]. Terms with Benjamini-corrected P value < 0.05 were listed as significant.

2.9. Expression Validation. We employed RT-qPCR, hierarchical clustering, and ROC analysis to validate the expression levels of drought-responsive rice miRNAs.

PrimeScript one-step miRNA cDNA Synthesis Kit (Takara) was used for reverse transcribing mature miRNAs according to the manufacturer's instructions. Universal primers were provided with the kit, and the miRNA-specific primers are given in Supplementary Table S1. SYBRH Premix Ex TaqII (Takara, Japan) was used to quantify mature miRNAs. The PCR amplification parameter was set as 94°C for 2 min, then 40 cycles of 95°C for 10 s, 60°C for 30 s, and 72°C for 1 min. The melting curves were adjusted as 95°C for 1 s and 40°C for 2 min. The 2-DDCT method was used to analyze the relative expression changes. The rice UBC and U6 RNAs were used as internal standards. ef1a was used as an internal control.

Hierarchical clustering was performed with Cluster 3.0. Heat map was visualized with Java TreeView 1.1.6. ROC (receiver operating characteristic) analysis was performed by SPSS 16.0.

2.10. RNA Ligase-Mediated 5′ RACE. Total RNA was separately extracted using TRIzol reagent and subjected to RLM-5′ RACE with 5′-Full RACE Kit (Takara, Japan) following the manufacturer's instructions. Briefly, 10 μg of total RNA was ligated to the RNA adapter followed by reverse transcription using the Oligo (dT) primer and superscript reverse transcriptase. PCR was performed with 5′ adaptor primers and 3′ gene-specific primers (Supplementary Table S2). The final PCR product was purified on 2% agarose gel and cloned into a pEASY-T1 Vector, and plasmid DNA was sequenced by Huada Biotech Co.

3. Results and Discussion

Dwindling water resources frequently challenge the quality and quantity of rice and cause high economic losses [14]. miRNAs are central to posttranscriptional gene regulation in a wide range of cellular processes including stress responses [6, 31]. Over the past decade, several drought-related miRNA families in rice have been identified by either microarray hybridization or RNA-seq [20, 21, 32]. Stress-responsive miRNAs regulate their target mRNAs by targeting the gene transcripts most of which are stress-associated functional proteins, e.g., transcription factors or hormones involved under drought stress [25]. Therefore, we hoped to uncover significant miRNA-mRNA relationships that are specifically responsive to drought stress [26].

In this study, a multistep approach combining microarray miRNA and mRNA expression profile and bioinformatics analysis was adopted to identify the drought-responsive miRNA-mRNA regulatory network. First, we simultaneously analyzed miRNA and mRNA expression profiles in the same rice leaf sample and identified 76 differentially expressed miRNAs and 2916 mRNAs in drought-stressed samples. Secondly, 13 deregulated miRNAs and their 58 target mRNAs (69 miRNA-mRNA pairs) were identified by a combination of PLS regression approach and sequence-based target prediction. Their differential expression and regulation were further validated experimentally by RT-qPCR and RLM-5′ RACE.

Most of the existing algorithms for miRNA target prediction are based on the sequence complementarity, and the results vary widely. A rule of thumb to address this problem is to intersect the results of multiple prediction algorithms to reduce the false-positive. However, this procedure improves accuracy at the cost of many bona fide targets and therefore leads to poor sensitivity. The research pipeline in this study combines the sequence features and the expression of miRNAs and mRNAs simultaneously. Thus, our method precludes artifacts and improves prediction accuracy, thereby facilitating a more targeted experimental validation.

3.1. Identification of Drought-Responsive miRNAs and mRNAs in Rice. Microarray-based miRNA and mRNA expression profiles for *Oryza sativa* under/without drought stress were simultaneously investigated. As a result, 76

Table 1: Drought-responsive miRNAs and their target genes predicted by PLS.

miRNA	Family	Response	Target no.	Targets predicted by PLS	Literature
osa-miR1430	MIR169	Down	3	MYB30, NFYA, UGE-4	
osa-miR156b-3p	MIR156	Down	11	49D11.4, LCAT3, NRAMP5, RPSM, SPL15, SPL16, SPL2, SPL3, SPL4, SPL9, NCED1	[21, 24]
osa-miR159a.2	MIR159	Down	3	49D11.4, MYB4, MYB39	[21, 24]
osa-miR164c	MIR164	Down	4	CML22, CUC, NAC22, TCP8	[35, 36]
osa-miR169d	MIR169	Down	6	CBF, DREB, HAP2D, NAAT2, NFYA, UGE-4	[32]
osa-miR172d-5p	MIR172	Down	7	AP2D2, AP2D23, AP2D4, ARF9, bZIP, GT43A, JMJ705	[21]
osa-miR319a-3p	MIR159	Down	5	49D11.4, MYB4, MYB59, TCP14, TCP7	[21]
osa-miR395a	MIR395	Down	4	ATPS3, P0651G05.2, RRF, URH2	[21]
osa-miR396g	MIR396	Down	9	6PGL 1, P0651G05.2, PMP22, RGA2, RHL1, RPM1, RRM_SF, TPT	[21]
osa-miR444b.1	MIR444	Down	3	MADS2, MADS27, UEL-1	[36]
osa-miR528-3p	MIR528	Up	5	AO, DCL1, DWARF-3, IAR, SODCC.1	[24]
osa-miR529a	MIR529	Down	4	ACOX4, NUS1, SPL2, SPL9	[21]
osa-miR812q	MIR812	Up	3	CIPK10, DWARF-3, ERF	

miRNAs (FDR < 0.05) were identified to be differentially expressed between drought-stressed and control leaves, corresponding to 38 rice miRNA families. The list of 76 differentially expressed miRNAs is provided in Supplementary Table S3. Of these, 9 were upregulated and 67 were downregulated. Using the same criteria, 2916 genes were identified to be differentially expressed.

3.2. PLS Reveals Significant miRNA-mRNA Associations. We first filtered possible miRNA target genes using miRNA target databases PMRD [33] and psRNATarget [34] and obtained 346 miRNA-mRNA pairs among 30 DE-miRNAs and 329 DE-mRNAs.

Next, we used the PLS regression method to remove false-positive discoveries among the 346 miRNA-mRNA pairs in order to find the most likely miRNA-mRNA associations.

The 30 DE-miRNAs were input as independent variables while the 329 DE-mRNAs as dependent variables for PLS regression. The association score b_{ij} was calculated for each miRNA-mRNA pair, and repeated bootstrapping (1000 times) was subsequently performed to identify significant associations. Repeated bootstrapping (1000 times) was performed to assess the statistical significance of the miRNA-mRNA relationship. The interactions with $b_{ij} > 0.8$, P value < 0.01, and FDR < 0.05 were regarded as significant.

This filtering step reduced considerably the number of potential candidate interactions, resulting in 69 significant miRNA-mRNA pairs among 13 DE-miRNAs and 58 DE-mRNAs, as listed in Table 1. Among them, 11 miRNAs were significantly downregulated and 2 miRNAs were upregulated. The maximum number of targets was 11 for osa-miR156b-3p.

Among the 13 differentially expressed miRNAs, some are known to respond to drought, including miR156b-3p [21], miR159a.2 [21], miR164c [35], miR169d [32], miR172d-5p [21], miR319a-3p [21], miR395a [21], miR396g [21], miR444b.1 [36], miR528-3p [24], and miR529a [21]. Two novel miRNAs (miR812q and miR1430) are reported for the first time to be involved in drought stress in rice, and their role remains unknown.

The regulatory edges in the miRNA-mRNA network well reflect the existing evidence described in the literature, indicating a high predictive value of our integrative method in identifying genuine miRNA-mRNA interactions.

Prominent drought-responsive miRNA targets were found to encode well-known stress-associated transcription factors or kinases. For example, MADS-box transcription factors targeted by miR444b.1. participate in diverse stress conditions, such as cold stress [37]. SBP-box transcription factor family members targeted by miR156b-3p and miR529a are involved in vegetative phase change induced by various environmental stimuli [38].

ATP-sulfurylase targeted by miR395a is involved in sulfate assimilation and allocation. miR169d targets NFYA regulates the expression of drought-responsive genes, e.g., GT and POD, and is crucial for plant adaptation to environmental stress [39]. miR319a targets two TCP family members (TCP7, 14) that inhibit cell proliferation as well as MYB4 and MYB59 mRNAs. miR164c targets NAC transcription factors, e.g., NAC22 and CUC, which determines the cellular levels of free auxin [40]. miR812q targets CIPK10, which interacts with CBL to trigger the CBL-CIPK-signaling pathway-modulating environmental stress tolerance [41].

3.3. Analysis of Drought-Responsive miRNA Coregulatory Network. Using the statistically significant PLS association score between the miRNA and mRNA, we built interaction networks out of the significant DE-miRNAs and their DE-mRNA targets. Edges are formed between miRNA-mRNA pairs with $b_{ij} > 0.8$. The resulting miRNA-mRNA regulatory network is essentially a bipartite graph including two disjoint sets of vertices, miRNAs and mRNAs. The network consists of 69 regulations between 13 miRNAs and 58 mRNAs.

Then we analyzed the out-degree and in-degree distribution of the network (Figures 2(b) and 2(c)). On average, each miRNA has 3–5 outgoing edges and is loosely connected with

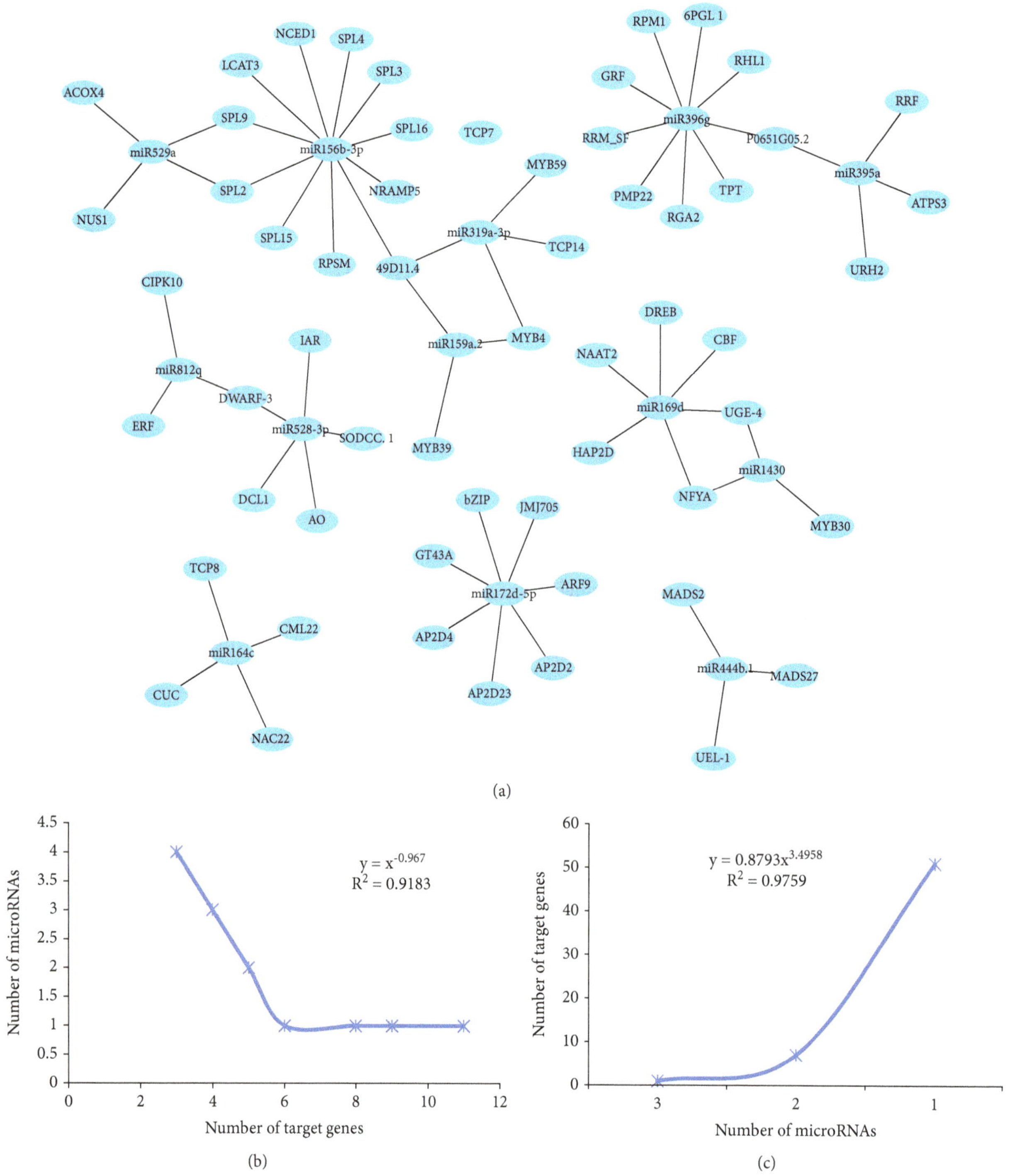

FIGURE 2: The layout of the miRNA-mRNA network and its structural features. (a) miRNA-mRNA network constructed by PLS. (b) Out-degree distribution of the miRNA-mRNA network. (c) In-degree distribution of the miRNA-mRNA network.

each other, with a few exceptions such as the highly conserved miR156b-3p with 11 target genes, most of which belong to the SPL family, and the miR172d-5p with 3 target genes belonging to APD family. The out-degree distribution fits the power law with a slope of −0.8399 and $R^2 = 0.9618$. The in-degree distribution reveals an exponential distribution with an exponent of −0.967 and $R^2 = 0.9183$. It is obvious that as with many biological networks, the miRNA-mRNA network is not random but features a core set of organizing principles in structure.

3.4. PLS Reveals miRNA-mRNA Coregulatory Modules. The PLS regression method successfully models the complex interrelationship of multiple miRNAs regulating one mRNA and identified miRNA-mRNA coregulatory modules, i.e., groups of miRNAs that regulate groups of mRNAs in

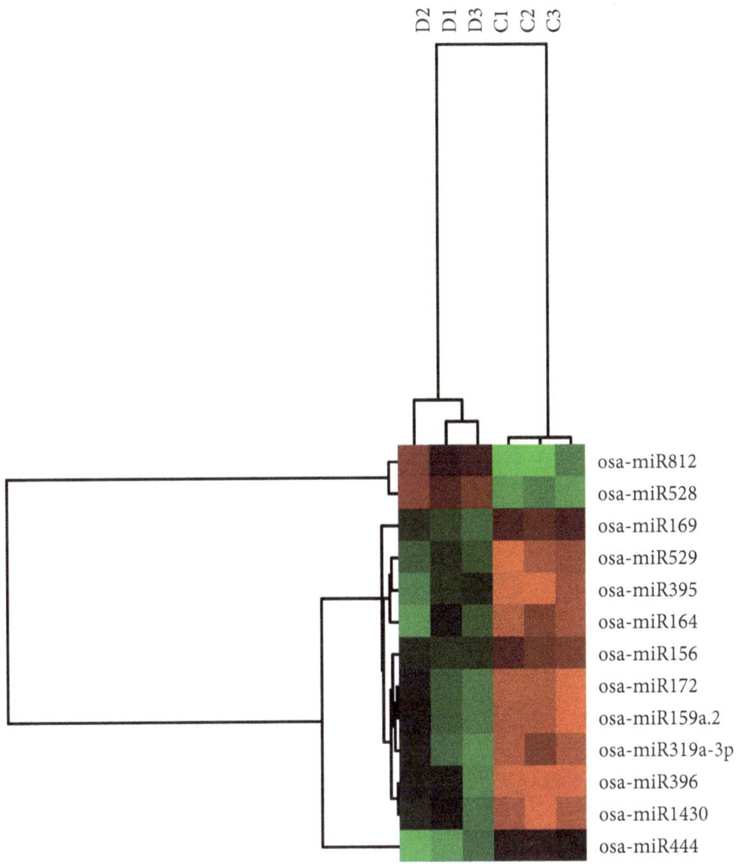

FIGURE 3: Heat map of the 13 DE-miRNAs in drought-stressed and control rice leaves.

concert. As shown in Figure 2(a), miRNAs regulate at least 3 targets, and 13.8% of mRNAs are coregulated by multiple miRNAs, indicating target multiplicity and miRNA cooperation in the drought response of rice.

One mRNA, 49D11.4 is common target of 3 significantly downregulated miRNAs, miR156b-3p, miR159a.2, and miR319a-3p. The coregulation of miR159 and miR319 was expected, since mature miR159 and miR319 have sequence similarity at 17 out of 21 nucleotides and belong to the same family [42]. The miRNAs within the same family are often derived from recent duplication events and functionally related. Likewise, miR529a showed 14 nucleotide similarities with miR156a-3p and is a shared common target of the SPL family (SPL2, 9) [43]. Moreover, miR1430 is similar to miR169d in sequence and shares the target of NFYA and UGE4 [44].

On the other hand, miRNAs within the same families also have exclusive targets. As shown in Figure 2(a), miR159 and miR319 share MYB targets, but miR159 fails to guide cleavage of TCPs as miR319. This case argues that a single nucleotide difference is sufficient to dictate miRNA target binding.

Interestingly, coregulation occurs also between phylogenetically unrelated miRNAs, i.e., miRNA belonging to different families, e.g., DWARF-3 by miR528-3p and miR8121 and P0651G05.2 by miR396g and miR395a, representing a stress-specific functional redundancy.

TABLE 2: ROC analysis of drought-responsive miRNAs.

miRNA	AUC	Sensitivity	Specificity	Accuracy
osa-miR1430	0.930	0.900	0.900	0.800
osa-miR156b-3p	0.978	0.900	1.000	0.900
osa-miR159a.2	0.908	0.800	1.000	0.800
osa-miR164c	0.902	0.900	0.800	0.700
osa-miR169d	0.945	0.850	0.950	0.800
osa-miR172d-5p	0.930	0.950	0.900	0.850
osa-miR319a-3p	0.950	0.950	0.950	0.900
osa-miR395a	0.957	0.850	1.000	0.850
osa-miR396g	0.913	0.800	0.950	0.700
osa-miR444b.1	0.870	0.900	0.850	0.750
osa-miR528-3p	0.890	0.800	0.900	0.700
osa-miR529a	0.931	0.850	0.950	0.800
osa-miR812q	0.986	0.900	0.950	0.850

Coordinate miRNA regulation has previously been demonstrated in Drosophila and mammals. let-7/lin-4 is the first miRNA pair to be experimentally verified in Drosophila [45]. miR-375/miR-124/let-7b were found to jointly regulate Mtpn in mammals [46]. In this study, we provide evidence for miRNA synergism in drought-stressed rice. It is proposed that synergistic gene regulation is less dependent on

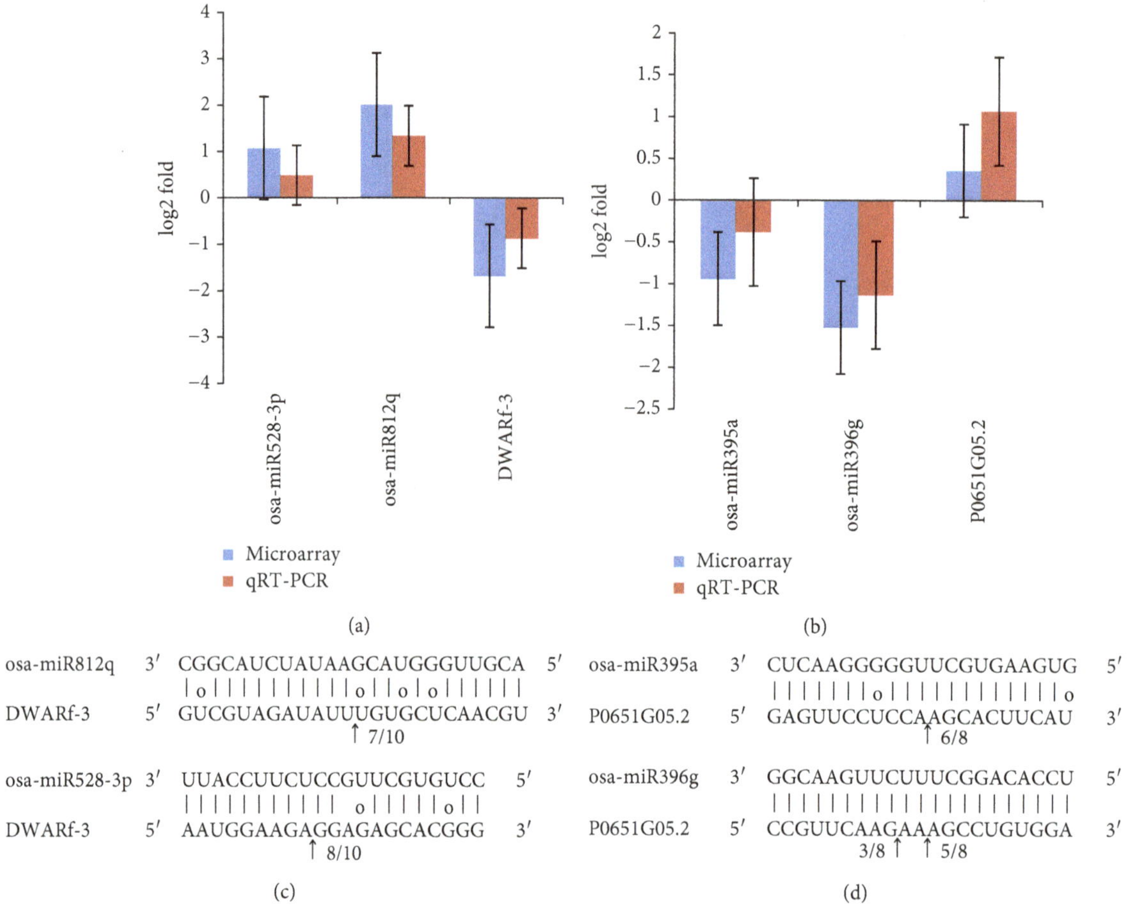

FIGURE 4: (a) Expression level of coregularory miR528-3p/miR812q and their target DWARF-3 by microarray and RT-qPCR. (b) Expression level of coregularory miR395a/miR396g and their target P0651G05.2 by microarray and RT-qPCR. The expression level is expressed as the mean fold changes of 3 replicates. Error bars depict the standard error ($n = 3$). (c) miR528-3p and miR812q cleavage sites on DWARF-3 identified by RLM-5′ RACE. (d) miR395a and miR396g cleavage sites on P0651G05.2 identified by RLM-5′ RACE. Arrows represent the 5′ termini of mRNA fragments, and the numbers beside them denote the cleavage frequency of cloned PCR products at the exact miRNA cleavage sites. The vertical dashes indicate matched RNA base pairs and circles represent GU mismatch whereas no dashes represent other types of mismatch.

miRNA abundance so that limited miRNAs are able to regulate more genes. Moreover, through synergism, miRNAs may exert reinforced repression effects and therefore enhance their regulatory efficacy. The identification of coregulatory miRNA modules would facilitate researchers to make more-informative choices of multiple mutant combinations to circumvent redundancy of miRNA regulation under drought stress.

3.5. Hierarchical Clustering and ROC Analysis. Hierarchical clustering and ROC were performed to validate the differential miRNA expression profiles. A heat map is illustrated in Figure 3.

The AUC (areas under the ROC curve), sensitivity, specificity, and accuracy for the 13 candidate miRNAs are listed in Table 2, and the ROC curves are provided as Supplementary Figure S1. ROC is a way of evaluating the performance of binary classifiers in classifying two classes, e.g., drought vs. control. A ROC curve is a plot of the true positive rates (sensitivity) vs. false-positive rates (1 – specificity). ROC curves can be used to evaluate the quality of classifiers for different discrimination thresholds. The closer the ROC curve to the upper left corner, the greater the AUC and the better the discrimination accuracy. The AUC for the 13 miRNAs are 0.870–0.986, with an average accuracy of 80.76%. The results showed that the candidate DE-miRNAs are able to differentiate drought-stressed samples from the controls. ROC analysis demonstrated that the miRNAs identified in this study are both sensitive and specific in sample discrimination, indicating that they are authentic miRNAs under drought stress.

3.6. RT-qPCR Validated the Expression Trends of Drought-Responsive miRNAs and Coregulated Target Genes. We conducted RT-qPCR analysis to validate the expression level of 13 drought-responsive miRNAs and their anticorrelation with coregulated target genes. The results were consistent with the microarray data (Supplementary Figure S2). The relative expression trends of all of the 13 miRNAs were confirmed as differentially expressed under drought stress and were negatively correlated with their target genes.

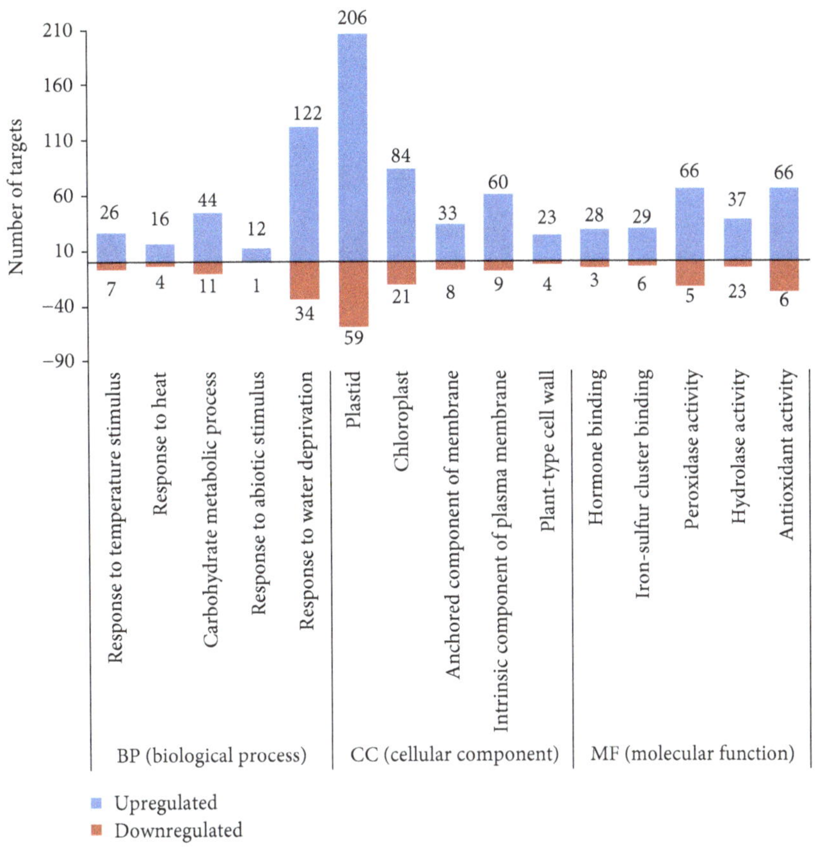

FIGURE 5: Enriched GO terms of drought-responsive miRNA targets.

3.7. RLM-5′ RACE Confirmed Direct Targets of Coregulatory miRNAs. To confirm the direct targets of coregulatory miRNAs, we performed RLM-5′ RACE to determine the exact miRNA cleavage sites on target genes. Two mRNAs (DWARF-3 and P0651G05.2) that are coregulated by miRNAs which do not belong to the same family were chosen for verification.

After sequencing the RLM-5′ RACE PCR products, we found that both DWARF-3 and P0651G05.2 had specific cleavage sites in the complementary miRNA sequences (Figures 4(c) and 4(d)). DWARF-3 was cleaved by miR528-3p and miR812q at positions 776 and 1150, respectively. Both cleavage sites are located at the positions 12–13 of miRNAs from the 5′ end. P0651G05.2 can be cleaved at position 486 by miR395a. The cleavage site was mapped to the paired miR395a sequence between the nucleotides 10 and 11. Moreover, P0651G05.2 could be simultaneously cleaved at the two sites by miR396g. The cleavage sites are located between nucleotides 10 and 11 and between 12 and 13, indicating that the cleavage sites of miR396g may be diverse.

The results of RLM-5′ RACE generally confirmed the coregulation of miR528-3p and miR812q on DWARF-3, as well as the coregulation of miR395a and miR396g on P0651G05.2, indicating a high reliability of the PLS analysis.

miR812q and miR528-3p were found to target auxin receptor F-box protein DWARF-3, which is essential for auxin-mediated signaling and plant development [47]. When auxin binds to DWARF-3, it promotes the ubiquitination and degradation of auxin transcriptional repressors, thereby activating auxin-responsive genes [48]. As a coreceptor for phytohormone, e.g., strigolactones and polar auxin, DWARF-3 has a pleiotropic effect on stalk height, leaf angle, and root growth in plants [49, 50]. Under drought stress conditions, upregulated miR812q and miR528-3p would decrease the expression of DWARF-3, and the miRNA-mediated DWARF-3 suppression would lead to attenuation of auxin signaling and inhibition of lateral root growth. We assume that the miR528-3p/miR812q-DWARF-3 coregulatory module might play a central role in development-related gene transcription and plant adaptation to drought stress.

In this study, we only validated two targets, DWARF-3 and P0651G05.2 by RACE experiment, because they are coregulatory hub nodes in the miRNA target network. Degradome sequencing is also a primary and effective tool to prove the miRNA targets. It can directly detect cleaved miRNA targets without predictions or overexpression. Degradome sequencing has been performed in rice by several studies and revealed endogenous small RNA targets under specific tissues or treatments [51, 52]. In the future, a large-scale target validation can be performed based on degradome sequencing data.

3.8. Functional Annotation of Drought-Stressed miRNAs. MiRNAs and their predicted target mRNAs were mapped to Gene Ontology and KEGG pathways to explore their biological significance.

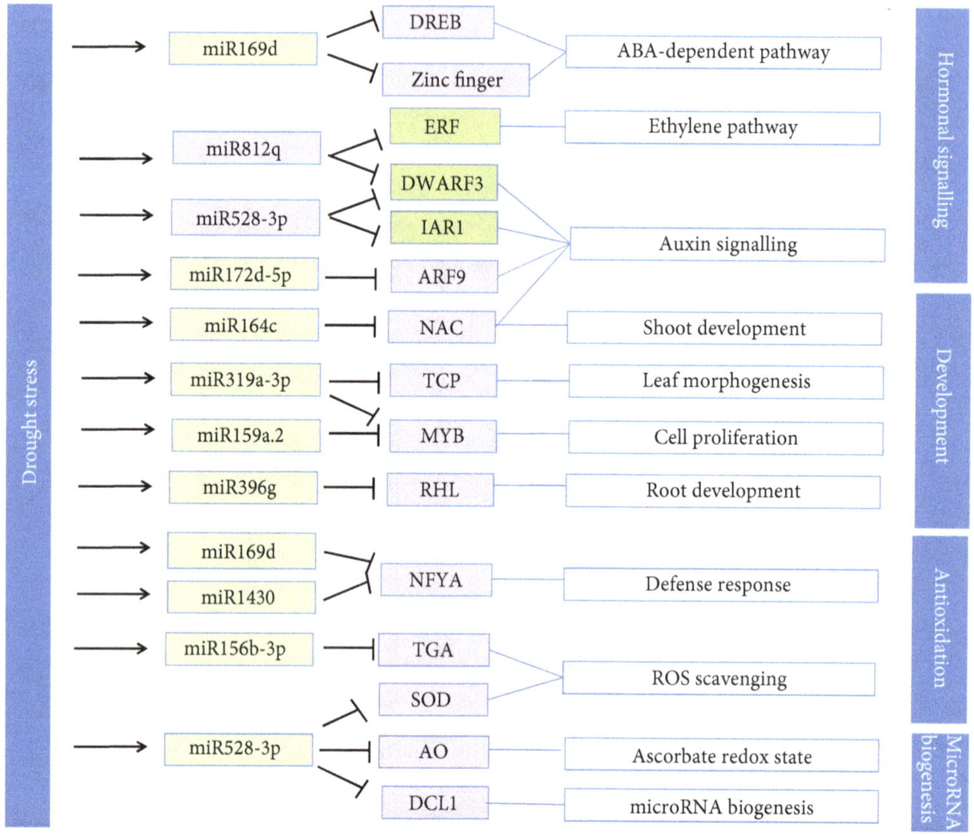

Figure 6: Schematic diagram of miRNA-regulated metabolic processes in rice under drought stress. Under drought stress, rice responds with the following physiological changes: (1) activating hormonal signaling pathways, (2) immobilizing fatty acid and downregulating starch and sucrose metabolism, (3) producing antioxidant compounds for detoxification, and (4) inhibiting miRNA biogenesis.

According to GO enrichment, the target of drought-responsive miRNAs belonged to five important biological processes (BP): response to temperature stimulus, response to heat, response to abiotic stimulus, carbohydrate metabolic process, and response to water deprivation. Enriched molecular function (MF) terms included hormone binding, iron-sulfur cluster binding, hydrolase activity, peroxidase activity, and antioxidant activity. Plastid, chloroplast, anchored component of membrane, intrinsic component of plasma membrane, and plant-type cell wall were most significant among the cellular components (CC). A detailed list of the enriched GO terms is given in Figure 5.

KEGG pathway analysis unraveled significant activation of hormone signaling, lipid and carbohydrate metabolism, and antioxidant defense under drought. The full list of significant pathways (adjusted P value < 0.05) and the number of differentially expressed genes enriched in each pathway are represented in Supplementary Table S4.

3.8.1. miRNA-Regulated Phytohormone Signaling. Plant hormone signal transduction represented the most significantly enriched pathway with 52 genes differentially expressed under drought. A number of miRNA-regulated phytohormones are known to participate in plant responses to abiotic stress, e.g., auxin, salicylic acid, ethylene, GAs, jasmonate acid, and abscisic acid.

Here we found that the expression of key genes in auxin synthesis and the ABA signaling pathway was induced under drought stress. For example, IAR1 transcripts are downregulated by increased miR528 under drought. IAR1 regulates auxin homeostasis by hydrolyzing an inactive form of auxin to release bioactive auxin [53]. The decreased IAR1 expression and hence decreased auxin level leads to the inhibition of adventitious rooting under drought. Several genes in the ABA signaling pathway were also activated under drought. ABA is a plant stress hormone produced under environmental stresses [54]. It is an apocarotenoid derivative produced via oxidative cleavage of carotenoids. Also, we found drought stress increases carotenoid synthesis, which suggests that carotenoid synthesis and the ABA signaling were cooperatively involved in the drought stress response.

3.8.2. Lipid and Carbohydrate Metabolism. Pathway analysis unraveled significant reduction of starch-sucrose metabolism. This includes a variety of sucrose synthases, amylase, and starch branching enzymes. Decreased starch and sucrose concentration is an adaptive feature under drought stress [55], since it allows plants to divert assimilates and energy intended for growth into protective responses for survival [54].

In the context of fatty acid metabolism, α-linolenic acid metabolism was significantly induced under drought.

α-linolenic acid is an unsaturated fatty acid in membrane lipids [56]. It enhances plant tolerance by maintaining membrane fluidity during stress [57].

3.8.3. miRNA-Mediated Antioxidant Defense. The immediate response of plants to stress is accumulation of reactive oxygen species (ROS). To remove the excessive ROS, the cell produces an antioxidative enzyme system, e.g., POD, AO, and SOD that serve as ROS scavengers.

In this study, we found that the production of ROS scavengers is regulated by miRNAs, as illustrated in Figure 6. For example, miR528 targets genes coding ascorbate oxidase, a central component of the hydrogen peroxide scavenging network. Ascorbate metabolism reduces oxidative damage by maintaining the ascorbate redox state in the form of apoplastic dehydroascorbate [58]. In addition, miR528 targets SOD transcripts, which catalyzes the dismutation of O_2^- and alleviates the damage from ROS. Drought also induces key enzyme genes involved in biosynthesis of several secondary metabolites, such as isoflavonoids and carotenoids which mitigate diverse stress in plants.

A simple model was constructed according to the function of miRNA and target genes under drought stress (Figure 6).

4. Conclusion

In conclusion, this study employed a multiple regression model for integrating paired miRNA and mRNA expression profiles. The integrated approach uncovered high-confidence reciprocally expressed miRNA target pairs, supporting the existence of a coregulatory network triggered by drought stress. Further validation of these hub miRNAs as coregulatory nodes may facilitate the development of novel molecular markers for breeding of rice cultivars with enhanced drought tolerance. In-depth analysis of coregulatory networks will reveal the regulatory role of miRNAs in a broader range of a posttranscriptional network under drought stress.

Abbreviations

ABA:	Abscisic acid
AUC:	Areas under the ROC curve
CIPK:	Calcineurin B-like protein interacting protein kinase
ERF:	Ethylene-responsive element-binding factor
GO:	Gene Ontology
GT:	Glutathione transferase
KEGG:	Kyoto Encyclopedia of Genes and Genomes
NFYA:	Alpha subunit of CCAAT binding NFY transcription factors
POD:	Peroxidase
RH:	Relative humidity
RLM-5′ RACE:	RNA ligase-mediated 5′ rapid amplification of cDNA ends
ROC:	Receiver operating characteristic
SOD:	Superoxide dismutase
SPL:	Squamosa promoter binding protein-like.

Conflicts of Interest

The authors declare that they have no conflicts of interest.

Acknowledgments

This work was supported by the National Natural Science Foundation of China [grant numbers 31770903, 31400712, and 21676173] and Qinglan Project of Jiangsu Province.

Supplementary Materials

Supplementary 1. Table S1: oligonucleotide primers used in RT-qPCR.

Supplementary 2. Table S2: the specific primer sequences for RLM-5′ RACE of target genes.

Supplementary 3. Table S3: differentially expressed miRNAs under drought stress.

Supplementary 4. Table S4: enriched KEGG pathways of drought-responsive miRNA targets.

Supplementary 5. Figure S1: ROC curves of 13 drought-responsive miRNAs.

Supplementary 6. Figure S2: validation of drought-responsive miRNAs and coregulated target genes by RT-qPCR. The expression level is expressed as the mean fold changes of 3 replicates. Error bars depict the standard error ($n = 3$).

References

[1] U. Ala, F. A. Karreth, C. Bosia et al., "Integrated transcriptional and competitive endogenous RNA networks are cross-regulated in permissive molecular environments," *Proceedings of the National Academy of Sciences of the United States of America*, vol. 110, no. 18, pp. 7154–7159, 2013.

[2] H. R. Lafitte, G. Yongsheng, S. Yan, and Z. K. Li, "Whole plant responses, key processes, and adaptation to drought stress: the case of rice," *Journal of Experimental Botany*, vol. 58, no. 2, pp. 169–175, 2007.

[3] B. Barnabas, K. Jager, and A. Feher, "The effect of drought and heat stress on reproductive processes in cereals," *Plant, Cell & Environment*, vol. 31, no. 1, pp. 11–38, 2007.

[4] T. Unver, D. M. Namuth-Covert, and H. Budak, "Review of current methodological approaches for characterizing microRNAs in plants," *International Journal of Plant Genomics*, vol. 2009, Article ID 262463, 11 pages, 2009.

[5] A. Kozomara and S. Griffiths-Jones, "miRBase: annotating high confidence microRNAs using deep sequencing data," *Nucleic Acids Research*, vol. 42, no. D1, pp. D68–D73, 2013.

[6] G. Akdogan, E. D. Tufekci, S. Uranbey, and T. Unver, "miRNA-based drought regulation in wheat," *Functional & Integrative Genomics*, vol. 16, no. 3, pp. 221–233, 2016.

[7] A. A. Covarrubias and J. L. Reyes, "Post-transcriptional gene regulation of salinity and drought responses by plant micro-RNAs," *Plant, Cell & Environment*, vol. 33, no. 4, pp. 481–489, 2010.

[8] P. Shuai, D. Liang, Z. Zhang, W. Yin, and X. Xia, "Identification of drought-responsive and novel Populus trichocarpa microRNAs by high-throughput sequencing and their targets using degradome analysis," *BMC Genomics*, vol. 14, no. 1, p. 233, 2013.

[9] Q. Liu, Y. C. Zhang, C. Y. Wang et al., "Expression analysis of phytohormone-regulated microRNAs in rice, implying their regulation roles in plant hormone signaling," *FEBS Letters*, vol. 583, no. 4, pp. 723–728, 2009.

[10] Y. Ding, Z. Chen, and C. Zhu, "Microarray-based analysis of cadmium-responsive microRNAs in rice (Oryza sativa)," *Journal of Experimental Botany*, vol. 62, no. 10, pp. 3563–3573, 2011.

[11] M. Hajyzadeh, M. Turktas, K. M. Khawar, and T. Unver, "miR408 overexpression causes increased drought tolerance in chickpea," *Gene*, vol. 555, no. 2, pp. 186–193, 2015.

[12] B. Li, Y. Qin, H. Duan, W. Yin, and X. Xia, "Genome-wide characterization of new and drought stress responsive micro-RNAs in Populus euphratica," *Journal of Experimental Botany*, vol. 62, no. 11, pp. 3765–3779, 2011.

[13] H. B. Cagirici, B. Alptekin, and H. Budak, "RNA sequencing and co-expressed long non-coding RNA in modern and wild wheats," *Scientific Reports*, vol. 7, no. 1, article 10670, 2017.

[14] B. A. Akpinar and H. Budak, "Dissecting miRNAs in wheat D genome progenitor, *Aegilops tauschii*," *Frontiers in Plant Science*, vol. 7, p. 606, 2016.

[15] B. Alptekin, P. Langridge, and H. Budak, "Abiotic stress miR-Nomes in the Triticeae," *Functional & Integrative Genomics*, vol. 17, no. 2-3, pp. 145–170, 2017.

[16] B. A. Akpinar, M. Kantar, and H. Budak, "Root precursors of microRNAs in wild emmer and modern wheats show major differences in response to drought stress," *Functional & Integrative Genomics*, vol. 15, no. 5, pp. 587–598, 2015.

[17] E. M. Fard, B. Bakhshi, R. Keshavarznia, N. Nikpay, M. Shahbazi, and G. H. Salekdeh, "Drought responsive micro-RNAs in two barley cultivars differing in their level of sensitivity to drought stress," *Plant Physiology and Biochemistry*, vol. 118, pp. 121–129, 2017.

[18] J. Aravind, S. Rinku, B. Pooja et al., "Identification, characterization, and functional validation of drought-responsive microRNAs in subtropical maize inbreds," *Frontiers in Plant Science*, vol. 8, p. 941, 2017.

[19] H. Budak and A. Akpinar, "Dehydration stress-responsive miRNA in *Brachypodium distachyon*: evident by genome-wide screening of microRNAs expression," *OMICS: A Journal of Integrative Biology*, vol. 15, no. 11, pp. 791–799, 2011.

[20] R. Sunkar, X. Zhou, Y. Zheng, W. Zhang, and J. K. Zhu, "Identification of novel and candidate miRNAs in rice by high throughput sequencing," *BMC Plant Biology*, vol. 8, no. 1, p. 25, 2008.

[21] L. Zhou, Y. Liu, Z. Liu, D. Kong, M. Duan, and L. Luo, "Genome-wide identification and analysis of drought-responsive microRNAs in Oryza sativa," *Journal of Experimental Botany*, vol. 61, no. 15, pp. 4157–4168, 2010.

[22] S. Balyan, M. Kumar, R. D. Mutum et al., "Identification of miRNA-mediated drought responsive multi-tiered regulatory network in drought tolerant rice, Nagina 22," *Scientific Reports*, vol. 7, no. 1, p. 15446, 2017.

[23] B. H. Cheah, S. Jadhao, M. Vasudevan, R. Wickneswari, and K. Nadarajah, "Identification of functionally important micro-RNAs from rice inflorescence at heading stage of a *qDTY4.1*-QTL bearing near isogenic line under drought conditions," *PloS One*, vol. 12, no. 10, article e0186382, 2017.

[24] B. Bakhshi, E. Mohseni Fard, N. Nikpay et al., "MicroRNA signatures of drought signaling in rice root," *PloS One*, vol. 11, no. 6, article e0156814, 2016.

[25] L. Giusti, E. Mica, E. Bertolini et al., "microRNAs differentially modulated in response to heat and drought stress in durum wheat cultivars with contrasting water use efficiency," *Functional & Integrative Genomics*, vol. 17, no. 2-3, pp. 293–309, 2017.

[26] L. Zhang, H. Xuan, Y. Zuo et al., "Topological characteristics of target genes regulated by abiotic-stress-responsible miRNAs in a rice interactome network," *Functional & Integrative Genomics*, vol. 16, no. 3, pp. 243–251, 2016.

[27] B. E. Michel and M. R. Kaufmann, "The osmotic potential of polyethylene glycol 6000," *Plant Physiology*, vol. 51, no. 5, pp. 914–916, 1973.

[28] V. Pihur, S. Datta, and S. Datta, "Reconstruction of genetic association networks from microarray data: a partial least squares approach," *Bioinformatics*, vol. 24, no. 4, pp. 561–568, 2008.

[29] X. Li, R. Gill, N. G. F. Cooper, J. K. Yoo, and S. Datta, "Modeling microRNA-mRNA interactions using PLS regression in human colon cancer," *BMC Medical Genomics*, vol. 4, no. 1, p. 44, 2011.

[30] D. W. Huang, B. T. Sherman, and R. A. Lempicki, "Systematic and integrative analysis of large gene lists using DAVID bioinformatics resources," *Nature Protocols*, vol. 4, no. 1, pp. 44–57, 2009.

[31] H. Budak, M. Kantar, R. Bulut, and B. A. Akpinar, "Stress responsive miRNAs and isomiRs in cereals," *Plant Science*, vol. 235, pp. 1–13, 2015.

[32] B. Zhao, R. Liang, L. Ge et al., "Identification of drought-induced microRNAs in rice," *Biochemical and Biophysical Research Communications*, vol. 354, no. 2, pp. 585–590, 2007.

[33] Z. Zhang, J. Yu, D. Li et al., "PMRD: plant microRNA database," *Nucleic Acids Research*, vol. 38, Supplement 1, pp. D806–D813, 2010.

[34] X. Dai and P. X. Zhao, "psRNATarget: a plant small RNA target analysis server," *Nucleic Acids Research*, vol. 39, Supplement 2, pp. W155–W159, 2011.

[35] Y. Fang, K. Xie, and L. Xiong, "Conserved miR164-targeted NAC genes negatively regulate drought resistance in rice," *Journal of Experimental Botany*, vol. 65, no. 8, pp. 2119–2135, 2014.

[36] F. Zhang, X. Luo, Y. Zhou, and J. Xie, "Genome-wide identification of conserved microRNA and their response to drought stress in Dongxiang wild rice (*Oryza rufipogon* Griff.)," *Biotechnology Letters*, vol. 38, no. 4, pp. 711–721, 2016.

[37] R. Lozano, T. Angosto, P. Gomez et al., "Tomato flower abnormalities induced by low temperatures are associated with changes of expression of MADS-box genes," *Plant Physiology*, vol. 117, no. 1, pp. 91–100, 1998.

[38] T. Usami, G. Horiguchi, S. Yano, and H. Tsukaya, "The more and smaller cells mutants of Arabidopsis thaliana identify novel roles for *squamosa promoter binding protein-like* genes

in the control of heteroblasty," *Development*, vol. 136, no. 6, pp. 955–964, 2009.

[39] R. Sunkar, Y. F. Li, and G. Jagadeeswaran, "Functions of microRNAs in plant stress responses," *Trends in Plant Science*, vol. 17, no. 4, pp. 196–203, 2012.

[40] A. Hasson, A. Plessis, T. Blein et al., "Evolution and diverse roles of the *cup-shaped cotyledon* genes in *Arabidopsis* leaf development," *The Plant Cell*, vol. 23, no. 1, pp. 54–68, 2011.

[41] Y. Xiang, Y. Huang, and L. Xiong, "Characterization of stress-responsive *CIPK* genes in rice for stress tolerance improvement," *Plant Physiology*, vol. 144, no. 3, pp. 1416–1428, 2007.

[42] S. T. Wang, X. L. Sun, Y. Hoshino et al., "*MicroRNA319* positively regulates cold tolerance by targeting *OsPCF6* and *OsTCP21* in rice (*Oryza sativa* L.)," *PloS One*, vol. 9, no. 3, article e91357, 2014.

[43] E. G. O. Morea, E. M. da Silva, G. F. F. e Silva et al., "Functional and evolutionary analyses of the miR156 and miR529 families in land plants," *BMC Plant Biology*, vol. 16, no. 1, p. 40, 2016.

[44] B. Zhao, L. Ge, R. Liang et al., "Members of miR-169 family are induced by high salinity and transiently inhibit the NF-YA transcription factor," *BMC Molecular Biology*, vol. 10, no. 1, p. 29, 2009.

[45] A. J. Enright, B. John, U. Gaul, T. Tuschl, C. Sander, and D. S. Marks, "MicroRNA targets in *Drosophila*," *Genome Biology*, vol. 5, no. 1, p. R1, 2003.

[46] A. Krek, D. Grün, M. N. Poy et al., "Combinatorial microRNA target predictions," *Nature Genetics*, vol. 37, no. 5, pp. 495–500, 2005.

[47] C. Pan, L. Ye, Y. Zheng et al., "Identification and expression profiling of microRNAs involved in the stigma exsertion under high-temperature stress in tomato," *BMC Genomics*, vol. 18, no. 1, p. 843, 2017.

[48] H. Bian, Y. Xie, F. Guo et al., "Distinctive expression patterns and roles of the miRNA393/TIR1 homolog module in regulating flag leaf inclination and primary and crown root growth in rice (*Oryza sativa*)," *The New Phytologist*, vol. 196, no. 1, pp. 149–161, 2012.

[49] L. H. Zhao, X. E. Zhou, W. Yi et al., "Destabilization of strigolactone receptor DWARF14 by binding of ligand and E3-ligase signaling effector DWARF3," *Cell Research*, vol. 25, no. 11, pp. 1219–1236, 2015.

[50] H. Chen, Z. Li, and L. Xiong, "A plant microRNA regulates the adaptation of roots to drought stress," *FEBS Letters*, vol. 586, no. 12, pp. 1742–1747, 2012.

[51] S. Dong, J. Zhang, D. Sun et al., "Identification of *Magnaporthe oryzae*-elicited rice novel miRNAs and their targets by miRNA and degradome sequencing," *European Journal of Plant Pathology*, vol. 151, no. 3, pp. 629–647, 2018.

[52] M. Zhou, L. Gu, P. Li et al., "Degradome sequencing reveals endogenous small RNA targets in rice (*Oryza sativa* L. *ssp. indica*)," *Frontiers in Biology*, vol. 5, no. 1, pp. 67–90, 2010.

[53] T. Li, H. Li, Y. X. Zhang, and J. Y. Liu, "Identification and analysis of seven H_2O_2-responsive miRNAs and 32 new miRNAs in the seedlings of rice (*Oryza sativa* L. ssp. *indica*)," *Nucleic Acids Research*, vol. 39, no. 7, pp. 2821–2833, 2011.

[54] J. K. Zhu, "Salt and drought stress signal transduction in plants," *Annual Review of Plant Biology*, vol. 53, no. 1, pp. 247–273, 2002.

[55] K. V. Padmalatha, G. Dhandapani, M. Kanakachari et al., "Genome-wide transcriptomic analysis of cotton under drought stress reveal significant down-regulation of genes and pathways involved in fibre elongation and up-regulation of defense responsive genes," *Plant Molecular Biology*, vol. 78, no. 3, pp. 223–246, 2012.

[56] S. K. Lenka, A. Katiyar, V. Chinnusamy, and K. C. Bansal, "Comparative analysis of drought-responsive transcriptome in Indica rice genotypes with contrasting drought tolerance," *Plant Biotechnology Journal*, vol. 9, no. 3, pp. 315–327, 2011.

[57] S. J. Prince, T. Joshi, R. N. Mutava et al., "Comparative analysis of the drought-responsive transcriptome in soybean lines contrasting for canopy wilting," *Plant Science*, vol. 240, pp. 65–78, 2015.

[58] L. Wu, Q. Zhang, H. Zhou, F. Ni, X. Wu, and Y. Qi, "Rice microRNA effector complexes and targets," *The Plant Cell*, vol. 21, no. 11, pp. 3421–3435, 2009.

PERMISSIONS

All chapters in this book were first published in IJG, by Hindawi Publishing Corporation; hereby published with permission under the Creative Commons Attribution License or equivalent. Every chapter published in this book has been scrutinized by our experts. Their significance has been extensively debated. The topics covered herein carry significant findings which will fuel the growth of the discipline. They may even be implemented as practical applications or may be referred to as a beginning point for another development.

The contributors of this book come from diverse backgrounds, making this book a truly international effort. This book will bring forth new frontiers with its revolutionizing research information and detailed analysis of the nascent developments around the world.

We would like to thank all the contributing authors for lending their expertise to make the book truly unique. They have played a crucial role in the development of this book. Without their invaluable contributions this book wouldn't have been possible. They have made vital efforts to compile up to date information on the varied aspects of this subject to make this book a valuable addition to the collection of many professionals and students.

This book was conceptualized with the vision of imparting up-to-date information and advanced data in this field. To ensure the same, a matchless editorial board was set up. Every individual on the board went through rigorous rounds of assessment to prove their worth. After which they invested a large part of their time researching and compiling the most relevant data for our readers.

The editorial board has been involved in producing this book since its inception. They have spent rigorous hours researching and exploring the diverse topics which have resulted in the successful publishing of this book. They have passed on their knowledge of decades through this book. To expedite this challenging task, the publisher supported the team at every step. A small team of assistant editors was also appointed to further simplify the editing procedure and attain best results for the readers.

Apart from the editorial board, the designing team has also invested a significant amount of their time in understanding the subject and creating the most relevant covers. They scrutinized every image to scout for the most suitable representation of the subject and create an appropriate cover for the book.

The publishing team has been an ardent support to the editorial, designing and production team. Their endless efforts to recruit the best for this project, has resulted in the accomplishment of this book. They are a veteran in the field of academics and their pool of knowledge is as vast as their experience in printing. Their expertise and guidance has proved useful at every step. Their uncompromising quality standards have made this book an exceptional effort. Their encouragement from time to time has been an inspiration for everyone.

The publisher and the editorial board hope that this book will prove to be a valuable piece of knowledge for researchers, students, practitioners and scholars across the globe.

LIST OF CONTRIBUTORS

Yueli Wang, Xiaoyan Li, Rongjuan Li, Ya Yang and Jie Du
Department of Echocardiography, Beijing Anzhen Hospital, Capital Medical University and Beijing Institute of Heart, Lung and Blood Vessel Diseases, Beijing, China

Kamila Borowczyk and Hieronim Jakubowski
Department of Microbiology, Biochemistry and Molecular Genetics, Rutgers-New Jersey Medical School, International Center for Public Health, Newark, NJ 07103, USA

Kamila Borowczyk
Department of Environmental Chemistry, Faculty of Chemistry, University of Łódź, 90-236 Łódź, Poland

Jacek Wróblewski
Institute of Bioorganic Chemistry, 71-704 Poznań, Poland

Jacek Wróblewski and Hieronim Jakubowski
Department of Biochemistry and Biotechnology, Poznań University of Life Sciences, 60-632 Poznań, Poland

Joanna Suliburska
Institute of Human Nutrition and Dietetics, Poznań University of Life Sciences, 60-632 Poznań, Poland

Noriyuki Akahoshi and Isao Ishii
Department of Health Chemistry, Showa Pharmaceutical University, Tokyo 194-8543, Japan

Omar Ramos-Lopez, Jose I. Riezu-Boj, Fermin I. Milagro, Marta Cuervo, Leticia Goni and J. A. Martinez
Department of Nutrition, Food Science and Physiology, and Center for Nutrition Research, University of Navarra, Pamplona, Spain

Jose I. Riezu-Boj, Marta Cuervo and J. A. Martinez
Navarra Institute for Health Research (IdiSNA), Pamplona, Spain

Fermin I. Milagro, Marta Cuervo and J. A. Martinez
CIBERobn, Fisiopatología de la Obesidad y la Nutrición, Carlos III Health Institute, Madrid, Spain

J. A. Martinez
Madrid Institute of Advanced Studies (IMDEA Food), Madrid, Spain

Milena Rizzo, Gabriele Berti, Monica Evangelista and Giuseppe Rainaldi
Non-coding RNA Laboratory, Institute of Clinical Physiology (IFC), CNR, Via Moruzzi 1, 56124 Pisa, Italy

Milena Rizzo and Giuseppe Rainaldi
Tuscan Tumor Institute (ITT), Via Alderotti 26/N, 50139 Firenze, Italy

Gabriele Berti
Instute of Life Science, Scuola Superiore Sant'Anna, Piazza Martiri delle Libertà 33, 56127 Pisa, Italy

Francesco Russo and Marco Pellegrini
Laboratory of Integrative Systems Medicine (LISM), Institute of Informatics and Telematics (IIT) and Institute of Clinical Physiology (IFC), CNR, Via Moruzzi 1, 56124 Pisa, Italy

Francesco Russo
Novo Nordisk Foundation Center for Protein Research, Faculty of Health and Medical Sciences, University of Copenhagen, Blegdamsvej 3b, 2200 Copenhagen, Denmark

Taiga Nishihori
Department of Blood and Marrow Transplantation, H. Lee Moffitt Cancer Center and Research Institute, Tampa, FL, USA

Kenneth Shain
Department of Malignant Hematology, H. Lee Moffitt Cancer Center and Research Institute, Tampa, FL, USA

Lucas F. Ribeiro
Department of Biology, Faculdade de Filosofia, Ciências e Letras de Ribeirão Preto, University of São Paulo, São Paulo, SP, Brazil

Liliane F. C. Ribeiro and Matheus Q. Barreto
Department of Biochemistry and Immunology, Faculdade de Medicina de Ribeirão Preto, University of São Paulo, São Paulo, SP, Brazil

Richard J. Ward
Department of Chemistry, Faculdade de Filosofia, Ciências e Letras de Ribeirão Preto, University of São Paulo, São Paulo, SP, Brazil

Jihye Kim, Minjae Yoo, Jimin Shin, Hyunmin Kim and Aik Choon Tan
Translational Bioinformatics and Cancer Systems Biology Laboratory, Division of Medical Oncology, Department of Medicine, School of Medicine, University of Colorado Anschutz Medical Campus, Aurora, CO 80045, USA

Jaewoo Kang
Department of Computer Science and Engineering, Interdisciplinary Graduate Program in Bioinformatics, Korea University, Seoul, Republic of Korea

Preethi Krishnan and Sambasivarao Damaraju
Department of Laboratory Medicine and Pathology, University of Alberta, Edmonton, Canada T6G 1Z2
Cross Cancer Institute, Edmonton, Canada T6G 1Z2

Runyu Jing, Shengzhong Feng, Yanjie Wei and Li He
Shenzhen Institutes of Advanced Technology, Chinese Academy of Sciences, Shenzhen 518055, China

Yu Liang
College of Chemistry, Sichuan University, Chengdu 610064, China

Yi Ran
Biogas Appliance Quality Supervision and Inspection Center, Biogas Institute of Ministry of Agriculture, Chengdu, Sichuan, China

Hongliang Wang, Wei Xing, Zhenglin Wang, Tiantian Lv, Shuzhen Guo and Wei Wang
College of Basic Medicine, Key Laboratory of Ministry of Education (Syndromes and Formulas), Key Laboratory of Beijing (Syndromes and Formulas), Beijing University of Chinese medicine, Beijing 100029, China

Hongliang Wang
School of Public Health, Capital Medical University, Beijing 100069, China

Shijie Tang
Key Laboratory of Systems Biology, CAS Center for Excellence in Molecular Cell Science, Innovation Center for Cell Signaling Network, Institute of Biochemistry and Cell Biology, Shanghai Institutes for Biological Sciences, Chinese Academy of Sciences, Shanghai 200032, China

Yan Wu and Jing Han
Institute of Chinese Medicine, Beijing University of Chinese medicine, Beijing 100029, China

Chun Li
Modern Research Center for Traditional Chinese Medicine, Beijing University of Chinese Medicine, Beijing 100029, China

Ruixin Zhu
Department of Bioinformatics, School of Life Sciences and Technology, Tongji University, Shanghai 200092, China

Wei-ni Ma, Ming-mei Zhou, Le Zhao and Fang Cen
Center for Chinese Medicine Therapy and Systems Biology, Shanghai University of Traditional Chinese Medicine, Shanghai 201203, China

Xiao-jun Gou
Central Laboratory, Baoshan District Hospital of Integrated Traditional Chinese and Western Medicine of Shanghai, Shanghai University of Traditional Chinese Medicine, Shanghai 201999, China

Fang Cen
Experiment Center of Teaching & Learning, Shanghai University of Traditional Chinese Medicine, Shanghai 201203, China

Ying Xu
Department of Physiology, Shanghai University of Traditional Chinese Medicine, Shanghai 201203, China

Hong-yi Shen
Research Center for Health and Nutrition, Shanghai University of Traditional Chinese Medicine, Shanghai 201203, China

Hongkun Zhao, Fu Xing and Jixun Guo
Key Laboratory of Vegetation Ecology, Ministry of Education, Institute of Grassland Sciences, Northeast Normal University, Changchun 130024, China

Hongkun Zhao, Guangxun Qi and Yingshan Dong
Institute of Crop Germplasms, Jilin Academy of Agricultural Sciences, Gongzhuling 136100, China

Yumin Wang, Xiaodong Liu, Cuiping Yuan and Yingshan Dong
Institute of Soybean Research, Jilin Academy of Agricultural Sciences, Changchun 130033, China

Jesús Espinal-Enríquez, Daniel González-Terán and Enrique Hernández-Lemus
Computational Genomics Division, National Institute of Genomic Medicine, 14610 México City, Mexico

Jesús Espinal-Enríquez
Centro de Ciencias de la Complejidad, Universidad Nacional Autónoma de México, 04510 México City, Mexico

Bahaeldeen Babikar Mohamed, Beenish Aftab, Muhammad Bilal Sarwar, Bushra Rashid, Zarnab Ahmad, Sameera Hassan and Tayyab Husnain
Center of Excellence in Molecular Biology, University of the Punjab, 87-West Canal Bank Road, Thokar Niaz Baig, Lahore 53700, Pakistan

Bahaeldeen Babikar Mohamed
National Center for Research (NCR), Ministry of Sciences and Telecommunications, Khartoum, Sudan

Ali Nabi Duman
Department of Mathematics and Statistics, King Fahd University of Petroleum and Minerals, Dhahran, Saudi Arabia

Harun Pirim
Department of Systems Engineering, King Fahd University of Petroleum and Minerals, Dhahran, Saudi Arabia

Guangxin Yuan, Yuhang Zhang and Guangyu Xu
College of Pharmacy, Beihua University, Jilin, Jilin 132013, China

Yu Bai
Pharmaceutical College, Jilin Medical University, Jilin, Jilin 132011, China

Shuang Zhao, Xuemei Zhang, Yang Liu, Meng Sun and Guohui Qi
Department of Forestry, Agricultural University of Hebei, Baoding 071000, China

Shuang Zhao
Cotton Research Institute, Hebei Academy of Agriculture and Forestry Sciences, Shijiazhuang 050000, China

Xuemei Zhang and Guohui Qi
Research Center for Walnut Engineering and Technology of Hebei, Lincheng 054304, China

Yanping Su
Department of Life Science, Normal University of Langfang, Langfang 065000, China

Yilan Chen
Baoding Forest Seedling Management Station, Baoding 071000, China

Jiajia Chen and Liangzhi Li
School of Chemistry, Biology and Materials Engineering, Suzhou University of Science and Technology, No. 1 Kerui Road, Suzhou 215011, China

Index

R

S

T

Z

www.ingramcontent.com/pod-product-compliance
Lightning Source LLC
LaVergne TN
LVHW060608210926
848485LV00005B/1389